Search for the Causes of Schizophrenia

Volume III

Edited by
H. Häfner and W.F. Gattaz

With 50 Figures

Springer-Verlag Berlin Heidelberg New York
London Paris Tokyo Hong Kong Barcelona Budapest

Prof. Dr. Dr. Dres. h.c. HEINZ HÄFNER
Prof. Dr. WAGNER FARID GATTAZ
Central Institute of Mental Health
P.O. Box 122120
68072 Mannheim, Germany

ISBN-13:978-3-642-79431-5 e-ISBN-13:978-3-642-79429-2
DOI: 10.1007/978-3-642-79429-2

Library of Congress Cataloging-in-Publication Data. (Revised for vol. 3) Search for the causes of schizophrenia. Papers from a conference held at the Forum of Science, Heidelberg University, Sept. 24–26, 1986, and organized by the Central Institute of Mental Health, Mannheim, and the Department of Psychiatry, University of Heidelberg. Vol. 2, edited by H. Häfner and W.F. Gattaz, presents papers from a second symposium held at the University of Heidelberg on May 24–26, 1989. Includes bibliographies and indexes. 1. Schizophrenia – Etiology – Congresses. I. Häfner, H. (Heinz), 1926– . II. Gattaz, W.F. (Wagner Farid), 1951– . III. Janzarik, W., 1920– . IV. Zentralinstitut für Seelische Gesundheit (Mannheim, Germany) V. Universität Heidelberg. Psychiatrische Klinik. RC514.S366 1987 616.89'82071 87-9504 ISBN-13:978-3-642-79431-5

Softcover reprint of the hardcover 1st edition 1995

Typesetting: Best-set Typesetter Ltd., Hong Kong
SPIN: 10480210 25/3130/SPS – 5 4 3 2 1 0 – printed on acid-free paper

Preface

This volume presents the contributions to the third symposium on the "Search for the Causes of Schizophrenia". The first symposium was organized by the Central Institute of Mental Health on the occasion of the 600th anniversary of the University of Heidelberg in 1986. Its objective was to give an overview of the rapid development in schizophrenia research in a situation characterized by an excessive variety of methods and topics. It was our intention, through substantial contributions to topical areas of schizophrenia research and through intensive discussions, to further the exchange of hypotheses and results and thus open up ways towards new concepts integrating different methodological approaches or towards new models of explanation.

The first symposium was a great success. The publication of the proceedings (Häfner et al. 1987) was received with enthusiasm and recommended in leading journals "as the best available state-of-the-art analysis of current research on the aetiology of schizophrenia". For this reason we held a second and a third symposium entitled "Search for the Causes of Schizophrenia" under the same auspices and at the same place in 1989 and 1993.

The variety and disparity of the disciplines and methods involved in schizophrenia research makes it essential to focus on certain issues, even if the symposium is meant to bring together the divergent branches of research for an exchange of experience. Therefore, we endeavoured to discuss the latest research progress by concentrating on issues promising further knowledge on the aetiology of schizophrenia. The speakers and discussants were leading experts engaged at the frontline of schizophrenia research: several speakers have already developed concepts for comprehensive aetiological models even beyond their own approaches. The most intricate but also interesting task in this context was for the invited discussants to critically analyse four or five contributions of partly loosely connected issues.

The concentration on a few focuses, which meant the disregard of other fields of schizophrenia research, was bound to meet with biased opinions. In "*Search for the Causes of Schizophrenia II*", we had taken up topics from the first symposium such as genetics, biological and psychophysiological aspects, vulnerability models, epidemiology and the course of schizophrenia and had pursued them up to the latest developments of that time. Extensive information was provided about recent research in analytical and genetical epidemiology and about the few methodologically outstanding national and transnational comparative course studies. Intensive and critical attention was paid to the vulnerability models; the descriptive genetics of schizophrenia was complemented by aspects of molecular genetics, biological approaches by brain anatomy and brain function.

The present volume of "*Search for the Causes of Schizophrenia III*" opens up new perspectives for schizophrenia research: On the assumption that, at the beginning, the symptoms and course of schizophrenia are not so greatly modified by secondary processes such as medication, environmental and personality factors, and on the assumption that antecedents of the psychosis can offer an approach to aetiological hypotheses, we focused in Part I on the precursors, onset and early course of schizophrenia. In this field, the British Child Development Study has provided substantial progress. This study of a representative birth cohort of the British population assessed in several cross sections during childhood and adolescence has for the first time provided generalizable data about the antecedents of schizophrenia on the basis of emotional, behavioural, cognitive and neuromotor deficits (see T.J. Crow, p. 3). The results of the high-risk studies (J. Parnas, p. 67), which are necessarily based on a selectively defined proportion of schizophrenic individuals, i.e. the children of schizophrenic mothers, could thus be complemented by epidemiologically founded results concerning the functional and developmental deficits of individuals who later develop schizophrenia.

These results have proved to be of theoretical importance for both genetic and exogenous developmental deficits. The variety and severity of the neurobehavioural deficits, above all in early-onset schizophrenia, starting before the age of 12, was demonstrated by Asarnow et al. (p. 87) with a multi-level approach. Altogether these results showed that the aetiological model of schizophrenia is no longer adequate without taking into account the neurodevelopmental aspects with their genetic and exogenous causes (R. M. Murray, p. 186). They also emphasized that a comprehensive or an exclusive interpretation of schizophrenia as a neurodevelopmental disorder is not or perhaps not yet feasible, since a great amount of schizophrenias break out without preceding symptoms of neurodevelopmental deficits.

The systematic investigation into onset and early course on the basis of a large first-episode sample and the explanation of the higher age of onset in women by the estrogen hypothesis (H. Häfner, p. 43; E.F. Walker, p. 21; L.E. DeLisi, p. 114) was also an important contribution towards generating new theories. The fact that the maximum of social consequences of schizophrenia already occur during the prodromal phase and are determined by the time when the disease breaks into the social biography, and thus by the age at onset, is not only of theoretical but also of practical importance (see H. Häfner, p. 43).

Part II, entitled "*Schizophrenia: Disease or Syndrome?*", deals with the complex relationships between brain function assessed by brain imaging and neuropathology and specific psychopathology of schizophrenia at the level of morphological, functional and regional brain abnormalities (P.F. Liddle, p. 150; E.C. Johnstone, p. 129). At these levels, the first steps are made towards elucidating the associations between localized brain function and psychopathology. They are firmly supported by M.R. Trimble (p. 141) with regard to the organic presentation of the schizophrenia syndrome or the comparison of symptomatic schizophrenia and its neuropathological causes with idiopathic schizophrenia. The issue of the genetics of schizophrenia was pursued by means of epidemiological family studies of sound methodological standard, coming up with the result of limited genetic heterogeneity (see W. Maier, p. 157).

Part III, entitled "*Brain, Receptors and Schizophrenia*", presents the associations between brain function, receptors and schizophrenia at the level of biochemical analyses, their molecular basis (P. Sokoloff, p. 195; W.F. Gattaz, p. 215), at the level of a 31p spectroscopic magnetic resonance study of the brain phospholipid metabolism (J.W. Pettegrew, p. 227) and with positron emission tomography (PET) studies of receptor ligands and regional changes in metabolism (F.-A. Wiesel, p. 252).

The discussants (L.E. DeLisi, R.E. Murray and F.A. Henn) performed the complex critical synthesis leading to integrative approaches and initiated the creative general discussion. Unfortunately, for reasons of expense, the general discussion could not be included in this volume.

"Search for the Causes of Schizophrenia III" was for the first time connected with the symposium "Schizophrenia – Psychotic Continuum or Distinct Entities" organized by A. Marneros (University of Halle, Germany), N.C. Andreasen (University of Iowa, USA) and M.T. Tsuang (Harvard Medical School, USA). The latter is also the third symposium in a series of internationally renowned research symposia. It was the intention of the organizers to hold the two symposia at the same place and in immediate succession so as to enable the exchange of experience between the participants of the two groups. This experiment was greatly estimated by all the participants. However, both groups had to put up with a limited time and programme compared to the preceding symposia. The proceedings of the symposium "Schizophrenia – Psychotic Continuum or Distinct Entities" will be published in a separate volume (Marneros et al. 1995).

Mannheim, September 1994

H. Häfner
W.F. Gattaz

Contents

Part III Brain, Receptors and Schizophrenia

List of Contributors

You will find the addresses at the beginning of the respective contributions.

Part I
Antecedents, Onset, and Early Course of Schizophrenia

Birth Cohort Study of the Antecedents of Psychosis: Ontogeny as Witness to Phylogenetic Origins

T.J. Crow[1], D.J. Done[2], and A. Sacker[1]

Introduction

Psychotic illness is common, and apparently universal in human society (Jablensky et al. 1992). What is its origin and why does it persist? These questions may be unanswerable if they are not recognised as related to more fundamental questions such as the characteristics which distinguish man from other primate species, and the nature of human individuality. It is arguable that the phenomena of psychotic illness provide important clues to the evolution of man.

Since Kraepelin (1919) it is conventional to distinguish between schizophrenic and manic-depressive psychoses. But although the association between form of illness and outcome is undoubted, and prototypical affective and schizophrenic illnesses are observed, great diversity in clinical manifestations and outcome is commonplace, and there is serious question whether a fundamental distinction between categories of psychotic illnesses can be drawn. For these reasons it is more realistic to regard the variations in psychosis as part of a continuum rather than discrete illnesses (Crow 1990b).

This conclusion influences the set we adopt in considering the problem of causation. In some way variation within the normal population is related to the abnormalities that present themselves as disease. The diversity of human personality and intellectual ability is relevant. One must assume either that the manifestations of illness depend upon an interaction between an extrinsic pathogen and the pre-existing psychological structure, or perhaps more parsimoniously, that the two are part of a single developmental process.

The epidemiology of psychosis emphasises the singularity of the problem. The WHO Ten Country study shows that when schizophrenia is defined restrictively and when steps are taken to identify all presenting cases, incidence rates are remarkably similar in populations in widely varying climatic, industrial and social environments (Jablensky et al. 1992). In this respect psychotic illness differs from most if not all other common illnesses. It appears invariant with respect to the environment; it may be intrinsic to human nature.

[1] Clinical Research Center, Division of Psychiatry, Northwick Park Hospital, Harrow, Middle Sex HA1 3UJ, UK

[2] Department of Psychology, University of Hertford, UK

H. Häfner and W.F. Gattaz (Eds.)
Search for the Causes of Schizophrenia, Vol. III

Age of onset is noteworthy. Onsets are rare before puberty but then rise steeply, and remain high throughout the reproductive period of life (Penrose 1991). This is paradoxical if the disease is genetic since it significantly impairs procreativity (MacSorley 1964). Onset itself requires an explanation since it cannot be attributed to an environmental precipitant. When illness occurs in two siblings it occurs at the same age and not at the same time (Crow and Done 1986).

Kraepelin (1919) drew attention to what is perhaps the most puzzling aspect of onset, the sex difference, well studied in recent years by a number of workers. Onsets are earlier in males than females by 3–4 years, and this difference cannot be accounted for by differences in the facility with which early symptoms are detected or patients admitted to hospital (Häfner et al. 1992; Lewine 1991). The sex difference in onset is perhaps the best established but least explained epidemiological fact about schizophrenia.

New light is cast upon onset by a cross-cultural comparison of case registers (Eaton et al. 1992). Early onset predicts likelihood of relapse, as also do male sex and single status. But when onset is taken into account neither sex nor marital status add significantly to prediction. A single variable concerned with timing an aspect of central nervous system development is a major determinant of the course of the disease.

But what is the significance of onset itself? Onset is conveniently defined as the point of hospital admission which reflects that a threshold of distress or behavioural disturbance has been passed. It may or may not correspond to the onset of psychotic symptoms, but it is well recognised that this event is often preceded by dysfunction that has lasted for months and sometimes years. In two recent studies duration of symptoms before onset was found to be a predictor of outcome following a first episode of illness (Crow et al. 1986; Loebel et al. 1992). Of more fundamental significance are precursors in childhood that are related in a systematic way to the form of psychosis in adult life. The timing and nature of such premonitory manifestations could provide clues to the identity of the functions that become deviant in psychotic disturbance.

That there are such precursors has been suggested by a number of retrospective studies of the childhood of individuals who later developed schizophrenia. Lane and Albee (1964) found that the intelligence scores in the second, sixth and eighth grades at school of children who became schizophrenic were significantly reduced compared to their siblings. Offord and Cross (1971) found low IQ in childhood to be a predictor of early age of onset and longer duration of hospitalisation, and Offord (1974) reported that the deficit in IQ and poor school performance relative to siblings applied particularly to males. Earlier Gittelman-Klein and Klein (1969) had identified a group of schizophrenic patients with "premorbid asociality" who also had academic problems and were likely to have a poor outcome. In a study of school records Watt (1978) described pre-schizophrenic girls in their early school years as emotionally unstable, introverted and passive, and the boys as emotionally unstable and disagreeable, becoming more so in the later school years.

Some follow-up studies have yielded comparable findings. Thus Hartmann et al. (1984) identified 24 individuals in the cohort of 500 delinquent and 500 "non-delinquent" boys that Glueck and Glueck (1950) had studied between that ages of 10 and 17 years 40 years earlier, who had been diagnosed as suffering from schizophrenia in adult life. The 24 pre-schizophrenics were matched for IQ, ethnicity, original grouping and age at testing with 48 non-schizophrenic controls. Five psychosocial indicators were identified as significant predictors of a schizophrenic outcome—lack of historicity (lack of sense of self continuing over time, poor goal direction), anxiety, difficulty in interpersonal relationships, lack of competency (developmental delays, poor schoolwork), and permeable boundaries (easily distracted, day dreaming, tangential thinking).

Amongst child guidance attenders Ambelas (1992) matched 18 boys who subsequently met DSM-III criteria for a diagnosis of schizophrenia for age, sex, and time and reason of referral with 18 control attenders. The index group were more likely to have been diagnosed as suffering from a mixed emotional and conduct disorder and less likely to have an adjustment reaction, were more likely to have had developmental (including speech, language and reading) and relationship problems, and had lower IQs.

Across these studies the picture that emerges is that individuals who are later destined to suffer from schizophrenic illnesses are distinguishable (at least on the basis of group means) from individuals not so predisposed 10 or 15 years before the onset of psychosis. The distinguishing features include academic performance and difficulties in establishing interpersonal relationships. There is at least a suggestion that deficits in IQ are a predictor of poor outcome. Questions that are unanswered are the time course of emergence of these impairments, the relationship between them, and the meaning of the sex difference.

United Kingdom Perinatal Mortality Survey and National Child Development Study

The Perinatal Mortality Survey included 98% of all births in England, Scotland and Wales registered during the week 3–9 March 1958. Four subsequent attempts to trace all members of the cohort to monitor physical, educational and social development (in 1965, 1969, 1974, and 1981) became known as the National Child Development Study (NCDS). There were 16980 individuals in the Perinatal Mortality Survey and 15398, 15303, 14761 and 12537 in the subsequent follow ups, at ages 7, 11, 16 and 23 years, respectively. Using the Mental Health Enquiry (a register of admissions to psychiatric hospitals and units that continued until 1986) we were able to identify all cases recorded in the NCDS who had been treated as inpatients for psychiatric reasons between 1974 and 1986, i.e., until the cohort were aged 28 years. From the casenote histories PSE Catego diagnoses (Wing et al. 1974) were derived to give groups of patients suffering from schizophrenia by broad ($n = 57$) or by narrow criteria (the presence of nuclear symptoms, $n = 40$), affective psychosis ($n = 35$) and neurosis ($n = 79$) (Done et al. 1991). These groups were compared either with the total sample (Done et al. 1991) or with a randomly selected 10% sample (Done et al. 1994a) of individuals within the NCDS population who had never been admitted to hospital for psychiatric reasons.

Pregnancy and Birth Complications

We assumed that the nature of prenatal and perinatal events relevant to the later development of psychiatric disease was the same as those responsible for stillbirth and neonatal death. Logistic regression was used to derive the probability of stillbirth or neonatal death for each cohort member (referred to as the "risk"). The risk of stillbirth and neonatal death for each of the groups identified through the Mental Health Enquiry was then compared with the sample as a whole (Table 1).

The risk was not significantly increased for any of the diagnostic groups. Nor did a further analysis in which the 30% of patients at greatest risk within each group reveal any evidence of the possible existence of an extreme subgroup with exceptionally high risk.

Some light on the literature concerning pregnancy and birth complications in relation to psychosis is cast by a subsequent analysis (Sacker et al. 1994b) of the data recorded in the Perinatal Mortality Survey.

Of 48 variables recorded concerning the pregnancy and delivery ten were found to show greater abnormality in pre-schizophrenics compared to controls ($p < 0.1$). Of these ten variables five (mother's weight being lower, fewer antenatal visits, delivery more likely to be attended by an untrained person, smoking during pregnancy and psychological problems) could be thought to relate to the mother's behaviour and/or personality. In a meta-analysis of the literature (Sacker et al. 1994a) it was concluded that many of the findings concerning the obstetric histories of pre-schizophrenics can be accounted for on the basis of differences in the lifestyle of their mothers.

Table 1. Estimated probability of stillbirth or neonatal death in groups diagnosed as psychotic (Done et al. 1991)

	Risk	Difference between means	95% confidence limits	p
Controls (n = 12 946)	0.0098			
Schizophrenia				
Narrow (n =35)	0.0111	0.0012	−0.00142–0.005	0.40
Broad (n = 57)	0.0112	0.0014	−0.0008–0.004	0.24
Affective psychosis (n = 32)	0.0125	0.0027	−0.0005–0.007	0.105
Neurosis (n = 75)	0.0103	0.0005	−0.001–0.0027	0.60

Prenatal Exposure to Influenza

By chance the mothers of the NCDS cohort were exposed to the 1957 influenza epidemic, and a record was made by the midwife at the time of birth based on the medical records and the mother's report. It was therefore possible to examine whether as has been suggested by Mednick et al. (1988) and O'Callaghan et al. (1991) exposure to influenza in the second trimester of pregnancy increases the risk of schizophrenia in the child. The numbers of mothers who suffered from influenza in each trimester together with the expected and observed numbers of their children who later developed schizophrenia (by broad criteria) are shown in Table 2.

No excess of children with schizophrenia born to mothers who suffered from influenza in the second trimester is observed. In the bottom row are the numbers of children predicted to suffer from schizophrenia on the basis of interpretations of population data by Mednick et al. (1988) and O'Callaghan et al. (1991). There is a highly significant discrepancy between the numbers predicted by the protagonists of the influenza hypothesis and the numbers actually observed.

Behavioural Assessments at the Ages of 7 and 11 Years

At the ages of 7 and 11 years the teacher of each subject in the NCDS cohort was asked to complete the Bristol Social Adjustment Guide (BSAG), a standardised psychometric test of social adjustment and maladjustment in which the teacher underlines which of some 200 descriptions of school behaviour are relevant to the child (Done et al. 1994a). From these assessments a number of core syndromes are identified, and these can be placed into two groups of over- and under-reaction. In over-reaction are included the syndromes hostility towards adults, hostility towards children, anxiety for acceptance by adults and children, inconsequential behaviour and restlessness, and in under-reaction are included unforthcomingness, withdrawal and depression as well as writing off adults. Two further categories comprise miscellaneous symptoms and miscellaneous nervous symptoms. The major descriptors of the items that make up these syndromes are listed in Table 3.

Comparisons were made between a group of patients with schizophrenia (narrow criteria, $n = 40$), with affective psychosis ($n = 35$), neurosis ($n = 79$),

Table 2. Maternal influenza in trimester

	n	1	2	3	χ^2	*P*
Total subjects	16 268	231	945	675		
Schizophrenia	57 Expected	0.8	3.3	2.4		
Broad criteria	Observed	0	3	4	0.04	NS
M/O'C predicted		0.4	28.3	1.3	17.6	<0.001

M/O'C, Mednick and O'Callaghan.

Table 3. Bristol social adjustment guide

Over-reaction
1. Anxiety for acceptance by adults
   - Either (i) Overly friendly or talking excessively to the teacher
     or
     (ii) Seeking to engage excessively/monopolise the teacher
2. Anxiety for acceptance by children
   - Buffoonery, overly brave, shows off, brags
3. Hostility towards adults
   - Includes (i) Variability of mood and performance when asked to do something
     (ii) Damages property, lies, uses bad language
     (iii) Steals, is aggressive
4. Hostility towards other children
   - A variety of criteria for being unpleasant towards other children
5. Inconsequential behaviour
   - Includes (i) Poor concentration, cannot stick at things
     (ii) Careless and untidy, lolls about, mischievous
6. Restlessness
   - Too restless in individual games or when working alone, lacks persistence

Under-reaction
1. Unforthcomingness
   - Timid and shy, e.g., does not initiate chat, reluctant to approach the teacher
2. Withdrawal
   - Distant, cut off from people, avoids communication
3. Depression
   - Includes (i) Mood variation, sometimes alert, sometimes lethargic and lacks interest
     (ii) Usually apathetic, lifeless, or miserable and depressed
4. "Writing off" adult values
   - Includes (i) Unwilling to work except when compelled to do so
     (ii) Suspicious, selfish, untrustworthy

Miscellaneous
1. Miscellaneous symptoms
   - Include (i) Immature and babyish, and plays with younger children, or gets bullied
     (ii) Truants
2. Miscellaneous nervous symptoms
   - Includes stutters, twitches, bites nails badly

and a 10% random sample of the NCDS populations assessed in the follow ups at 7 and 11 years never admitted to a psychiatric facility. Reports on clinically recognised emotional maladjustment were made by the medical officer who interviewed cohort members and had access to information obtained in an interview with the parents. There was no significant increase in emotional maladjustment in any of the patient groups.

On the BSAG ratings differences are present at the age of 7 and 11 years which cast light on the nature and development of the psychotic disease process.

At the age of 7 pre-schizophrenic subjects are rated as more anxious and hostile towards adults and children and more likely to engage in inconsequential behaviours (Fig. 1), each of these differences being significant at the 1% level or less. These differences are largely confined to males. Moreover with the exception of the category of miscellaneous symptoms the differences between patients and controls are confined to the component of over-reaction; there are no differences with respect to unforthcomingness and withdrawal, and the difference for depression is significant (for males) only at the 10% level.

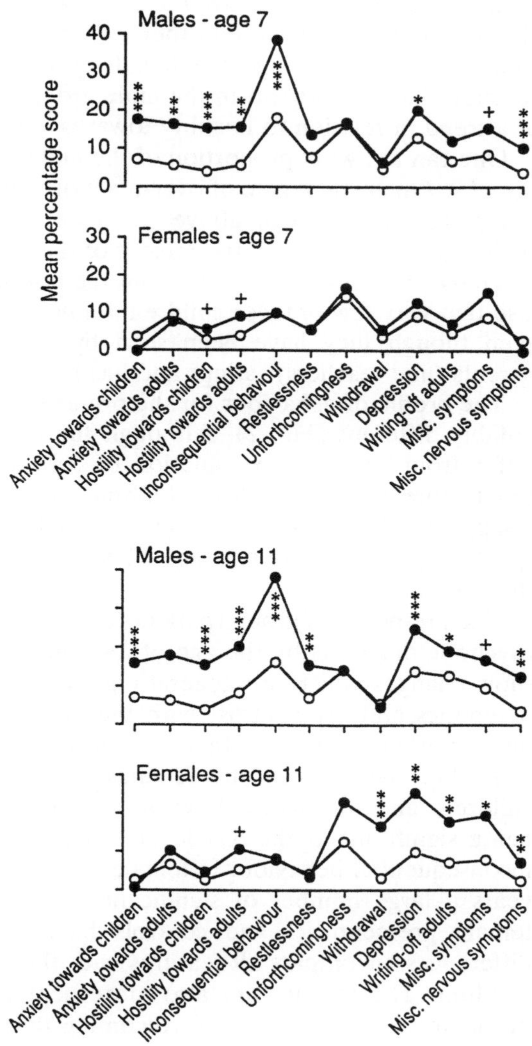

Fig. 1. BSAG profile in pre-schizophrenics. $^{+}$, $p < 0.1$; *, $p < 0.05$; **, $p < 0.01$; ***, $p < 0.005$. Note that p values based on Mann-Whitney U tests for each syndrome are uncorrected for multiple testing. We have previously (Done et al. 1994a) examined profile differences across core syndromes, the significance values for those comparisons incorporating a correction for multiple testing

The changes therefore are sex-specific and relate to those components of behaviour that one might be inclined to relate to positive rather than to negative symptoms.

At the age of 11 years the picture has changed (Fig. 1). Pre-schizophrenic females, as a group, are now rated as more withdrawn and depressed than controls, and as more likely to write off adult values, each of these differences being significant at the 1% level or less (see legend to Fig. 1 for note regarding significance levels). Males are rated as significantly more likely to be depressed, although not withdrawn, and there is a trend with respect to writing off adult values. Even at this age the female group shows no evidence of the over-reaction that was present in the male group 4 years earlier, with the exception of a trend in relation to hostility towards adults.

One can ask what proportion of the male group are rated as abnormal at the age of 7. On an aggregate of over-reaction ratings 50% of the pre-schizophrenic group were rated 13 or above, whereas only 10% of the comparison group achieved such a rating. Fifty percent of the boys (and presumably nearly all the females) are being rated well within the normal range. It should be recollected also that very few of these children are being assessed as emotionally disturbed even though they have been seen by a medical officer who has available parents (and teachers) ratings and has been asked to address this question.

The profile of behaviour for the pre-affective psychotic boys at age 7 years is notably different (Fig. 2). Although the overall profile does not significantly differ from controls, examination of core syndromes reveals that they were rated more hostile to other children and more restless. Both features have disappeared by the age of 11 years. Females are rated as slightly more restless ($p < 0.1$) than controls at age 11 years, but are otherwise unremarkable in their profiles.

The pre-neurotic group (patients admitted to psychiatric hospital with a non-psychotic diagnosis by the age of 28 years) are notable in that at age 11 they show changes which are generally greater than those present in pre-affective psychotics and, by contrast with the pre-schizophrenics, these deviations are more marked in females than in males (Fig. 3). At age 7 the differences are small but the pre-neurotic females are more anxious and hostile to other children, and the males show more anxiety toward adults, these differences being significant at the 5% level. There is a trend ($p < 0.1$) towards more inconsequential behaviour in females, and restlessness in both sexes. At age 11 years a large number of significant ($p < 0.01$ or less) differences between female pre-neurotics and controls have emerged. For over-reaction these differences encompass all features with the exception of anxiety toward other children. The females are also more depressed and withdrawn (less so than pre-neurotics) but not less forthcoming than controls. The males differ from controls only in being more depressed.

Therefore in several respects these individuals who will be admitted with a non-psychotic illness to a psychiatric unit before they are aged 28 years resemble those who will be admitted with a diagnosis of schizophrenia, but the changes are present (and have emerged at a slightly later stage of development)

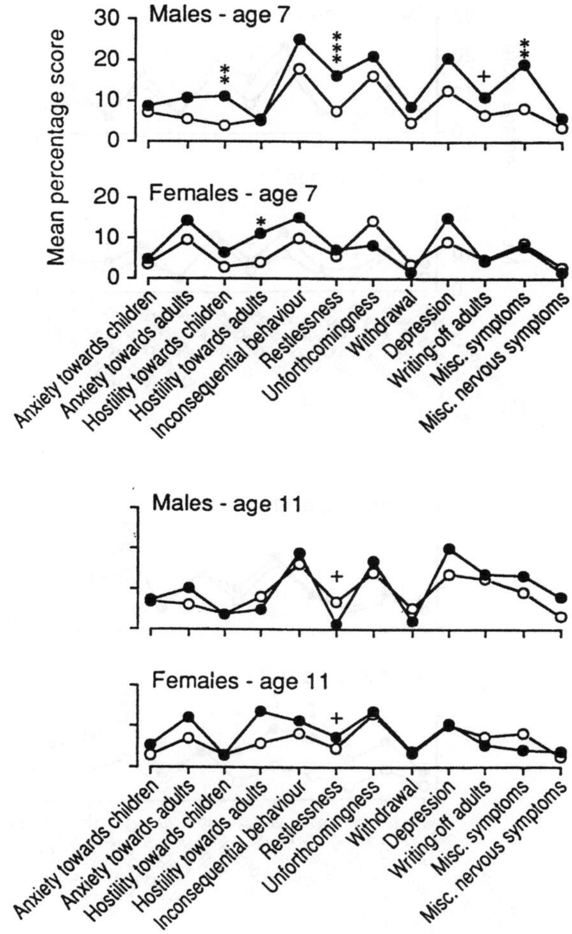

Fig. 2. BSAG profile in pre-affectives aged 7 (for significance values see legend to Fig. 1)

in females rather than in males. Most of these patients acquire a diagnosis of depression. Whether they will later suffer from a psychotic illness, either affective or paraphrenic, at a later age cannot be determined from our database because the Mental Health Enquiry did not continue beyond 1986.

The major changes over time and their relationships between the groups are summarised in terms of total BSAG ratings in Fig. 4. Here it can be seen that for males the pre-schizophrenic group is well separated from the controls at ages 7 and 11, while the pre-affective psychotics apparently become more normal and the pre-neurotics are not distinguished from the controls at either age. For the females the pre-affective psychotics show no gross abnormality at either age, but the pre-schizophrenic and pre-neurotic groups become more abnormal between these ages.

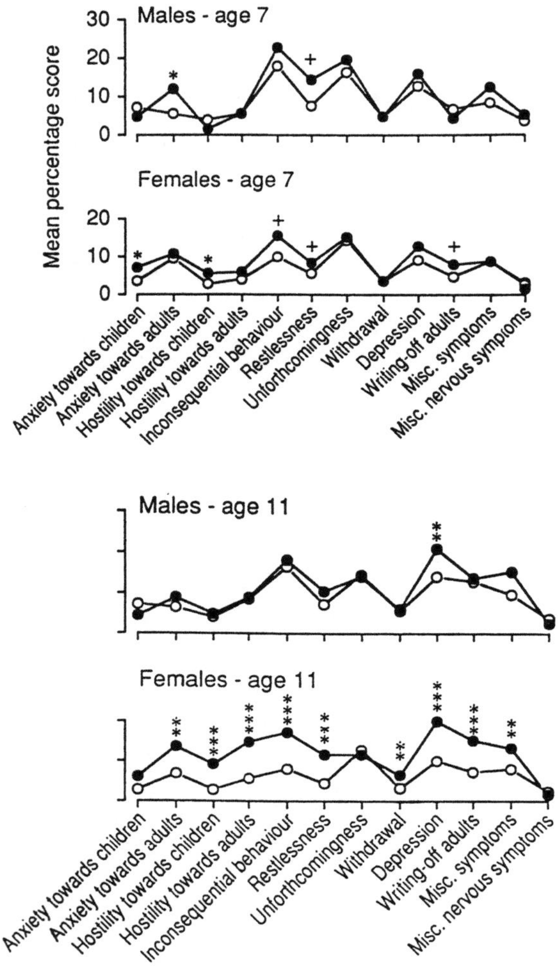

Fig. 3. BSAG profile in pre-neurotics aged 11 (for significance values see legend to Fig. 1)

Since in various respects the pre-schizophrenic group differs from the NCDS population as a whole – for example they show academic impairments (see below) and were more likely to have been in care, experienced parental separation or loss, or to have been referred to a specialist for emotional problems – the question arises whether the behavioural anomalies could be secondary to these differences. However, in a covariance analysis (Done et al. 1994a) which took account of these variables and also included social class the difference between pre-schizophrenics and controls remained significant ($F_{(1,1470)} = 6.3$, $p = 0.01$) after these adjustments. Therefore the group differences in behaviour do not appear to be a consequence of adverse social or educational experience.

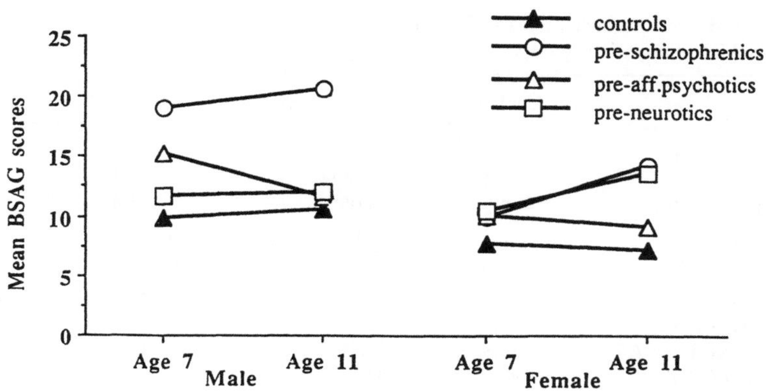

Fig. 4. Changes in BSAG profiles between 7 and 11 years

Academic Performance

Assessments of reading and mathematical ability were made in the NCDS cohort at the ages of 7, 11 and 16 years, and a test of general ability that includes an assessment of verbal and non-verbal intelligence was administered at the age of 11 (Done et al. 1994b). The tests used differed at the different ages – at age 7 a word recognition test (the Southgate group reading test) was used to assess reading skill, while at the later ages this was assessed by a reading comprehension test. Mathematical skill was assessed by tests appropriate to each age. Test performance was correlated between ages with coefficients of between 0.64 and 0.79 for reading and between 0.48 and 0.76 for mathematical skill, the lower coefficient being for the relationship between ages 7 and 16 and the higher between ages 11 and 16 in each case.

The comparisons between the groups at the three ages are summarised in Table 4.

The striking feature of the findings is the widespread nature of the impairments. The deficits are greatest in the pre-schizophrenic group followed by the pre-neurotics and least in the pre-affectives, although the greater significance of the differences between pre-neurotics and controls may in part be attributable to the greater sample size in this group.

Of potential interest are sex differences and trends over time. Although there was a trend toward a greater impairment in performance on the GAT in male pre-schizophrenics, and the impairment was greater in verbal than non-verbal ability, the ANOVA group by sex by ability interaction did not reach the 5% level of significance. Scores on reading and mathematics attainments were standardised with respect to normal controls at each age. Deficits at age 7 years (with the schizophrenic group 1.1 SD below the mean for controls) were as great as at later ages, suggesting that the impairments reflect a constitutional limitation rather than the emergence of a disease process. For the reading test

Table 4. Performance on tests of reading and mathematical skill at the ages of 7, 11 and 16 years in psychiatric in-patient groups compared to controls

	Controls	Pre-schizophrenic	Pre-affective	Pre-neurotic
AGE 7 (*n*)	1520	33	31	67
Reading	23.3	18.4***	20.3*	22.3
Mathematics	5.2	3.9**	4.4*	4.6*
AGE 11 (*n*)	1446	29	29	70
Reading	16.0	12.6**	14.4	13.6***
Mathematics	16.8	11.6**	13.4*	12.4***
Verbal GAT	22.4	16.9**	19.5*	17.5***
Non-verbal GAT	20.9	16.3***	17.5*	17.5***
Total GAT	43.2	33.2***	37.0*	35.0***
AGE 16 (*n*)	1217	22	21	56
Reading	25.3	21.2**	23.8	23.2*
Mathematics	12.6	10.2	11.6	9.8**

performance there were no significant age by sex interactions for the pre-schizophrenic and pre-affective groups (although there was a trend for the schizophrenic males to perform worse at younger ages), but for the pre-neurotic group there was a significant age by sex interaction, the males improving and the females deteriorating with time. On the mathematics tests the deficits were greatest in the pre-schizophrenic group, somewhat less in the pre-neurotics, and least (0.2 SD) in the pre-affectives; there were no significant age by sex interactions.

The deficits on objective tests can be matched against ratings made by parents, teachers and individuals themselves at the various ages. Thus when questioned when the probands were aged 7 parents did not rate their pre-schizophrenic children as significantly abnormal with respect to their acquisition of speech by the age of 2, as ever having stammered or stuttered, or as having had unintelligible speech, but according to the teachers' assessments at this age these children were poor on oral and reading ability and mispronounced words. At age 11 the pre-schizophrenic children were rated (at the 2% level of significance or less) by teachers as having speech difficulties, an imperfect grasp of English, mispronouncing words and having difficulties with number work and the use of books. At the age of 16 years the pre-schizophrenics were rated ($p < 0.001$) as poor on English ability, as having problems with reading and spelling, and again as mispronouncing words, and these individuals rated themselves as having problems with reading.

Overall the picture that emerges is of a group of individuals who are significantly impaired across a range of intellectual abilities from an early age, these impairments having a particular impact on their ability to communicate either in written or oral format. The sex differences in academic ability are less striking than in the case of the disturbances of behaviour.

Physical and Neurological Development

No significant differences were observed between patients and controls with respect to height or weight at the ages of 7, 11 or 16 years or with respect to head circumference assessed at the age of 7 years. The timing of puberty was assessed with ratings of pubic hair and genital and breast development at the age of 11, and pubic hair, the onset of menstruation and age of the voice breaking made at the age of 16. These assessments did not distinguish psychiatric patients from controls.

A number of aspects of neurological function were recorded (by medical officers) at each age, and some of these separated the pre-schizophrenics from controls. For example at age 7 patients were more likely to have been recollected by parents as having been wet by day at age 3 years and by night at age 5 years. At 7 they were also recorded ($p < 0.01$) as showing poor coordination and were rated as poor on a simple test of vision for both left and right eyes. By age 11 vision is recorded as normal and on indices of motor coordination (walking backwards, heel-to-toe standing) they were not now impaired. At this age they were still more likely than controls to be recorded as incontinent. At age 16 years although coordinated in heel-to-toe standing, and hopping on either foot they were rated as significantly ($p < 0.01$) clumsy.

Summary of Salient Findings

The findings with respect to birth complications and the 1957 influenza epidemic give no support to claims that these factors contribute significantly to the aetiology of psychosis. The absence of abnormalities of physical development fail to support theories, e.g., of Kretschmer (1921), that psychosis is associated with significant anomalies of somatic growth, and the relative normality of indices of puberty adds no empirical support to hypotheses, e.g., that of Saugstad (1989), that there are substantial deviations in the timing of endocrine changes.

The changes in behaviour suggest that psychosis reflects a disturbance of selected aspects of central nervous system function that are time-dependent and in certain respects sex-specific. Thus the schizophrenics-to-be are hostile and anxious towards other children and adults and show inconsequential behaviours at ages 7 and 11. But the differences at age 7 are more marked in males (50% of whom are rated as deviant as the most deviant 10% of the school population), and there is little evidence of the social withdrawal that has sometimes been regarded as characteristic of pre-schizophrenic children. By the age of 11 the changes in the pre-schizophrenics are more striking and include (particularly in females) those items of under-reaction, e.g. withdrawal, that might be thought to correspond to the asociality which is often a precursor as well as a sequel to a first episode of schizophrenic psychosis.

The affective-psychotics-to-be (but not the schizophrenics) are rated restless at age 7 (but not 11) years, and the schizophrenics-to-be (but not the depressives) are rated depressed at age 11. A surprising finding is the emergence of a range of abnormalities (more striking and, with the sole exception of depression confined to, females) in the neurotic group. Uncertainty concerning the nature of the disorder in these patients who have required admission for non-psychotic illness by the age of 28 makes these changes difficult to interpret: some patients it may be suspected will go on to develop affective or other psychoses at a later age.

Academic impairments are present in each of the psychiatric groups, but are uniformly more severe in the pre-schizophrenics. In this group language and reading difficulties are present in both sexes from the age of 7 years. While not abnormal on a number of indices of coordination at ages 11 and 16 years and vision tested at 11 years schizophrenics-to-be are slow to develop continence and show poor coordination and vision at age 7 years, and are rated clumsy at 16.

Commentary

In several respects our findings are in agreement with those of Watt's (1978) school record study of schizophrenic patients and normal controls: marked sex differences were present, the boys showing a pattern of irritability, disagreeableness and defiance of authority, while the girls were insecure, shy and participated less in groups. As in the present study the changes were more striking in the males; our study supports Watt's conclusion that "over-inhibition" is characteristic of female rather than male pre-schizophrenics. But the two studies suggest differences in the time sequence of development of changes in the two sexes. Watt found that schizophrenic girls were rated lower on an "extraversion" factor in the kindergarten years through to grade 6 (12 years), and this difference remained through grades 7–12 (13–18 years), whereas boys increased their separation from controls on the factors of "disagreeableness" between these two sets of ratings; he concluded there was greater evidence of progression of deviance in males than in females. Our findings, resting on measures at two specific ages, suggest that deviation is certainly present (mainly as over-reaction) in males as a group as early as 7 and is more widespread (and includes depression) at age 11, whereas the changes in females (that include withdrawal and depression) present at age 11 are of borderline significance at age 7. "Progressive deviance" could be used as a term to describe the changes in both males and females, but the sequence in either case is different.

The range of deviations in behaviour and intellectual function that are present many years before onset casts light on the nature of psychotic illness. It is unlikely, although further work is required, that the disturbances in behaviour can be accounted for in terms of the intellectual deficits or vice versa. In some sense it seems that psychosis is a disorder of both intellect and social ability, and in this respect our findings resemble those of Watt (1978), Hartmann et al. (1984) and Ambelas (1992). That anomalies of personality or social ability are present, and differ between the psychoses, long before psychotic symptoms adds credence to the concept of Kretschmer (1921) that there is a relationship between personality variation in the general population and the varieties of psychosis. According to Kretschmer there is an association between schizophrenia and the schizothymic or introverted personality on the one hand, and between manic-depressive illness and the cyclothymic or extraverted personality on the other.

This view can be contrasted with the categorical or disease entity view of psychotic illness according to which disease onset represents the intrusion into psychic life of a pathological process that is independent of the existing personality structure. Such a concept is sometimes ascribed to Emil Kraepelin, but it is interesting that towards the end of his life Kraepelin came to a view of psychotic illness not far distant from that of Kretschmer. Thus in 1920 Kraepelin wrote "... we must seriously consider how far the phenomena on which we normally base our diagnosis really do afford insight into the basic pathological process. While it may be admitted that this procedure is generally valuable there is a fairly extensive area in which such criteria are lacking.... This is understandable if we assume that the affective and schizophrenic forms of mental disorder do not represent the expression of particular pathological processes, but rather indicate the areas of our personality in which these processes unfold." Such a concept raises the question of the nature of personality variation within the normal population. Kraepelin went on "if ... we try to relate the syndromes of insanity to particular stages of personality, we have few hypotheses on which to base our arguments. If these attempts are to be more than tentative gropings we must trace the phenomena of our inner life back to their roots in the psyche of the child, of primitive man, and of animals." The relationship between personality and psychosis suggested by our findings is perhaps more compatible with the disorder being seen in terms of a lifetime persepective, as Kraepelin appears to be suggesting rather than in relation to a static personality type as Kretschmer's concept is often interpreted.

A clue to the roots of psychosis in personality structure is the sex difference in age of onset, perhaps the best established but least explained fact about the illness. Onsets are earlier in males, a difference which applies to the first symptom as well as to the time of admission. The sex difference in the personality deviations that precede psychosis indicates the areas of psychological function in which the origins of psychosis may be found, and these include particularly the ability to interact with others, whether parents or contemporaries. Of potential relevance is the finding that the personality dimension of cooperativeness is one for which a substantial sex difference has been identified (Cloninger et al. 1991).

One can envisage that there are trajectories of development that are determined by genetic factors acting early in ontogeny. Such genes must be variable between individuals and the range of variation presumably differs between the sexes. The male brain shows a greater mean asymmetry than the female brain, and it is plausible that this accounts for the small mean discrepancy in the pattern of intellectual abilities between the sexes – females performing better on tests of verbal fluency and males on tests of spatial ability. This asymmetry presumably is determined by a gene(s) that acts early in development and it is conceivable that this action could also account for sex differences in personality traits (e.g., sociability), and perhaps in the age of onset of psychosis (Crow 1993).

Characteristics of the human species are a long period of maturation and a complex social life, and these phenomena may well be related, the complexities of social interaction being a selective factor in the delay of sexual maturity. The human being, as Bolk (1926) first suggested, is a neotenous species, the prolongation of foetal characteristics into adult life endowing them with a brain that is enlarged relative to other primates and retains its plasticity for a greater proportion of the life span. Such a brain confers an increased capacity to deal with conspecifics (e.g., through the evolution of language), and for this purpose the ability to take account of the dispositions and intentions of other individuals – or as it has been labelled, to develop a "theory of mind" – is an adaptive advantage. Deficits in this ability are at the heart of a number of psychiatric syndromes including some that are related to psychosis (Brothers 1990; Frith 1992). Complex social behaviour requires a well-developed concept of the self and its relation to others. Increasing hostility to others culminating in paranoid delusions can be seen as one type of abnormality in this relation, and first rank symptoms with loss of the psychological boundary between the self and the external world, can be seen as another.

But if there is variation in the several aspects – intelligence, language and social ability – of psychological function, and particularly if the variation is genetic in origin, one must ask how does this variation arise, and how is it maintained in the population? Only when this more general question has been answered will it be possible to explain the persistence of the genes that predispose to psychosis in the face of a fertility disadvantage, the problem addressed by evolutionary theories of the origin of psychosis (Crow 1993).

Timing is central to a number of these related problems. Within the spectrum of psychotic illness early onset is a predictor of poor outcome. Males generally have an earlier onset and worse outcome but as Eaton et al. (1992) have shown when age of onset is taken into account sex no longer has an influence. A single genetic factor that has a range (overlapping but with a small mean difference in males and females) of influence on the development of an aspect of CNS function, could account both for the premorbid deviations in behaviour and the sex difference in onset of psychosis.

But what is the critical rate-determining factor? A number of findings in morphological studies are compatible with the concept that in schizophrenia there is an arrest of development of cerebral asymmetry (Crow 1990a) there may also be a reduction in total cortical mass. Perhaps these two findings are related and there is a trade-off between brain size and asymmetry. Perhaps various combinations of size and asymmetry are associated with different cognitive and personality structures, and this diversity is maintained by the selective factors that have increased brain size and retarded its maturation. The extremes of this variation may be expressed as the cognitive and behavioural deviations that we recognise as psychosis.

References

Ambelas A (1992) Preschizophrenics: adding to the evidence. Br J Psychiatry 160:401–404

Bolk L (1926) Das Problem der Menschwerdung. Fischer, Jena

Brothers L (1990) The social brain: a project for integrating primate behavior and neurophysiology in a new domain. Concepts Neurosci 15:27–51

Cloninger CR, Svrakic DM, Przybeck TR (1991) The tridemensional personality questionnaire: US normative data. Psychol Rep 69:1047–1057

Crow TJ (1990a) Temporal lobe asymmetries as the key to the etiology of schizophrenia. Schizophr Bull 16:433–443

Crow TJ (1990b) Nature of the genetic contribution to psychotic illness – a continuum viewpoint. Acta Psychiatr Scand 81:401–408

Crow TJ (1993) Sexual selection, Machiavellian intelligence and the origins of psychosis. Lancet 342:594–598

Crow TJ, Done DJ (1986) Age of onset of schizophrenia in siblings: a test of the contagion hypothesis. Psychiatry Res 18:107–117

Crow TJ, Macmillan JF, Johnson AL, Johnstone EC (1986) A randomised controlled trial of prophylactic neuroleptic treatment. Br J Psychiatry 148:120–127

Done DJ, Crow TJ, Johnstone EC, Sacker A (1994a) Childhood antecedents of schizophrenia and affective illness: social adjustment at ages 7 and 11. Br Med J (in press)

Done DJ, Crow TJ, Johnstone EC, Sacker A (1994b) Childhood antecedents of schizophrenia and affective illness: intellectual performance at ages 7 and 11 (to be published)

Done DJ, Johnstone EC, Frith CD, Golding J, Shepherd PM, Crow TJ (1991) Complications of pregnancy and delivery in relation to psychosis in adult life: data from the British perinatal mortality survey sample. Br Med J 302:1576–1580

Eaton WW, Mortenson PB, Herrman H et al (1992) Long-term course of hospitalisation for schizophrenia: I. Risk for hospitalisation. Schizophr Bull 18:217–228

Frith CD (1992) The cognitive neuropsychology of schizophrenia. Erlbaum, Hove

Gittelman-Klein R, Klein DF (1969) Premorbid asocial adjustment and prognosis in schizophrenia. J Psychiatr Res 7:35–53

Glueck S, Glueck E (1950) Unravelling juvenile delinquency. Commonwealth Fund, New York

Häfner H, Riecher-Rössler A, Maurer K, Fätkenheuer B, Löffler W (1992) First onset and early symptomatology of schizophrenia. Eur Arch Psychiatr Clin Neurosci 242:109–118

Hartmann E, Milofsky E, Vaillant G, Oldfield M, Falke R, Ducey C (1984) Vulnerability to schizohrenia: prediction of adult schizophrenia using childhood information. Arch Gen Psychiatry 41:1050–1056

Jablensky A, Sartorius N, Ernberg G et al (1992) Schizophrenia: manifestations, incidence and course in different cultures. A World Health Organization ten country study. Psychol Med Suppl 20:1–97

Kraepelin E (1919) Dementia Praecox and Paraphrenia (translated by R M Barclay, facsimile edition published in 1971). Krieger, New York

Kraepelin E (1920) Die Erscheinungsformen des Irreseins (translated by H Marshall as: Patterns of mental disorder. In: Hirsch SR, Shepherd M (eds) Themes and variations in European psychiatry. Wright, Bristol, pp 7–30, 1974). Z Gesamte Neurol Psychiatr 62:1–29

Kretschmer E (1921) Körperbau und Charakter. Springer, Berlin

Lane EA, Albee GW (1964) Early childhood intellectual differences between schizophrenic adults and their siblings. J Abnorm Soc Psychol 68:193–195

Lewine RRJ (1991) Ontogenetic implications of sex differences in schizophrenia. In: Walker EF (ed) Schizophrenia, a life-course developmental perspective. Academic, New York, pp 195–211

Loebel AD, Lieberman JA, Alvir JMJ, Mayerhoff DI, Giesler SH, Szymanski SR (1992) Duration of psychosis and outcome in first-episode schizophrenia. Am J Psychiatry 149:1183–1188

MacSorley K (1964) An investigation into the fertility rates of mentally ill patients. Ann Hum Genet 27:247–256

Mednick SA, Machon RA, Huttunen MO et al (1988) Adult schizophrenia following prenatal exposure to an influenza epidemic. Arch Gen Psychiatry 45:189–192

O'Callaghan E, Sham P, Takei N et al (1991) Schizophrenia after prenatal exposure to the 1957 A2 influenza epidemic. Lancet 337:1248–1250

Offord DR (1974) School performance of adult schizophrenics, their siblings and age mates. Br Med J 125:12–19

Offord DR, Cross LA (1971) Adult schizophrenia with scholastic failure of low IQ in childhood. Arch Gen Psychiatry 24:431–436

Penrose LS (1991) Survey of cases of familial mental illness. Eur Arch Psychiatry Neurol Sci 240:315–324

Sacker A, Done DJ, Crow TJ (1994a) Antecedents of schizophrenia and affective illness: a meta-analysis of obstetric complications in children born to schizophrenic parents (to be published)

Sacker A, Done DJ, Crow TJ Golding J (1994b) Antecedents of schizophrenia and affective illness: obstetric complications (to be published)

Saugstad LF (1989) Age at puberty and mental illness: towards a neurodevelopmental aetiology of Kraepelin's endogenous psychoses. Br J Psychiatry 155:536–544

Watt NF (1978) Patterns of childhood social development in adult schizophrenics. Arch Gen Psychiatry 35:160–165

Wing JK, Cooper JE, Sartorius N (1974) The measurement and classification of psychiatric symptoms. Cambridge University Press, Cambridge

Antecedents of Schizophrenia: Moderating Effects of Development and Biological Sex

E.F. Walker, J. Weinstein, K. Baum, and C.S. Neumann

Introduction

When the manifest symptoms of a disorder vary predictably with age, we are alerted to the potential role of maturational processes in the moderation of symptom expression. In the case of schizophrenia, it is well established that there is a developmental period, namely, late adolescence/early adulthood, during which florid psychotic symptoms are most likely to emerge (Gottesman 1991; Loranger 1984). This suggests that CNS maturational events during this period serve to trigger the expression of these symptoms. But, for many patients, the manifestation of psychotic symptoms is preceded by dysfunction in a variety of domains, including motor, interpersonal and cognitive functions (Walker 1991; Watt et al. 1984a). Thus constitutional vulnerability to schizophrenia influences behavior long before the onset of clinical symptoms.

Similarly, it has become apparent that there are sex differences in schizophrenia. Most consistently reported are a later age at onset, more favorable prognosis, greater responsivity to neuroleptic medication, more affective symptoms, and greater susceptibility to tardive dyskinesia in female patients (Goldstein and Tsuang 1990; Häfner et al. 1992; American Psychiatric Association 1992). It also appears that there is an increase in the risk for onset among females, but not males, in the 40–50-year age range (Häfner et al. 1992). These findings have led several writers to conclude that biological sex alters the expression of vulnerability to schizophrenia (Lewine 1988; Seeman and Lang 1990). Thus, like maturational level, sex may be associated with CNS characteristics that have implications for the behavioral manifestations of the neural substrate underlying schizophrenia.

We have undertaken a research program that is aimed at elucidating the nature and course of developmental precursors of schizophrenia. The ultimate goal is to use the findings as a basis for generating hypotheses about the neurodevelopmental underpinnings of the disorder. In this chapter, we describe some of the findings and explore their implications for conceptualizations of the neurodevelopmental process(es). In particular, we consider how sex and age differences in the premorbid behavior and illness course of patients might shed light on the nature of the neuropathology.

Department of Psychology, Emory University, Atlanta, GA 30322, USA

H. Häfner and W.F. Gattaz (Eds.)
Search for the Causes of Schizophrenia, Vol. III

Premorbid Behavioral Problems

Retrospective, follow-back and high-risk research designs have been employed in the study of childhood behavioral precursors of schizophrenia. Although these studies have used diverse measures, the findings are consistent in showing a higher rate of behavior problems in preschizophrenic subjects (Offord and Cross 1969; Walker 1991; Watt et al. 1979). However, each of these methods has unique limitations that could influence the validity of the findings they yield (Walker et al. 1991). *Retrospective* studies, which rely on the recollections of an informant, may be subject to biases produced by the informant's awareness of the subject's psychiatric outcome. *High-risk studies*, which examine the development of biological offspring of schizophrenic parents, may be focusing on a unique subgroup, given that most schizophrenic patients do not have a biological parent with the disorder (Gottesman 1991). Moreover, only a fraction (10%–15%) of high-risk subjects is expected to develop schizophrenia (Gottesman 1991), and adult psychiatric outcome data have not yet been acquired on most high-risk samples. Although *follow-back studies* of adult patients may be focusing on more representative patient samples, they have relied on archival sources, notably school records (e.g., Offord and Cross 1969; Watt 1978), that contain very limited information on behavior.

Given the differences among these research paradigms, we must be concerned with the possibility that the findings they yield will reflect method artifacts. Thus a critical question is whether there is evidence of convergence among the findings from these paradigms. In order to address this question, we recently examined data on childhood behavior from two sources; one is a prospective study of biological (i.e., "high-risk") offspring of schizophrenic parents, and the other is a retrospective study of adult schizophrenic patients. The same child behavior rating scale, the Achenbach Child Behavior Checklist (Achenbach 1982; Achenbach and Edelbrock 1981), was used in both studies, so that the pattern of results could be more readily compared.

In the prospective study, behavioral data on the subjects were collected at two time points (for a detailed description of the study, see Walker et al. 1989a). At the first assessment, the subjects averaged 9.70 (SD = 2.35) years of age. The second assessment was conducted 1 year later. We focused on this preadolescent developmental period because it may be a critical one with respect to the prodromal signs of schizophrenia. There were 30 offsprings of schizophrenic parents (17 males and 13 females) and 51 subjects from families with no mental illness (25 males and 25 females). The families of the high-risk subjects were of low socioeconomic status (most parents were unemployed or in blue-collar positions), and the normal comparison group was selected to be comparable with respect to demographic factors.

The Achenback Child Behavior Checklist was completed by parents at both assessments, and scores on two general behavioral dimensions, "*internalized*" behavior problems and "*externalized*" behavior problems, were computed for each subject. *Externalized* behavioral problems are those that involve a failure of behavioral inhibition, and include physical and verbal aggression and hyperactivity. In contrast, the *internalized* behavior problems scale includes items such as social-withdrawal, anxiety and depression. (These two behavioral dimensions were previously identified with factor analyses applied to data from large standardization samples (Achenbach and Edelbrock 1981).)

The mean scores on these two scales, by sex and diagnostic group, are presented in Figs. 1 and 2. Given previously reported sex differences in the premorbid adjustment of schizophrenic patients, diagnostic group comparisons were conducted within sex. As shown in Fig. 1, high-risk males had the highest scores on the *externalizing* factor at both time points, with increasing elevations with age. The *externalized* behavior problem score for high-risk males was significantly greater at time 2 when compared to same-sex controls [$t(41) = 1.59$, $p = 0.05$]. Although the high-risk females also showed higher scores on this scale at time 2 than did the same-sex comparison group, the elevation was not significant.

The pattern of group differences on the *internalizing* scale was quite different from the *externalizing* scale. Figure 2 illustrates that the high-risk females showed the highest rate of *internalizing* behavior problems, differing significantly from same-sex controls at time 2 [$t(37) = 2.63$, $p < 0.01$]. When compared to same-sex controls, high-risk males showed more *internalized* problems at assessment time 1 [$t(41) = 1.65$, $p < 0.05$] and time 2 [$t(41) = 3.19$, $p < 0.05$].

This pattern of sex and diagnostic group differences in childhood behavior is similar to that reported by John et al. (1982) in their study of subjects from the Danish High-Risk Project. Specifically, they found that high-risk males who

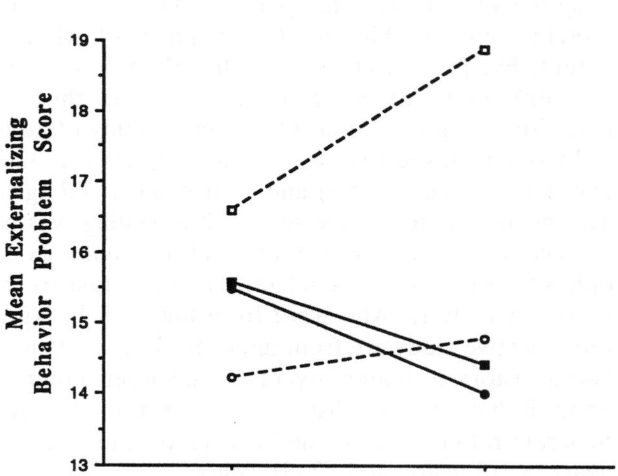

Fig. 1. Mean externalized behavior problem scores by time of assessment, parental diagnostic group, and sex. *Solid Squares*, male offspring of normals; *open squares*, male offspring of schizophrenics; *solid circles*, female offspring of normals; *open circles*, female offspring of schizophrenics

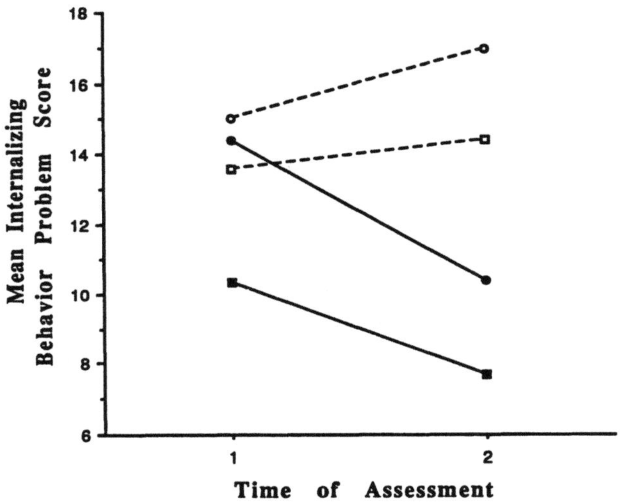

Fig. 2. Mean internalized behavior problem scores by time of assessment, parental diagnostic group, and sex. *Solid squares*, male offspring of normals; *open squares*, male offspring of schizophrenics; *solid circles*, female offspring of normals; *open circles*, female offspring of schizophrenics

later developed schizophrenia were more disruptive *and* socially withdrawn, whereas females were primarily withdrawn.

However, as mentioned above, the behavior problems manifested by the offspring of schizophrenic parents may differ from those typically shown by preschizophrenic children, the majority of whom do not have a disturbed parent. Further, psychiatrically disturbed parents may not be reliable sources of information. In order to cross-validate this pattern, we used the same behavior rating scale in a retrospective study of adult schizophrenic patients.

In our retrospective study, the subjects are 32 schizophrenic patients (25 males and seven females) and their healthy siblings (16 males and 14 females). The nearest-in-age, same-sex healthy sibling was selected as the comparison subject. (In cases where there was no same-sex sibling, the nearest-in-age, opposite sex sibling was selected.) In contrast to the high-risk subjects in the prospective study, who were from low SES backgrounds, the subjects of this retrospective study are from upper SES, intact families. Parents, none of whom had a history of major psychiatric disorder, completed the Achenback Childhood Behavior Checklist for their psychiatrically ill and healthy children. Separate ratings were made for each of four age periods; birth to 4 years, 4–8 years, 8–12 years, and 12–16 years.

The ratings on the *externalizing* scale are shown in Fig. 3. As can be seen, the scores are much lower than those obtained with the high-risk sample, presumably reflecting the lower rate of externalizing problems in children of higher SES families. However, the pattern of sex and diagnostic group differences is highly similar to that found in the prospective study. Preschizophrenic

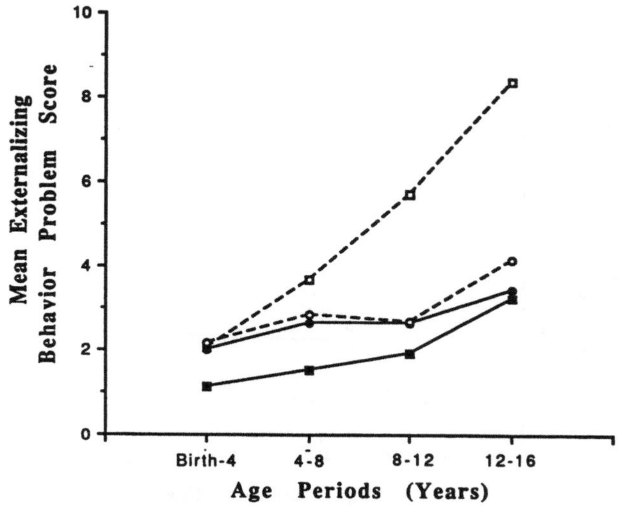

Fig. 3. Mean externalized behavior problem scores by age period, diagnostic group, and sex. *Solid squares*, male sibling controls; *open squares*, preschizophrenic males; *solid circles*, female sibling controls; *open circles*, preschizophrenic females

males are characterized by an increase in *externalizing* behaviors, with significantly higher scores than same-sex healthy siblings at all but the birth to 4-year age period [4–8 years, $t(38) = 2.07$, $p < 0.03$; 8–12 years, $t(38) = 2.43$, $p < 0.01$; 12–16 years, $t(38) = 2.21$, $p < 0.02$]. In contrast, preschizophrenic females did not differ from same-sex healthy siblings in *externalized* scores.

As shown in Fig. 4, preschizophrenic males also manifest an increase in *internalizing* behaviors when compared to healthy siblings of the same sex. The differences were significant at 4–8 [$t(38) = 3.18$, $p < 0.01$], 8–12 [$t(38) = 2.83$, $p < 0.01$] and 12–16 years [$t(38) = 2.97$, $p < 0.01$]. But, overall, preschizophrenic females show the highest rate of *internalizing* behaviors, differing from same-sex controls at 8–12 [$t(19) = 2.33$, $p < 0.02$] and 12–16 years [$t(19) = 2.54$, $p < 0.01$]. These findings mirror those reported in follow-back studies of the school records of schizophrenic patients (Watt 1978; Watt et al. 1979).

The convergent pattern of findings across methods lends support to the assumption that both male and female preschizophrenic subjects manifest childhood behavior problems, with the two sexes showing different profiles. Both the preschizophrenic and high-risk females showed increases in *internalizing*, but not *externalizing*, behavior problems. For the preschizophrenic and high-risk males, there were significant increases in both types of behavior problems. In some respects, it appears that vulnerability to schizophrenia produces an exaggeration of the modal sex differences in childhood behavior problems that have been reported in the literature (Eme 1979). Thus, biological sex appears to moderate the behavioral expression of vulnerability before the onset of clinical symptoms.

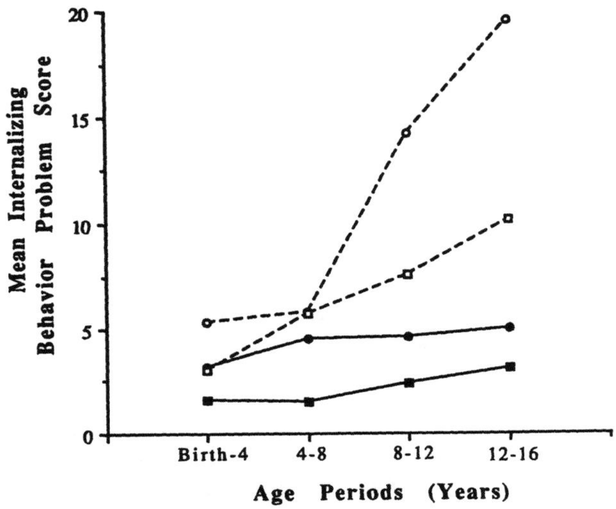

Fig. 4. Mean internalized behavior problem scores by age period, diagnostic group, and sex. *Solid squares*, male sibling controls; *open squares*, preschizophrenic males; *solid circles*, female sibling controls; *open circles*, preschizophrenic females

Facial Expressions of Emotion in Preschizophrenic Children

The assumption that behavioral signs of vulnerability are observable in childhood is also supported by the findings of our recently published study of facial expressions of emotion in preschizophrenic children and their healthy siblings (for details of the study, see Walker et al. 1993). In this study, subjects' facial expressions in childhood home movies were coded. The films spanned from infancy through 15 years of age. Analyses of data on positive ("joy") facial expressions, shown in Fig. 5, revealed a significant interaction between sex and diagnostic group. Compared to same-sex healthy siblings, the preschizophrenic females showed significantly less positive emotion beginning in infancy and extending to adolescence (i.e., at all age periods). The same comparisons for males revealed no effect of diagnostic group. Comparisons within diagnostic group revealed that the preschizophrenic females also showed less positive emotion than preschizophrenic males. With respect to negative emotional expressions (Fig. 6), both male and female preschizophrenics showed elevations relative to their healthy siblings; preschizophrenic males differed significantly from same-sex siblings at the birth to 4, 8–12 and 12–16-year periods, and preschizophrenic females at the 12–16-year period.

Comparison of the age trends in the data on facial expressions and parental ratings of behavior problems reveals some divergence. The diagnostic group differences in facial expressions are most pronounced in the birth to 4-year and 12–16-year periods. In contrast, the behavioral data suggest a linear increase in behavior problems with age. Thus, the diagnostic group differences in facial

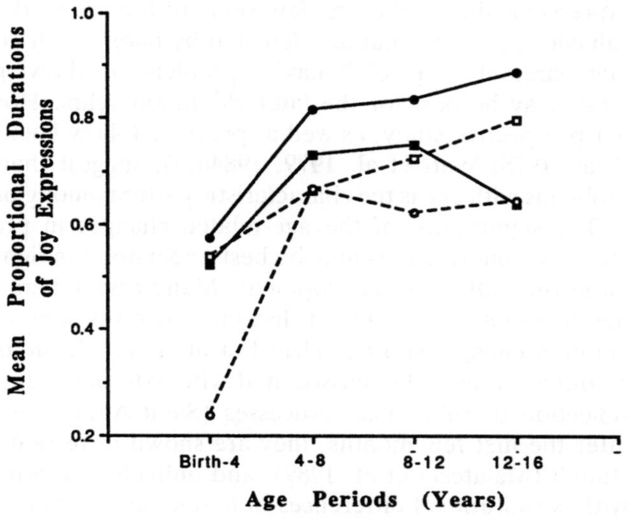

Fig. 5. Rate of positive facial expressions by age period, diagnostic group, and sex. *Solid squares*, male sibling controls; *open squares*, preschizophrenic males; *solid circles*, female sibling controls; *open circles*, preschizophrenic females

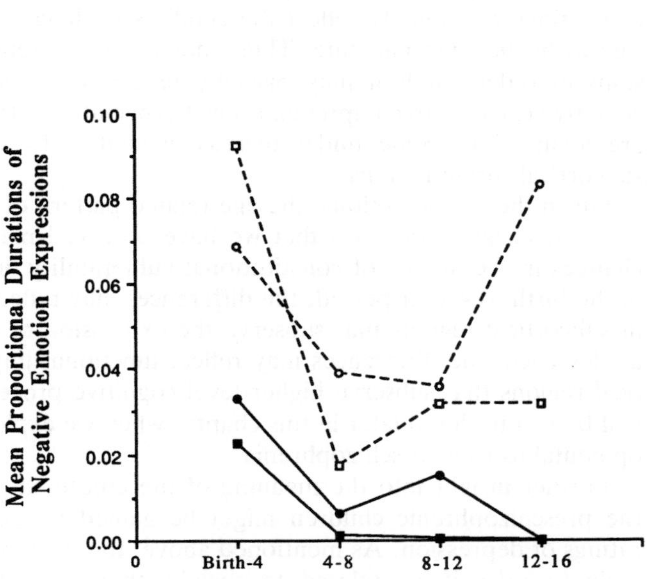

Fig. 6. Rate of negative facial expressions by age period, diagnostic group, and sex. *Solid squares*, male sibling controls; *open squares*, preschizophrenic males; *solid circles*, female sibling controls; *open circles*, preschizophrenic females

expressions during the first few years of life are evidently not accompanied by bahavior problems that are detected by parents. Of course, it might be argued that parental recall of behavior problems is showing a recency effect; their recall may be best for the later childhood years. However, the findings from our prospective study, as well as previous follow-back studies of school records (Watt 1978; Watt et al. 1979, 1984a,b), suggest that an increase in behavior problems with age is the characteristic pattern among preschizophrenic children.

The significance of the age-related changes in preschizophrenic children's affective abnormalities may be best understood in light of the theories and data on normal affective development. Many researchers in the field of emotional development have assumed that the processes subserving facial expressions of emotion change with age (Izard et al. 1984). In infancy, facial expressions of emotion are initially unassociated with external events and may be primarily a reflection of subcortical processes (Saint-Anne Dargassies 1982). Gradually, after the first few months, they are shown in response to social and nonsocial stimuli (Malatesta et al. 1989), and individual differences appear to be linked with constitutional differences in temperament (Goldsmith and Campos 1986). Thus, early facial expressions may be relatively direct manifestations of internal state. But with age, the child acquires knowledge about social display rules and expectations (Cole 1985; Saarni 1979). By about 5 years of age, children begin to manifest the capacity to modify facial expressions in accordance with perceived social expectations. Thus higher level cognitive processes come into play in determining emotional displays. Beyond this development period, facial expressions presumably reflect the child's knowledge of social expectations as well as his/her internal state. Thus, individual differences in the facial expressions of older children may partially reflect differences in the higher-level, cognitive capacity to comprehend social expectations. In sum, facial expressions are assumed to come under increasing control by cortical, as opposed to subcortical, brain regions.

Given these assumptions, the age-related pattern of diagnostic group differences in facial expressions that we have observed may reflect developmental changes in the impact of constitutional vulnerability on behavior. Specifically, in the birth to 4-year period, the differences may reflect functional impairment in subcortical regions that subserve the expression of temperament. Later, in adolescence, the differences may reflect functional impairment in the neocortical regions that subserve higher level cognitive processes. These possibilities will be reconsidered later in this chapter when we explore possible neurodevelopmental aspects of schizophrenia.

Further insight into the meaning of the emotional abnormalities shown by the preschizophrenic children might be gained by examining the behavioral ratings of depression. As mentioned above, the *internalized* behavior problems scale includes items related to social-withdrawal, anxiety, and depression. The depression rating may be especially helpful in elucidating gender differences; it may shed some light on the facial expression data presented above, and help explain how such gender differences influence the expression of schizophrenia.

Based on prior research (Goldstein and Tsuang 1990), we hypothesized that the preschizophrenic females would have higher scores on the depression-related item compared to the preschizophrenic males and the healthy sibling controls. Figure 7 shows mean depression scores as a function of diagnosis, gender, and age.

As shown in Fig. 7, the preschizophrenic females show dramatic increases in depression as a function of age. The three-way interaction (diagnostic group × gender × age) was significant [Wilks Lambda(3,48) = 0.785, $p < 0.01$]. The preschizophrenic females had significantly higher depression scores than the preschizophrenic males at 12–16 years [$F(1,27) = 7.18$, $p < 0.01$]. There was also a (nonsignificant) trend for preschizophrenic females to have higher depression scores than the preschizophrenic males at 8–12 years [$F(1,27) = 3.08$, $p = 0.08$]. The preschizophrenic femals had higher depression scores than their same-sex controls at 8–12 years [$F(1,15) = 6.97$, $p < 0.02$], and at 12–16 years [$F(1,15) = 14.40$, $p < 0.01$]. Notably, the preschizophrenic males had higher depression scores than their same-sex controls at the 8–12-year age period [$F(1,35) = 4.91$, $p < 0.05$], and there was a similar (nonsignificant) trend at 12–16 years [$F(1,35) = 3.62$, $p = 0.06$].

The overall pattern of findings indicate that by ages 12–16, and to some degree ages 8–12, the female preschizophrenic subjects are exhibiting more depression than the male preschizophrenic subjects, as well as the healthy sibling controls. Because the depression score is a component of the internalizing behavior problems dimension, it is not surprising that this pattern of findings is quite similar to the previous findings regarding *internalizing* behavior problems (see Fig. 4). However, the current findings suggest that, within the internalizing behavior problems dimension, the depression rating is contributing

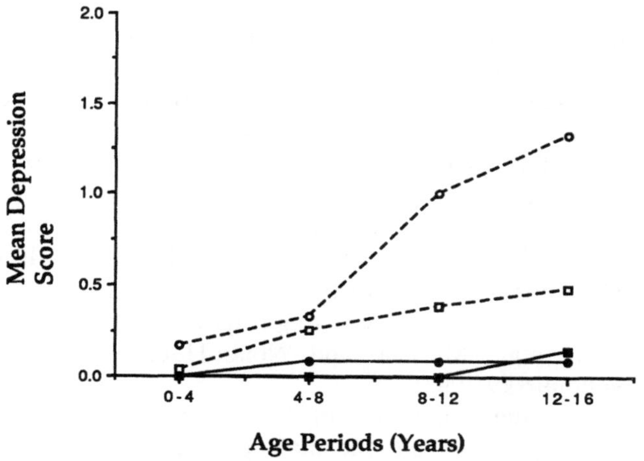

Fig. 7. Mean depression scores by age period, diagnostic group, and sex. *Solid squares*, male sibling controls; *open squares*, preschizophrenic males; *solid circles*, female sibling controls; *open circles*, preschizophrenic females

significantly to the discrimination between the female and male preschizophrenic subjects. Such a finding is consonant with research on adult schizophrenic patients; compared to schizophrenic males, schizophrenic females have more affective symptoms (Goldstein and Tsuang 1990).

Furthermore, the current findings help in understanding the gender differences regarding positive ("joy") facial expressions reported above (see Fig. 5). That the preschizophrenic females displayed significantly less positive emotion than their same-sex controls and the preschizophrenic males may be, in part, a function of their increased levels of depression. Also, it appears that the preschizophrenic females' elevations of negative emotion expressions at 12–16 years, relative to their same-sex controls, closely parallels the increase in depression at this age period (see Figs. 6, 7). Compared to the preschizophrenic females, the scenario for the preschizophrenic males is not as straight forward; the age-related increase in positive emotion expressions in preschizophrenic males is paralleled by an increase in depression. However, the preschizophrenic males do show elevated rates of negative emotion expressions at various age periods, as well as a modest elevation in depression, compared to their same-sex controls. Thus, it appears that the correspondence between facial expression and overt behavior is different for males and females. What may be more helpful in understanding depression in the male preschizophrenic subjects is their high level of *externalized* behavior problems.

Recall that the preschizophrenic males had high rates of externalizing (e.g., aggression and delinquency) behavior problems, although they also showed some *internalizing* problems (withdrawal, anxiety, and depression). The preschizophrenic females had the highest rate of internalizing behavior problems, but much lower *externalizing* scores. It could be hypothesized that the elevated rate of *externalized* behavior problems in preschizophrenic males (see Figs. 1, 3) reflects a "masculine" response style to internal dysphoric states. In contrast, it may be the case that the high rate of *internalized* behavior problems (other than depression, e.g., withdrawal) in preschizophrenic females reflects a "feminine" response style to internal dysphoric states.

These proposed gender-related response styles to depression are speculative, but there is support for such styles in the literature. Nolan-Hoeksema (1987) proposed a response style theory to address the sex differences in depression. According to the theory, women are more likely to engage in ruminative responses to depression and this response style contributes to women's greater susceptibility to prolonged periods of depression. In contrast, men's response style tends to involve distracting responses to depression. Laboratory studies have shown that the distracting response style is effective in reducing depressed mood (e.g., Morrow and Nolan-Hoeksema 1990; Nolan-Hoeksema and Lyubomirsky 1992).

Given the assumptions of the response style theory (Nolan-Hoeksema 1987), we can propose that the male preschizophrenics' response to dysphoria, or effort to avoid it, is reflected in their high rate of *externalized* behavior problems. Given that the *externalized* behavior problems dimension is composed of variables related to aggression and delinquency, it may be these behavioral tendencies that constitute the male preschizophrenic subjects' distracting response style. Thus, the preschizophrenic males' response to dysphoric internal states may be to act out. In contrast, the female preschizophrenics response style corresponds with their high rate of internal behavior problems. In particular, it appears that the female preschizophrenics show increases in withdrawn behaviors compared to the male preschizophrenics.

Sex differences in the rate of depression and in its overt behavioral manifestations may have both short-term and long-term prognostic implications. Presuming, as suggested by Nolen-Hoeksema (1987), that depression in women is accompanied by a ruminative, introspective cognitive style, then women with schizophrenia may be better at self-monitoring and self-regulation of symptoms. Further, the tendency to express dysfunction in the form of *internalized*, rather than *externalized*, behavior problems may deem female schizophrenic patients more clinically manageable than males. Thus, sex differences in schizophrenia may be partially determined by cognitive factors.

Neuromotor Precursors of Schizophrenia

Numerous studies have documented motor deficits in schizophrenic patients (Manschreck 1986) as well as high-risk children (Fish 1987; Hans and Marcus 1991). It thus appears that motor dysfunction is another manifestation of vulnerability to schizophrenia. Further, although deficits in reaction time and psychomotor coordination are apparent across the life span in schizophrenia (Manschreck 1986), movement abnormalities, specifically spontaneous dyskinesias (i.e., nondrug-induced dyskinesias), appear to increase in prevalence and severity with age in schizophrenic patients (Casey and Hansen 1984; Reiter 1926). Thus, elderly schizophrenic patients, especially female patients, manifest spontaneous dyskinesias at a higher rate than age-matched normals.

In a recent report, we presented the results of our study of neuromotor precursors of schizophrenia (E. Walker, T. Savoie and D. Davis 1993, unpublished manuscript). As in the study of facial expressions of emotion, we used childhood home movies as the source of information on motor functions. Preschizophrenic subjects showed a higher rate of neuromotor deficits than the normal and psychiatric comparison subjects. However, the diagnostic group differences were only significant in the birth to 2-year age period. Among the individual items that differentiated the preschizophrenic subjects from controls were: associated reactions, abnormal posturing of the hand, choreoathetoid movements, and hypotonicity. This "dyskinetic syndrome" has been observed to occur in some infants who have sustained pre- or peri-natal injury and tends to improve with age (Knobloch and Pasamanick 1974; Saint-Anne Dargassies 1982; Wolf et al. 1983).

In a previous study of high-risk children, Fish (1987) and Fish et al. (1992) also found an increase in motor abnormalities that was most pronounced in the first 2 years of life. Combined with our findings, it appears that motor dysfunction, including dyskinesias, are more pronounced during infancy in preschizophrenic children. Further, it is of interest to note that spontaneous dyskinesias are also more prevalent and pronounced later in the life course of schizophrenic patients (Casey and Hansen 1984; Reiter 1926). This suggests that CNS maturational changes are altering the expression of dysfunction in the domain of motor behavior.

In contrast to the findings on premorbid behavior problems and facial expressions, we found no sex differences in neuromotor functions. An *absence* of sex differences in neuromotor functions, including limb movement abnormalities, was also reported by Hertzig and Birch (1968) in a large sample of psychotic adolescents, most with schizophrenic disorders. Thus, sex may not be related to neuromotor deficits in preschizophrenic children and very young patients. However, it should be noted that there is evidence that there are sex differences in the rate of spontaneous and neuroleptic-induced oral–facial dyskinesias in middle-aged and elderly schizophrenic patients, with females showing a higher rate (American Psychiatric Association 1992; Casey and Hansen 1984). To date, the neural basis for this is unknown. There is evidence, however, that oral–facial and limb movement abnormalities constitute independent syndromes (Gureje 1988; Kidger et al. 1980) and that they are associated with a relative excess of D_2 and D_1 dopamine (DA) receptor activity, respectively (Waddington 1989). Thus the balance of activation of these two DA receptor subtypes may determine the behavioral consequence.

Cognitive Deficits in Preschizophrenic Subjects

It is well established that schizophrenic patients manifest cognitive deficits relative to normals and patients with affective disorders (for a review, see Aylward et al. 1984). Within the past 2 decades, a number of high-risk and follow-back studies have demonstrated that these deficits precede the onset of the illness. In a meta-analysis of the research data, Aylward et al. (1984) found that there was reliable evidence of performance deficits on standardized intelligence tests in preschizophrenic children beginning in the early elementary school years. Further, these deficits appeared to be more pronounced in preschizophrenic males.

Recent prospective studies have revealed deficits in more subtle aspects of cognitive function, specifically sustained attention (Erlenmeyer-Kimling et al. 1993). It is possible this attentional deficit is responsible for the more global cognitive performance deficits shown by preschizophrenic subjects.

Summary: Life-Course Developmental Changes in Schizophrenia

Our findings, and those of other investigators, suggest several *modal* developmental changes in the premorbid course in schizophrenia: (a) there is a gradual increase in the manifestation of behavior problems with age, although the nature of the problems varies as a function of sex: (b) more subtle abnormalities in affective behavior may be apparent as early as infancy and subsequently become more pronounced in late childhood; (c) neuromotor abnormalities are present throughout the premorbid period, but are more pronounced in infancy; and (d) cognitive performance deficits have been detected by the early elementary school years, although they may exist prior to that period.

Research on the postmorbid course of schizophrenia has revealed significant variability among patients; some show dramatic improvement after their first and only illness episode, whereas other show a chronic, deteriorating course following their first episode. The modal trend is toward improvement over time, especially in the positive symptoms of psychosis (Bridge et al. 1978; McGlashan and Fenton 1993). But, as previously stated, both spontaneous *and* drug-induced dyskinesias tend to worsen as patients age.

Based on these and other findings, the following general conclusions are offered:

1. The neuropathology is congenital and influences behavior long before the onset of clinical symptoms.
2. The neuropathology is in a region of the brain that is involved in multiple behavioral functions; motor, cognitive and socioemotional.
3. The neuropathology is in a region of the brain that is not *essential* to these functions (i.e., it does not eliminate functions), but rather influences their quality.
4. The neuropathology is modulated in its expression by development across the life span, but anatomical and/or biochemical changes that typically occur in late adolescence/early adulthood trigger the expression of psychotic symptoms.

Age and Sex as Moderating Factors in Schizophrenia

Effects of Age and Sex on the CNS

It is well established that both age and biological sex influence CNS characteristics. Although our knowledge of these influences is still rudimentary, the research data are rapidly accumulating. The findings indicate that brain structure and function are affected.

To date, research has documented age-related changes in patterns of brain metabolism, EEG activity, myelination, catecholamine receptors, and circulating neurotransmittors. Table 1 summarizes findings from studies that have examined the relation between age and various aspects of the CNS. Because changes in CNS structure and function tend to be most pronounced in infancy, adolescence and advanced age, the findings are presented within these three developmental periods.

Several points about these findings are worthy of note. First, there is a developmental change in the relative activation of various cortical regions, with the motor region being most activated in infancy, and the frontal and limbic

Table 1. Developmental changes in CNS structure/function

Infancy
– Regional cortical metabolism: sensorimotor regions most activated, prefrontal and limbic areas least activated (Chugant 1992; Chugani and Phelps 1986; Chugani et al. 1987)
– Myelination incomplete in all cortical regions, except sensorimotor (Gibson 1991)
– Cortical synaptic density increasing (Huttenlocher 1979, 1990)
– Cortico-cortical connections minimally developed (Thatcher et al. 1987)
Late adolescence/early adulthood
– Regional cortical metabolism: sensorimotor region least activated, prefrontal and limbic regions most activated (Chugani 1992; Chugani and Phelps 1986; Chugani et al. 1987)
– Myelination of association and limbic cortex completed (Konner 1991; Benes 1989)
– Synaptic pruning completed (Huttenlocher 1979, 1990)
– Cortico-cortical connections refined and enhanced (Thatcher et al. 1987)
– Peak in DA activity (McGeer and McGeer 1981; Ulloa-Aguirre et al. 1990)
Aging
– Regional cortical metabolism: general decline in activation, with sensorimotor regions relatively spared (Obrist 1980; Rapaport 1991)
- Decrease in cortical synaptic density, with relative sparing of motor regions (Huttenlocher 1979, 1990; Winblad et al. 1985)
– Decrease in DA activity (Carlsson and Winblad 1976)
– Decrease in DA receptor density, especially D_2 receptors (Farde et al. 1988; Wong et al. 1984; Morgan et al. 1987)

regions showing increased metabolic activity in early adulthood. Then, with advanced age, there is a decline in cortical metabolic activity, with relative sparing of the motor regions of the cortex. Second, there are significant age-related changes in DA activity and DA receptors. DA receptor density, especially DA D_2 receptors, decline with age, whereas DA activity appears to peak in early adulthood. Third, there is evidence that maturational processes that serve to refine cortico-cortical and limbic circuit interconnections, such as synaptic pruning and the myelination of certain brain regions (especially limbic), extend well into adolescence.

When compared to the literature on CNS development, there has been relatively little research on sex differences in human brain structure and function. The two findings that have been most consistently reported are greater brain weight and greater hypothalamic volume in males than in females (for a review, see Witelson 1991). We are not aware of any reports documenting significant sex differences in brain metabolism, myelination, synaptic density, neuronal density, neurotransmitter receptor density or neurotransmitter concentrations. One recent study showed a trend toward greater serotonin $5HT_2$ receptors on platelets in adolescent girls relative to boys (Biegon and Greuner 1992), but further research is needed to confirm this. However, numerous studies on animals have revealed sex differences in the morphology and function of neurotransmitter systems (Reisert and Pilgrim 1991), so there is reason to believe similar differences exist in humans. Of greatest potential relevance to schizophrenia are the reports of sex differences in DA transmission in animals. Female rats show lower levels of DA in the striatum and hypothalamus when compared to males, but greater release of striatal DA in response to amphetamine or estrogen (for a review, see Reisert and Pilgrim 1991). There is also evidence of sex differences in DA cell morphology, with male rats having larger DA neuron cell bodies in the diencephalon and females having a larger number of DA neurons in the mesencephalon. Recently Häfner et al. (1991) demonstrated that estrogen reduced the behavioral effects of dopamine in rats, and provided evidence to suggest that this effect is mediated by a reduction in DA D_2 receptor sensitivity. These findings are consistent with the proposal that estrogens have an antidopaminergic property that serves to reduce the severity of schizophrenia in women (Seeman and Lang 1990).

Speculations on the Neural Mechanisms Subserving Age and Sex Differences in Schizophrenia

The critical question raised by the findings on age and sex differences in CNS characteristics is which, if any, of these factors are influencing the expression of vulnerability to schizophrenia. In this section, we will speculate on some of these factors, beginning with the assumption that DA overactivity is involved in the neuropathology of schizophrenia.

Dopamine and Neural Circuitry Malfunction in Schizophrenia

Although a thorough discussion of the literature linking DA overactivity with schizophrenia is beyond the scope of this chapter, recent reviews document the continued viability of this assumption (Davis et al. 1991; Weinberger 1987). As knowledge of the DA system has accumulated, the focus on subtypes of DA receptors, especially D_1 and D_2, has intensified. Some researchers have reported an excess of D_2 DA receptors in the striatum in schizophrenic patients (Cross et al. 1981; Hess et al. 1987; Toru et al. 1988). Others have reported an increase in the ratio of D_2 to D_1 receptors in this region of the brain (Joyce et al. 1988). Thus, there is evidence that the functional DA overactivation may be mediated by abnormalities in receptors.

Several recent hypotheses about the neural substrate for schizophrenia have proposed DA overactivation in limbic circuitry (Csernansky et al. 1991; Swerdlow and Koob 1987; Weinberger 1987), and Davis et al. (1991) suggest this is specifically due to DA D_2 overactivation. Swerdlow and Koob offer one of the most detailed models, proposing that the excess activation of DA receptors in the "limbic" striatum results in a phasic interruption of transmission in the thalamocortical feedback loop. They suggested that this culminates in an inability to filter inappropriate cognitive or emotional processes and to maintain attentional focus. The behavioral consequence is psychotic symptomatology.

As Swerdlow and Koob (1987) point out, a functional excess of DA activity in the striatum has also been posited as the neural substrate for dyskinesias. Recently, Alexander et al. (1990) proposed, more specifically, that limb dyskinesias involve a relative overactivation of the inhibitory DA pathway to the striatum which is part of the "motor circuit" linking the striatum with the motor cortex. It has also been shown that the inhibitory effects of DA on striatal neurons are primarily mediated by D_2 DA receptors (Gerfen 1992). Thus, a relative excess of D_2 receptor activity has been implicated in the motor circuit as the basis for dyskinesias, and in the limbic circuit as the basis for psychotic symptoms. In both cases, the locus of this excess receptor activity is presumed to be the striatum, dorsal and ventral, respectively.

Neurodevelopment of Schizophrenia

Turning back to the issue of developmental changes in the behavioral expression of schizophrenia, we now consider how normal maturational processes in the CNS might be implicated. The available data on postnatal CNS maturation suggest that cortical regions are sites of significant change. Further, although cortical regions vary in the functions they subserve, most appear to have important connections with striatal regions. Thus, it has been proposed that there are functionally segregated "basal ganglia-thalamocortical circuits" that constitute parallel substrates for motor, oculomotor, prefrontal, and limbic cortical functions (Alexander et al. 1990). Given the developmental changes in the activation of these cortical regions, concomitant changes in the corresponding neural circuitry would be expected.

If, as the developmental data suggest, the motor circuit is maximally activated (relative to other cortical circuits) in infancy, then abnormalities in DA transmission in the striatum will be most likely to be expressed in the domain of motor functions during this period. This may explain the relative predominance of motor abnormalities in preschizophrenic infants, and the gradual tendency for these abnormalities to become less pronounced as the child gets older. Conversely, the gradual increase in observable behavior problems as the preschizophrenic child matures may reflect the maturation of cortical regions, frontal and limbic, that play a role in interpersonal behaviors. Subsequently, in adolescence/early adulthood, it appears that the limbic circuit attains maturity. This may account for the modal age-at-onset of psychotic symptoms in young adulthood. Finally, with advanced age, maturational changes in the brain may reduce limbic circuitry activation, thereby reducing the severity of positive psychotic symptoms. Movement abnormalities, in contrast, may become more pronounced in old age due to the relative sparing of motor regions of the cortex and the reduction of modulating influences on movement from other cortical regions that appear to undergo greater deterioration.

Sex Differences

As mentioned earlier, previous research suggests that schizophrenia in females has a more benign course and is more responsive to standard neuroleptic treatment. It has been suggested that this may be due to DA antagonistic effects of estrogen. This notion is consistent with the assumption that excess D_2 receptor activation occurs in schizophrenia, and with the finding that estrogen reduces D_2 receptor sensitivity in animals. Thus, as suggested by Häfner et al. (1992), the later illness onset and more favorable prognosis of schizophrenia in females may reflect the effects of estrogen on D_2 receptor activity.

However, some of the findings on sex differences in schizophrenia are not readily explained by the anti-dopaminergic effects of estrogen. If spontaneous limb movement abnormalities are subserved by excess D_2 receptor activation in the motor circuit, then postpubescent female schizophrenics should also show a lower rate of spontaneous limb dyskinesias than their male counterparts. To date, there is no evidence to support this, although few studies have addressed the issue.

As previously mentioned, there is evidence suggesting that oral-facial dyskinesias are due to a relative excess in DA D_1 activation, and that this type of dyskinesia increases with age and is more pronounced in female schizophrenic patients. These findings are consistent with the assumption that estrogens have a selective effect on D_2 receptors, reducing their sensitivity (Häfner et al. 1992). However, if estrogens are responsible for sex differences in the occurrence of oral-facial dyskinesia, we would expect to see significant age-related changes in rates of dyskinesias in female patients; in other words, these dyskinesias should increase in postpubescence and decrease with menopause. There are no data demonstrating such an effect.

Also, the findings of sex differences in affective symptoms and premorbid behavioral problems are not readily explained on the basis of estrogen's effect on DA transmission. It may be that some of the observed sex differences in schizophrenia are best explained on the basis of nonhormonal factors. As Reisert and Pilgrim (1991) point out, some sexual dimorphisms in the nervous systems of animals appear to be under primary genetic control and are not influenced by sex steroids. Presuming this is also true of humans, the behavioral differences that we, and other research groups, have observed in very young preschizophrenic children may reflect the interaction between sex-linked differences in brain morphology and neuropathological processes. Thus, sex differences in the behavioral expression of schizophrenia may be a consequence of both hormonal *and* neuroanatomical factors, with hormonal influence predominating between pubescence and menopause.

Conclusion

Our understanding of the longitudinal course of schizophrenia has expanded rapidly. However, many important questions about age and sex differences in the expression of schizophrenia remain unanswered. For example, how closely tied are changes in gonadal steroids and changes in symptom expression? Because both age and sex have implications for CNS functioning, elucidating their impacts on the expression of dysfunction in schizophrenia may inform us about the neural substrate for the disorder. At the present time, our efforts are hampered by the paucity of data on normative sex and age effects on the CNS. Technological advances in the study of brain structure and function will undoubtedly enhance our knowledge of these effects. As this knowledge base increases, opportunities for refining our models of neurodevelopmental processes in schizophrenia will also increase. In the interim, findings on life-course changes in the behavioral expression of schizophrenia must be taken into consideration as we develop models of the underlying neuropathology.

Our findings also indicate a need for further research on the inter-relations among deficits in schizophrenia. In particular, we need to know more about the relations between various forms of motor dysfunction and clinical symptoms, both cross-sectionally and longitudinally. The literature suggests that there are subtypes of movement abnormality that are the result of different neurotransmission imbalances. If this is confirmed, then the study of motor dysfunction in schizophrenia may prove highly informative with respect to neuropathological processes. This holds for both spontaneous and neuroleptic-induced motor dysfunction.

Acknowledgement. This work was supported by a research grant (MH46496) and a Research Scientist Development Award (MH00876) to the first author from the National Institute of Mental Health.

References

Achenbach TM (1982) Child behavior checklist. University of Vermont, Burlington, Vermont

Achenbach TM, Edelbrock CS (1981) Behavioral problems and competencies reported by parents of normal and disturbed children aged 4 through 16. Monogr Soc Res Child Dev 46(1):188

Alexander GE, Crutcher MD, DeLong MR (1990) Basal ganglia-thalamocortical circuits: parallel substrates for motor, oculomotor, "prefrontal" and "limbic" functions. Prog Brain Res 85:119–145

American Psychiatric Association (1992) Tardive dyskinesia: a task force of the American Psychiatric Association. American Psychiatric Association, Washington DC

Aylward E, Walker E, Bettes B (1984) Intelligence in schizophrenia: meta-Analysis of the research. Schizophr Bull 10:430–459

Benes FM (1989) Myelination of cortical hippocampal relays during late adolescence. Schizophr Bull 15:585–593

Biegon A, Greuner N (1992) Age-related changes in serotonin $5HT_2$ receptors on human blood platelets. Psychopharmacology 108:210–212

Bridge TP, Cannon HE, Wyatt RJ (1978) Burned-out schizophrenia: evidence for age effects on schizophrenic symptomatology. J Gerontol 33:835–839

Buchsbaum MS (1990) The frontal lobes, basal ganglia, and temporal lobes as sites for schizophrenia. Schizophr Bull 16:379–389

Carlsson A, Winblad B (1976) Influence of age and time interval between death and autopsy on dopamine and 3-methoxytyramine levels in human basal ganglia. J Neural Transm 38:271–276

Carpenter WT, Buchanan RW, Kirkpatrick B, Tamminga C, Thaker G, Breier A (1992) The neuroanatomy of the deficit syndrome (abstract). Schizophr Res 6:166

Casey DE, Hansen TE (1984). Spontaneous dyskinesias. In: Jeste, DV, Wyatt RJ (eds) Neuropsychiatric movement disorder. American Psychiatric Press, Washington DC, pp 68–95

Chugani HT (1992) Development of regional brain glucose metabolism in relation to behavior and plasticity. In: Dawson G, Fischer K (eds) Human behavior and the developing brain. Guilford, New York
Chugani HT, Phelps ME (1986) Maturational changes in cerebral function in infants determined by ¹8FDG positron emission tomography. Science 231:840–843
Chugani HT, Phelps ME, Mazziotta JC (1987) Positron emission tomography study of human brain functional development. Ann Neurol 22:487–497
Cole PM (1985) Display rules and the socialization of affective displays. In: Sivin G (ed) The development of expressive behavior. Academic, New York, pp 269–290
Cross AJ, Crow TJ, Owen F (1981) 3H-Flupenthixol binding in post-mortem brains of schizophrenics: evidence for a selective increase in dopamine D2 receptors. Psychopharmacology 74:122–124
Csernansky JG, Murphy GM, Faustman WO (1991) Limbic/mesolimbic connections and the pathogenesis of schizophrenia. Soc Biol Psychiatry 30:383–400
Davis, KL, Kahn, RS, Ko, G, Davidson, M (1991) Dopamine in schizophrenia: a review and reconceptualization. Am J Psychiatry 148:1474–1486
Eme R (1979) Sex differences in childhood psychopathology. Psychol Bull 86:574–592
Erlenmeyer-Kimling L, Cornblatt BA, Rock D, Roberts S, Bell M, West A (1993) The New York High-Risk Project: anhedonia, attentional deviance, and psychopathology. Schizophr Bull 19:141–153
Farde L, Wiesel FA, Nordstrom AL, Sedvall G (1988) PET examination of human D-1 and D-2 characteristics. Psychopharmacology 96[Suppl]:79
Feinberg, I (1982) Schizophrenia: caused by a fault in programmed synaptic elimination during adolescence? J Psychiatr Res 17:319–334
Fish B (1987) Infant predictors of the longitudinal course of schizophrenic development. Schizophr Bull 13:395–410
Fish B, Marcus J, Hans SL, Auerbach JG, Perdue S (1992) Infants at risk for schizophrenia: sequelae of a genetic neurointegrative defect. Arch Gen Psychiatry 49:221–235
Gerfen CR (1992) The neostriatal mosaic: multiple levels of compartmental organization in the basal ganglia. Annu Rev Neurosci 15:285–320
Gibson K (1991) Myelination and behavioral development. In: Gibson K, Peterson A (eds) Brain maturation and cognitive development. De Gruyter, New York, pp 29–64
Goldsmith H, Campos J (1986) Fundamental issues in the study of early temperament. In: Lamb M (ed) Advances in developmental psychology, vol 4, chap 6. Erlbaum, Hillsdale
Goldstein JM, Tsuang MT (1990) Gender and schizophrenia: an introduction and synthesis of findings. Schizophr Bull 16:179–184
Gottesman II (1991) Schizophrenia genesis. Freeman, New York
Gureje O (1988) Topographic subtypes of tardive dyskinesia in schizophrenic patients aged less than 60 years. J Neurol Neurosurg Psychiatry 51:1525–1530
Häfner H, Behrens A, de Vry J, Gattaz WF, Loffler W, Maurer K, Riecher-Rossler A (1991) Warum erkranken Frauen spater an Schizophrenie? Erhohung der Vulnerabilitatsschwelle durch Ostrogen. Nervenheilkunde 10:154–163
Häfner H, Riecher-Rossler A, Maurer K, Fatkenheuer B, Loffler W (1992) First onset and early symptomatology of schizophrenia: a chapter of epidemiological and neurobiological research into age and sex differences. Eur Arch Psychiatry Clin Neurosci 242:109–118
Hans S, Marcus J (1991) Neurobehavioral development of infants at risk for schizophrenia. In: Walker EF (ed) Schizophrenia: a life-course developmental perspective. Academic, New York, pp 35–53
Hertzig ME, Birch HG (1968) Neurologic organization in psychiatrically disturbed adolescents: a comparative consideration of sex differences. Arch Gen Psychiatry 19:528–537
Hess EJ, Bracha HS, Kleinman JE, Creese I (1987) Dopamine receptor subtype imbalance in schizophrenia. Life Sci 40:1487–1497
Huttenlocher PR (1979) Synaptic density in human frontal cortex – developmental changes and effects of aging. Brain Res 163:195–205

Huttenlocher PR (1990) Morphometric study of human cerebral cortex development. Neuropsychologia 28:517–527
Izard E, Kagan J, Zajonc R (1984) Emotion, attention and temperament. Emotion, cognition and behavior. Cambridge University Press, Cambridge
John RS, Mednick SA, Schulsinger F (1982) Teacher reports as predictor of schizophrenia and borderline schizophrenia. J Abnorm Psychol 91:399–413
Joyce JN, Lexow N, Bird E, Winokur A (1988) Organization of dopamine D1 and D2 receptors in human striatum: Receptor autoradiographic studies in Huntington's disease and schizophrenia. Synapse 2:546–557
Kidger T, Barnes TR, Trauer T, Taylor PJ (1980) Subsyndromes of tardive dyskinesia. Psychol Med 10:513–520
Kinney DK, Woods BT, Yurgelun-Todd D (1986) Neurologic abnormalities in schizophrenic patients and their families: II. Neurologic and psychiatric findings in relatives. Arch Gen Psychiatry 43:665–668
Knobloch H, Pasamanick B (1974) Developmental diagnosis. Harper and Row, New York
Konner M (1991) Universals of behavioral development in relation to brain myelination. In: Gibson K, Peterson A (eds) Brain maturation and cognitive development. De Gruyter, New York, pp 181–224
Kornhuber J, Konradi C, Mack-Burkhardt F, Reiderer P, Heinsen H, Beckmann H (1989) Ontogenesis of monoamine oxidase-A and -B in the human brain frontal cortex. Brain Res 499:81–86
Lewine RRJ (1988) Gender and schizophrenia. In: Nasrallah HA (ed) Handbook of schizophrenia, vol 3. Elsevier, Amsterdam, pp 379–397
Loranger A (1984) Sex differences in age at onset of schizophrenia. Arch Gen Psychiatry 41:157–161
Malatesta CZ, Haviland JM (1982) Learning display rules: the socialization of emotion expression in infancy. Child Development 53:991–1003
Malatesta CZ, Culver C, Tesman JR, Shepard B (1989) The development of emotion expression during the first two years of life. Monogr Soc Res Child Dev 54:1
Manschreck T (1986) Motor abnormalities in schizophrenia. In: Nasrallah H, Weinberger DR (eds) Handbook of schizophrenia, vol 1: the neurology of schizophrenia. Elsevier, New York, pp 65–96
Marcus J (1974) Cerebral functioning in offspring of schizophrenics: a possible genetic factor. Int J Ment Health 3:57–73
Marcus J, Hans SL, Mednick S, Schulsinger F, Michelsen N (1985) Neurological dysfunctioning in offspring of schizophrenics in Israel and Denmark: a replication analysis. Arch Gen Psychiatry 42:753–761
McGeer PL, McGeer EG (1981) Neurotransmitters in the aging brain. In: Darrison AM, Thompson RH (eds) The molecular basis of neuropathology. Arnold, London, pp 631–648
McGlashan TH, Fenton WS (1993) Subtype progression and pathophysiologic deterioration in early schizophrenia. Schizophr Bull 19:71–84
Morgan DG, May PC, Finch CE (1987) Dopamine and serotonin systems in human and rodent brain: effects of age and neurodegenerative disease. J Am Geriatr Soc 35:334–345
Morrow J, Nolan-Hoeksema S (1990) Effects of responses to depression on the remediation of depressive affect. J Pers Soc Psychol 58:519–527
Nolan-Hoeksema S (1987) Sex differences in unipolar depression: evidence and theory. Psychol Bull 101:259–282
Nolan-Hoeksema S, Lyubomirsky S (1993) Self-perpetuating properties of dysphoric rumination. J Pers Soc Psychol 65(2):339–349
Obrist ED (1980) Cerebral blood flow and EEG changes associated with aging and dementia. In: Busse EW, Blazer DG (eds) Handbood of Geriatric Psychiatry. Van Nostrand Reinhold, New York, pp 83–101
Offord DR, Cross LA (1969) Behavioral antecedents of adult schizophrenia: a review. Arch Gen Psychiatry 21:267–284

Rapaport SI (1991) Positron emission tomography in Alzheimers disease in relation to disease pathogenesis: a critical review. Cerebrovasc Brain Metab Rev 3:297–335
Reisert I, Pilgrim C (1991) Sexual differentiation of monoaminergic neurons: genetic or epigenetic? Trends in Neurosci 14:468–473
Reiter PJ (1926) Extrapyramidal motor-disturbances in dementia praecox. Acta Psychiatr Neurol Scand 1:287–309
Rinne JO, Lonnberg P, Marjamaki P (1990) Age-dependent decline in human brain dopamine D_1 and D_2 receptors. Brain Res 508:349–352
Saarni C (1979) Children's understanding of display rules for expressive behavior. Dev Psychol 15:424–429
Saint-Anne Dargassies S (1982) The neuromotor and psychoaffective development of the infant. Elsevier, Amsterdam
Seeman MV, Lang M (1990) The role of estrogens in schizophrenia gender differences. Schizophr Bull 16:185–194
Suzuki H, Shimohira M, Iwakawa Y, Nagatsu T (1990) Developmental change of dopamine B-hydroxylase activity in cerebrospinal fluid of epileptic and non-epileptic children. J Neural Transm 80:225–230
Swerdlow NR, Koob GF (1987) Dopamine, schizophrenia, mania, and depression: toward a unified hypothesis of cortico-striato-pallido-thalamic function. Behav Brain Sci 10:197–245
Thatcher RW, Walker RA, Giudice S (1987) Human cerebral hemispheres develop at different rates and ages. Science 236:1110–1113
Toru M, Watanabe S, Shibuya H, Nishikawa T, Noda K, Mitsushio H, Ichikawa H, Kurumaji A, Takaskhima M, Mataga N, Ogawa A (1988) Neurotransmitters, receptors and neuropeptides in post-mortem brains of chronic schizophrenic patients. Acta Psychiatr Scand 78:121–137
Ulloa-Aguirre A, Tarraso J, Mendez JP, Garza-Flores J, Diaz-Sanchez V, Perez-Palacios G (1990) Changes in the responsiveness of prolactin secretion to dopaminergic blockade and TRH stimulation throughout sexual maturation in men. Psychoneuroendocrinology 15:279–286
Waddington JL (1989) Functional interactions between D-1 and D-2 dopamine receptor systems: their role in the regulation of psychomotor behaviour, putative mechanisms, and clinical relevance. J Psychopharmacol 3:54–63
Walker E (ed) (1991) Schizophrenia: a life-course developmental perspective. Academic, New York
Walker E, Emory E (1983) Infants at risk for psychopathology: offspring of schizophrenic parents. Child Dev 54:1269–1285
Walker E, Downey G, Bergman A (1989a) The effects of parental psychopathology and maltreatment on child behavior: a test of the diathesis-stress model. Child Dev 60:15–24
Walker E, Downey G, Nightingale N (1989b) The nonorthogonal nature of risk factors. J Prim Prevent 9:143–163
Walker E, Davis D, Gottlieb L (1991) Charting the developmental trajectories leading to psychopathology. In: Cicchetti D, Toth S (eds) Developmental psychopathology, vol 3. Rochester University Press, Rochester
Watt NF (1978) Patterns of childhood social development in adult schizophrenics. Arch Gen Psychiatry 36:160–165
Watt NF, Fryer JH, Lewine RJ, Prentky RA (1979) Toward longitudinal conceptions of psychiatric disorder. In: Maher BA (ed) Progress in experimental personality research. Academic, New York
Watt NF, Anthony EJ, Wynne L, Rolf TE (1984a) Children at risk for schizophrenia. Cambridge University Press, London
Watt NF, Grubb TW, Erlenmeyer-Kimling L (1984b) Social, emotional, and intellectual behavior at school among children at high risk for schizophrenia. In: Watt NF, Anthony J, Wynne LC, Rolf JE (eds) Children at risk for schizophrenia. Cambridge University Press, London, pp 212–226
Weinberger D (1987) Implications of normal brain development for the pathogenesis of schizophrenia. Arch Gen Psychiatry 44:660–669

Willemse J (1986) Benign idiopathic dystonia with onset in the first year of life. Dev Med Child Neurol 28:355–360.
Winblad B, Hardy J, Backman L, Nilsson LG (1985) Memory function and brain biochemistry in normal aging and in senile dementia. Ann NY Acad Sci 444:255–268
Witelson SF (1991) Neural sexual mosaicism: sexual differentiation of the human tempro-parietal region for functional asymmetry. Psychoneuroendocrinology 16:131–153
Wolf PH, Gunnoe CE, Cohen C (1983) Associated movements as a measure of developmental age. Dev Med Child Neurol 25:417–429
Wong DF, Wagner HN Jr, Dannals RF, Links JM, Frost JJ, Ravert HT, Wilson AA, Rosenbaum AE, Gjedde A, Douglass KH, Petroms KA, Folstein JD, Toung JKT, Burns HD, Kuhar MJ (1984) Effects of age on dopamine and serotonin receptors measured by positron emission tomography in the living human brain. Science 226:1393–1396

Onset and Early Course of Schizophrenia

H. Häfner, K. Maurer, W. Löffler, S. Bustamante, W. an der Heiden, A. Riecher-Rössler, and B. Nowotny

Introduction

When, with what symptoms and how schizophrenia really begins, and how this beginning can be distinguished from already existing deficits or antecedents are all interesting questions. They arise following the papers of Crow et al. and Walker et al. on the antecedents of psychosis. Kraepelin (1909) and Bleuler (1911) had already observed that the disease rarely begins without non-specific symptoms in advance. Kraepelin (1909) mentioned, for instance, "small changes in emotional life, irritability, loss of interest, overactivity and poor concentration." DeLisi et al. (1986) and Huber and Gross (1989) recently expressed the expectation that the elucidation of the prodromal phase might provide a new understanding of the psychopathology and etiology of schizophrenia.

Nevertheless, this effort faces certain problems: prodromal symptoms generally occur prior to the first psychiatric contact and, being non-specific, cannot be assessed by prospective population studies. Therefore, they are beyond direct clinical and epidemiological observation.

Attempts at Assessing the Prodromal Symptoms of Relapse

Prodromal symptoms gained new significance when intermittent neuroleptic therapies were introduced. Early insight in prodromal symptoms should enable the patient to anticipate an imminent relapse and to resort to coping or to therapeutic intervention in time. As a matter of fact, in a previous study (Thurm and Häfner 1987) we found that early signs of relapse could be identified by 78% of the 37 German outpatients examined by us who had a mean duration of disease of 7 years from first admission. Kumar et al. (1989), in India, using the same methodology, came up with similar results (87% of the outpatients).

Several authors have retrospectively compared the patients' and partly also their relatives' statements concerning prodromal symptoms prior to relapse and periods free from relapse for the same patients or for matched patient groups without relapse (Herz et al. 1982; Marder et al. 1984, 1987; Subotnik and

Central Institute of Mental Health, P.O. Box 12 21 20, 68072 Mannheim, Germany

H. Häfner and W.F. Gattaz (Eds.)
Search for the Causes of Schizophrenia, Vol. III

Nuechterlein 1988). They found significant differences proving the association between prodromal symptoms and subsequent relapse and their high frequency of up to 90% for psychotic episodes (Herz and Melville 1980). The study of the prodromal symptoms of relapse in schizophrenic patients undergoing continuous treatment opens up the possibility of a prospective and standardized assessment of prodromi and of the early course of later psychotic episodes.

When Does Schizophrenia Begin?

The elucidation of the prodromal phase of the first schizophrenic episode aims at further objectives. Unlike the study of relapse, it has some advantages: at the beginning of the disease the influence of environmental factors on symptomatology and course is presumably small, and at this stage the contamination of prodromal symptoms of the psychosis with neuroleptic side effects of the therapy is still likely to play a minor role – 60% of our first-episode sample (n = 232) had not yet received any pharmacotherapy and 80% had not received neuroleptic drugs until first admission. The greatest disadvantage as compared with the study of relapse, however, is the inevitability of retrospective assessment (Häfner and Maurer 1993a).

Methodological Requirements

Three difficulties must be overcome: (a) the questionable insight of acute schizophrenics in the disorder; (b) the memory deficits in retrospective assessment; and (c) the distinction between non-specific symptoms of onset and premorbid traits of schizophrenia or symptoms of other diseases.

As the instruments available were not adequate for this purpose, we developed a structured interview, the Instrument for the Retrospective Assessment of the Onset of Schizophrenia (IRAOS; Häfner et al. 1990, 1992) on the basis of internationally approved instruments. It is employed for the assessment of prodromal, negative and positive symptoms, environmental and individual social changes, and treatment data. The timing is performed by means of a time grid using anchor events. The information is derived from patients and independently from significant others as well as from available case notes.

Sample

Our study sample comprised 232 patients with a first psychotic episode (84% of a sample of 276 first admissions interviewed by PSE within the first 2 weeks; 12–59 years of age) with a diagnosis of schizophrenia (broad definition: ICD-9 295, 297, 298.3, 298.4) from a population of approximately 1.5 million. For a detailed description and analysis of the sample see Häfner et al. (1993).

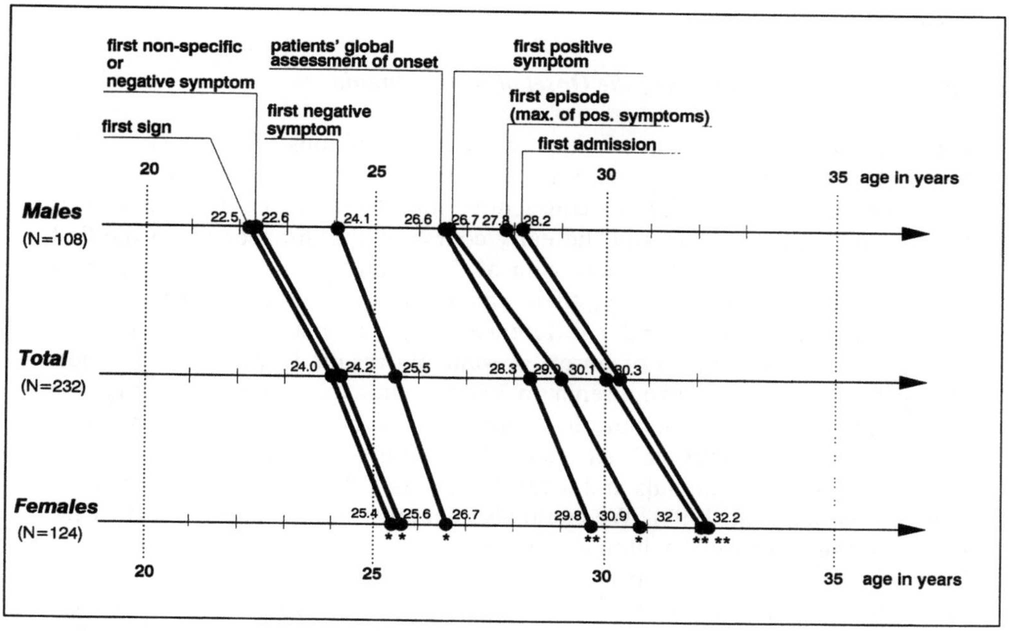

Fig. 1. Mean age values at seven definitions of onset until first admission – first episode sample of broad-definition schizophrenia (n = 232). * = $p \leqslant 0.05$; ** = $p \geqslant 0.01$

Figure 1 presents the age at first admission and six definitions of disease onset for both males and females based on the IRAOS interviews of the patients. If a more restricted definition of the diagnosis is used such as ICD 295 or the Catego diagnosis: PSE S+, then no essential changes are observed either in age at first onset or in the significant difference between the sexes.

Operationalization of the Phases of Onset

To reduce the inevitable risk of a contamination of prodromal symptoms by schizophrenia-independent symptoms, the three earliest definitions were operationalized according to a phase model:

1. *Non-specific symptoms* were included only if continuously present from their occurrence until index admission.
2. *Negative symptoms* were included only if they were either continuous or recurring.
3. *Positive symptoms* were considered in any case, even if they occurred only once.

The *first episode* was operationalized as the point when the *maximum number of positive symptoms* was reached *for the first time* irrespective of the time of first admission, which is also determined by disease-independent factors.

Results

Age at Different Definitions of the Onset of Schizophrenia

Figure 1 shows the mean age values seven for definitions of onset in females and males from the first sign of a mental disorder until first admission. These definitions can also be taken as consecutive milestones in the development of the first psychotic episode and the early course of the disorder up to the first hospitalization. The parallel succession of the mean age values for both sexes supports, though only roughly, their validity down to the earliest and less precise definitions. The gender difference in age at first onset, which is significant for each definition of onset, presumably results from a raised vulnerability threshold for schizophrenia in women due to estrogen reducing the sensitivity of D_2 receptors in the brain. We were able to demonstrate a structural effect of chronic estrogen substitution in experiments with ovariectomized newborn animals and a functional effect beginning later with adult animal (Häfner et al. 1991). We could also demonstrate the applicability of the neurohormonal model to human schizophrenics by showing the correlation between measures of schizophrenic, paranoid, and unspecific symptoms on the one hand and estrogen levels in the course of the menstrual cycle of acutely schizophrenic women on the other hand (Riecher-Rössler et al. 1992). Depressive symptom scores did not change in association with the level of estrogen secretion, neither in the index group of acute schizophrenic women nor in a parallelized non-schizophrenic control group of acutely hospitalized female patients mainly with a diagnosis of unipolar depressive disorder (Riecher-Rössler et al. 1994).

We compared the mean age values for five definitions of onset (except for the maximum of positive symptoms and for first admission, which are not reasonable to compare) in all cases where they were available from the three independently rated sources – patient IRAOS, close-relative IRAOS, and case-notes IRAOS (Table 1). While all the mean age values based on the estimates of patients and relatives showed no significant differences – varying between 0 and 0.7 years – and the global estimates of age at first onset and age at the first positive symptom were highly consistent across all three sources, the age at the earliest sign of a disorder and, above all, at the earliest negative symptom were recorded in the case notes with some delay. The difference from the patients' estimates was 1.4 and 1.6 years and from the relatives' estimates 0.8 and 0.9 years (Table 1). This result is plausible on the grounds that the soundest changes – e.g., the first positive symptom or the onset of the psychotic episode – may well be observed quickly by others, whereas the less sound negative or non-specific symptoms, which may be recognized almost at the same time by the patient's close relatives, are generally registered in the records somewhat later due at least partly to more distant or later observation.

Table 1. Mean age values and standard deviations in years at different points prior to first hospital admission: comparison of IRAOS data from patient and close-relative interviews and from the analysis of case notes

		Patients		Close relatives		Case notes		Significance
	(*n*)	Age	Sd	Age	Sd	Age	Sd	
Age at earliest sign of a schizophrenic disorder	140	23.3	10.7	23.9	9.9	24.8	11.0	*
Age at first negative or non-specific symptom	137	23.7	10.8	24.3	9.9	25.0	11.1	ns
Age at earliest negative symptom	117	24.1	9.7	24.7	9.1	25.6	10.1	***
Age at subjective assessment of onset of the disorder	121	28.0	9.7	28.0	9.6	28.0	9.3	ns
Age at earliest positive symptom	109	28.6	10.5	28.7	10.3	28.9	10.3	ns

* = $p < 0.1$.
** = $p < 0.05$.
*** = $p < 0.01$.
Operationalizations of onset are based on a phase model. Symptoms taken into account: non-specific symptoms only if continuously present until index admission, negative symptoms only if continuous or recurrent, positive symptoms in any case, even if they occurred only once.

Age Distribution of Onset

When the distribution of onset for males and females together is depicted by 5-year age groups across the entire range until the age of 59 for three definitions, namely earliest sign of the disease, first positive symptom, and first episode [gender differences in the type of onset and the early symptomatology have been analyzed elsewhere Häfner et al. 1993], then the predominantly early beginning of schizophrenia becomes obvious (Fig. 2). The maximus of positive symptoms, i.e., the culmination of the first psychotic episode, is reached before the age of 30 in 59% and before the age of 20 in 9% of cases.

Referring to the earliest sign of the disease, the proportion of patients afflicted before the age of 20 increases to 41%. In ten cases (4%) the disorder began before the age of 10. When the distribution of age at first onset is depicted only for the onset of the first sign of the disorder (Fig. 3), another important difference between the sexes emerges: schizophrenia starts earlier on average in males and reaches its peak between 15 and 25 years of age, whereas in females it occurs only between 20 and 30 years of age and shows a less steep curve. In addition, females exhibit a second peak in the age group 45–49 years, which differs at the 0.05 level from the number of first episodes in males in the same age group as well as from the value expected by the age trend.

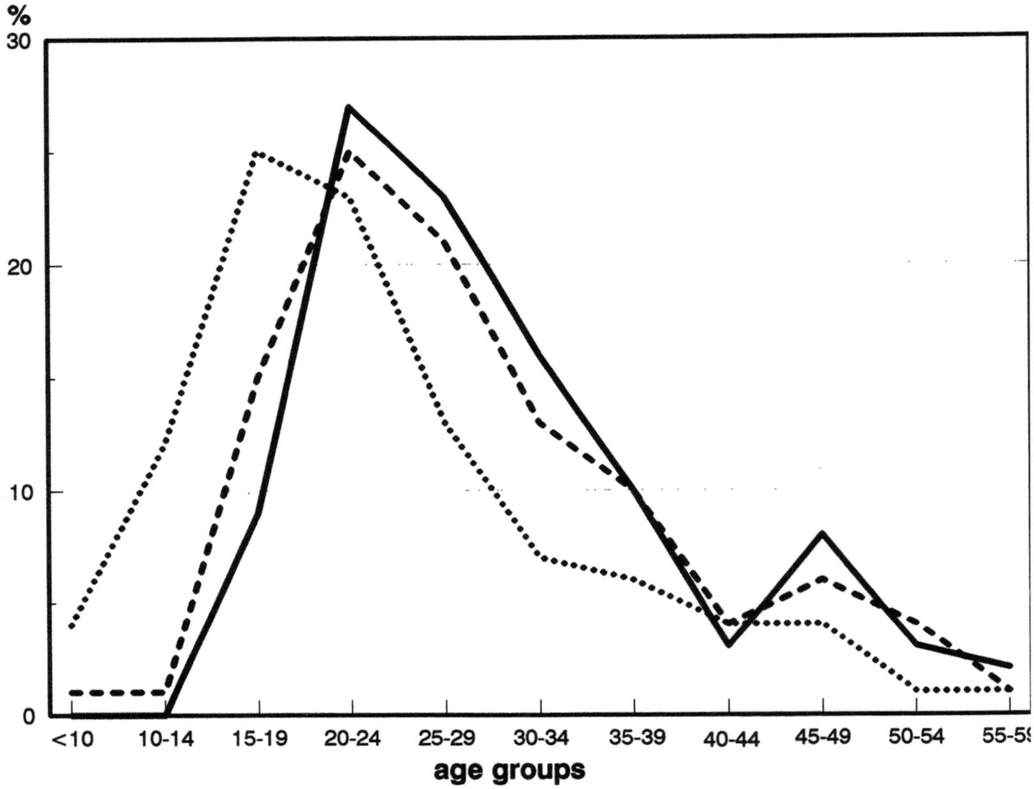

Fig. 2. Age distribution at onset of the prepsychotic (prodromal) and the psychotic prephase and of the culmination of the first psychotic episode ABC first episode sample, broad diagnostic definition; ICD-9–295, 297, 298.3, 298.4. Age ≤60 years, both sexes together (n = 232). *Dotted line*, earliest sign of a schizophrenic disorder; *broken line*, first occurrence of positive symptom; *solid line*, maximum number of positive symptoms

We interpreted these phenomena according to our estrogen hypothesis as a consequence of the decreasing neuromodular effect of estrogen on D_2 receptors in the brain and thus a fading of the protective function with the falling estrogen level in the pre-menopause. The increasing vulnerability which is presumably associated with it may lead to the outbreak of schizophrenia among those women disposed to schizophrenia who had until then been relatively protected by the higher level of estrogen.

With What Symptoms Does Schizophrenia Begin?

When the most frequent symptoms cited by the patients are listed in rank order (Table 2), then the results are in close agreement with the findings of Docherty et al. (1978), Herz and Melville (1980), Heinrichs and Carpenter (1985),

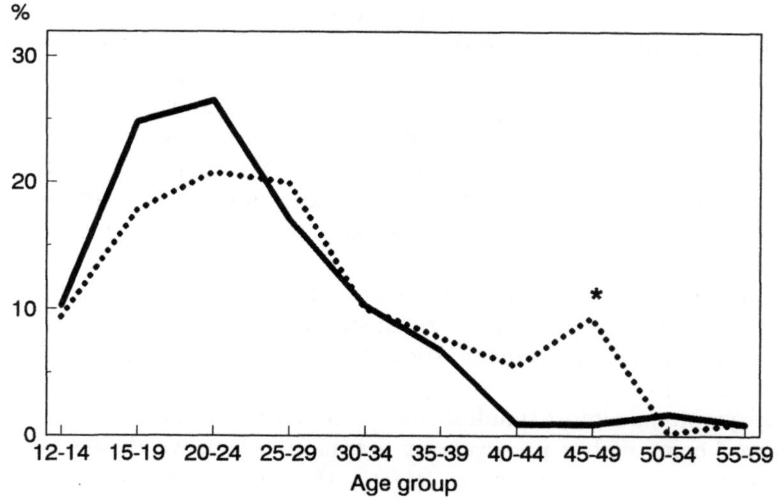

Fig. 3. Sex-specific age distributions in the early course of schizophrenia of a broad definition (ICD-9–295, 297, 298.3, 298.4) at earliest sign of mental disorder – Mannheim, Heidelberg, Rhein-Neckar-District, eastern Palatinate. *Solid line*, males, $n = 117$; *dotted line*, females, $n = 131$. $^{*}p < 0.05$

Table 2. The ten most frequent earliest signs of the disorder[a] (independent of the course) cited by the patients

	Total ($n = 232$)		Males ($n = 108$)		Females ($n = 124$)		P
	(n)	(%)	(n)	(%)	(n)	(%)	
Restlessness	43	19	16	15	27	22	
Depression	43	19	16	15	27	22	
Anxiety	42	18	18	17	24	19	
Trouble with thinking and concentration	38	16	21	19	17	14	
Worrying	35	15	10	9	25	20	*
Lack of self-confidence	29	13	11	10	18	15	
Lack of energy, slowness	28	12	9	8	19	15	
Poor work performance	26	11	13	12	13	10	
Social withdrawal, distrust	24	10	9	8	15	12	
Social withdrawal, communication	24	10	9	8	15	12	

[a] Evaluation based on closed questions, multiple counting possible.
All signs tested for sex differences.
$* = p \leqslant 0.05$.

Thurm and Häfner (1987), Hirsch and Jolley (1989), Tarrier et al. (1991), and Hamera et al. (1992), which were partly based on the study of prodromal phases and early symptoms of relapse and showed mainly unspecific, affective symptoms such as nervousness and restlessness, depression and anxiety as ranking highest. In our first-episode sample the three top-ranking symptoms are also restlessness, depression, and anxiety. Positive symptoms do not appear among the ten most frequent first symptoms. Negative symptoms, however, such as poor concentration, lack of energy and slowness, which are often associated with cognitive and social impairment, are in fact listed. The items "poor work performance" and "social withdrawal" reflect the early occurrence of socially disabling impairments. Only the symptom "worrying" is more frequent in women. No gender differences can be found among the other symptoms.

Having categorized the first symptoms by the traditional clinical dimensions of "positive," "negative," or "unspecific" (Table 3), we find that 6.5% of patients with a broad diagnostic definition of schizophrenia start with positive, 73.3% with negative or non-specific symptoms, and 20.3% with both categories of symptoms almost simultaneously, i.e., within 1 month. Schizophrenia therefore begins with a prepsychotic prephase, which we define as the prodromal phase, in three quarters of cases, whereas a beginning only with positive symptoms seldom occurs. Significant gender differences in the first symptoms of schizophrenia are not found.

How Does Schizophrenia Start?

In this question both temporal and contextual aspects are involved. The traditional temporal aspect is the speed or acuity of beginning operationalized by the time span between occurrence of the earliest sign of the disorder and the culmination of the first episode (Table 4). We have shown (Zedlick et al. 1993) that, in spite of being an important prognostic indicator, there is much confusion in the literature regarding the definition and assessment of the type of onset. We defined acute onset as up to 1 month, subacute as more than 1 month up to 1 year, and insidious onset as over 1 year from the occurrence of the first sign of the disorder until the culmination of the first episode, i.e., the maximum number of psychotic symptoms. Two thirds of the schizophrenic patients had a prephase lasting more than 1 year before the first episode, only about 18% had an acute onset, and about 15% a subacute onset.

Table 3. Symptom categories at the onset of a schizophrenic disorder

	Total (n = 232)		Males (n = 108)		Females (n = 124)	
	(n)	(%)	(n)	(%)	(n)	(%)
Positive startes	15	6.5	8	7.4	7	5.7
Mixed starters	47	20.3	24	22.2	23	18.6
Negative or non-specific starters	170	73.3	76	70.4	94	75.8

Sex differences ns. See note to Table 1.

Table 4. Type of onset (acute, subacute, insidious)

	Total (n = 232)		Males (n = 108)		Females (n = 124)	
	(n)	(%)	(n)	(%)	(n)	(%)
Acute (≤1 month)	41	17.7	20	18.5	21	16.9
Subacute (>1 month–1 year)	34	14.7	12	11.1	22	17.7
Insidious (>1 year)	157	67.7	76	70.4	81	65.3

Sex differences NS. See note to Table 1.

The accumulation of positive, negative, and non-specific symptoms defined according to simple sum scores of PSE and SANS items – 17 positive, 13 negative, and 35 non-specific – over 15 years until 1 year before the first episode and during the last 12 months before the first episode shows an exponential increase of all three symptom dimensions (Fig. 4). We found that symptom development is similar in males and females. Positive symptomatology occurs with a time lag of several years, but overtakes the negative during the psychotic episode. This implies that additive rather than successively alternating phase models – for instance, Conrad's (1958), Docherty et al. 's (1978), and Klosterkötter's (1992) models – are likely to give an adequate picture of the onset of schizophrenia, particularly as the increase of the earlier occurring negative and non-specific symptoms continues undiminished with the increasing number of psychotic symptoms.

With the accumulation of negative plus non-specific and of positive symptoms translated into a simple model, two prephases of a beginning psychosis can be discerned (Fig. 5): the prepsychotic (prodromal) phase, which almost three quarters of our sample (73.3%) presented, lasts on average 5.0 years until the first positive symptoms; the positive phase until the maximum of psychotic symptoms covers only 1.1 years. In some cases these are followed by a short latency phase (on average 0.2 years) until first admission. Thus, in those 73.3% of schizophrenics who show a prodromal phase before the onset of the psychosis, a mean period of 6.3 years passes from the occurrence of the first symptom until first admission, i.e., the entire prephase of the schizophrenic disorder.

An analysis of the interdependence of positive and negative symptom scores based on the 17 positive and 13 negative symptoms (irrespective of the non-specific symptom dimension) again depicted across 15 annual cross-sections until 1 year and across 12 monthly cross-sections during the last year before hospital admission mainly shows highly significant correlations – except for the 3rd year – with low to medium coefficients from year 11 befor first admission (Fig. 6). These findings do not fully agree with the assumed independence of the two symptom dimensions proposed by the dichotomy models of Crow (1980) and Murray and Lewis (1987). Andreasen, who had originally assumed a negative correlation between positive and negative symptom dimensions

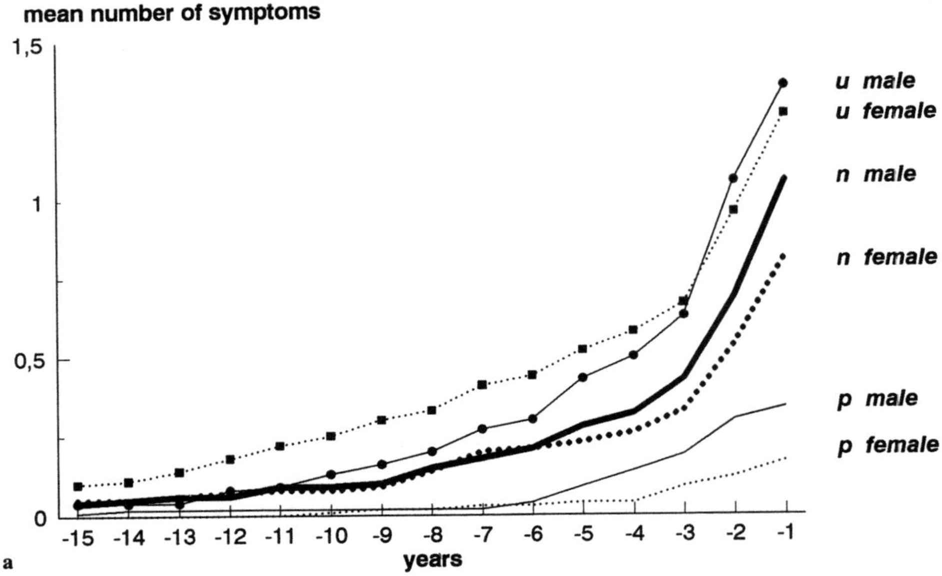

a

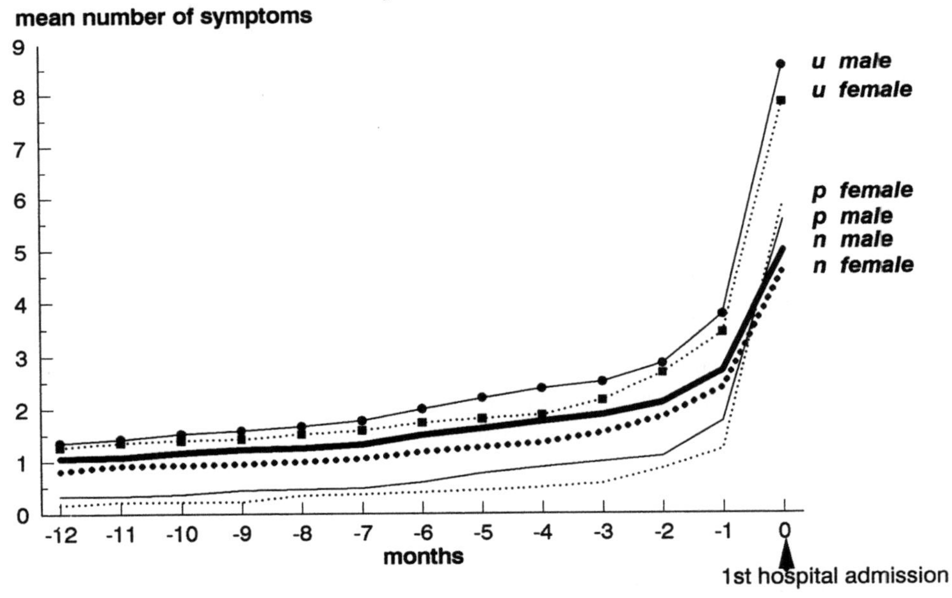

b

Fig. 4a,b. Cumulative numbers of positive (p), negative (n), and unspecific (u) symptoms of onset of schizophrenia until first hospital admission for schizophrenia (males = 108; females = 124)

(Andreasen and Olsen 1982) which would correspond to an alternating course of positive and negative symptom measures, has in the meantime come to regard dependence as possible (Andreasen et al. 1991). The majority of studies on the later course, however, find no dependence and only a few reveal a low

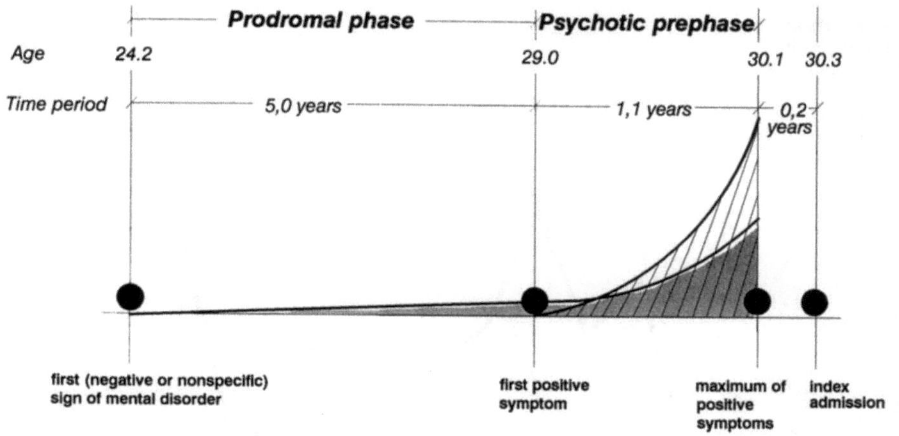

Fig. 5. Prephases of schizophrenia from first sign of mental disorder to first admission $n = 232$; 108 males, 124 females (both sexes together). *Hatched area*, positive symptoms; *shaded area*, negative symptoms

non-significant positive correlation between the two symptom dimensions (review: Häfner and Maurer 1993b).

There are possible explanations for this difference in our findings from the literature on the course of schizophrenia. On the one hand, environmental factors may gain more and differential influence on the symptomatology in the course of time, for example, by increasing the negative symptoms through social deprivation in- and outside mental hospitals (Wing 1978; Carpenter et al. 1991). Another reason is more evident: in irregularly recurring diseases, cross-sectional assessments are bound to assess a mixture of different stages of the disease. Usually 15%–20% of schizophrenic patients are in a psychotic episode at the time of assessment (Häfner and an der Heiden 1986; Biehl et al. 1986), the others are in various stages up to the free interval. If negative and positive symptom dimensions vary differently with the phases, then high positive correlations cannot be expected. A sample of patients in their first psychotic episode, however, presents an artificial homogenization of disease stages, which can also result in greater homogeneity of the prephases and the early course. Since significant correlations were found over almost the whole prephase despite the shortcomings of retrospective assessment, a functional or causal association between the leading negative and the psychotic symptom dimensions of schizophrenia seems more likely, at least for this part of the course, than a dichotomy model.

The Later Course – 2 Years After First Admission

We examined a representative cohort of schizophrenics admitted for the first time ($n = 133$) from our sample in four cross-sections within a period of up to 2

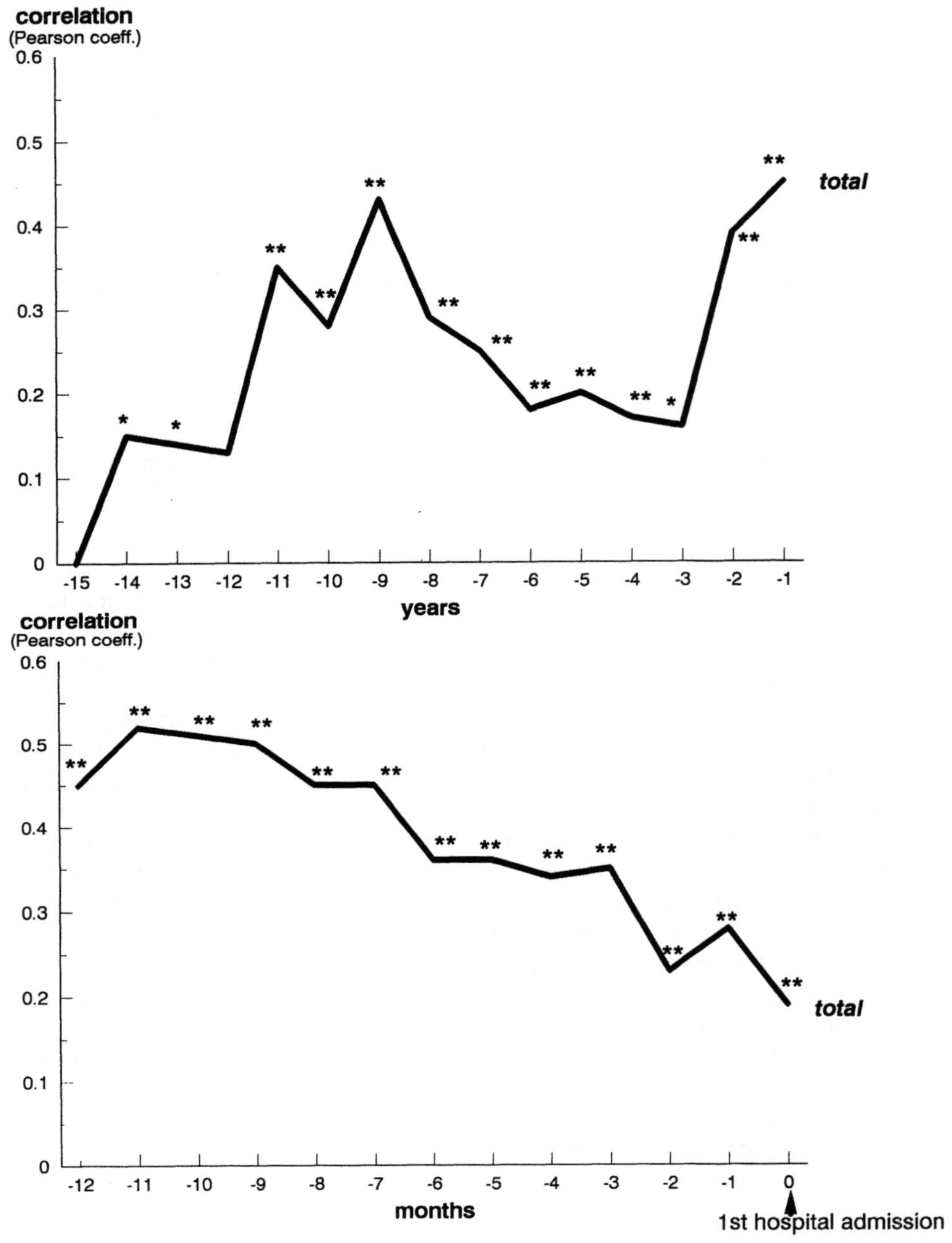

Fig. 6. Correlations between positive and negative symptoms of onset of schizophrenia (n = 232)

years. Operationalized on the basis of factor scores built by 13 negative and 12 positive PSE items, significant positive correlations between the symptom scores are found just as in the prephase, except for the timepoint of first admission (Fig. 7). The PSE interviews were carried out within the first 14 days after first admission in a period of rapidly changing symptomatology through intensive treatment, and therefore it was probably rather unlikely at this moment to expect interdependence of the two symptom dimensions. This overall result, which differs from the findings of most longitudinal studies that reveal independence or only slight correlation of the two symptom dimensions, is presumably due to the still effective yet limited homogenization after first admission, as are the clearly positive correlations in the prephase of the disease.

In two representative first admission cohorts studied prospectively over 2 and 5 years, Pogue-Geile and Harrow (1985) and Biehl et al. (1986) found a reduction of the negative symptom scores after the acute episode followed by stable mean values. We were able to confirm this trend. Figure 8 demonstrates that negative symptoms do not occur for the first time or more often after the first psychotic episode of the acute course, as Kraepelin had originally assumed. As a matter of fact, they mostly increase before or together with the positive symptoms, reach their peak in the acute episode, and then decrease again in a course almost parallel to that of the psychotic symptoms (see Fig. 4b). But they often decrease only down to a residual value of varying size, which can then be called the residual state. The proportion of patients with negative symptoms at the time of assessment is also comparatively stable, while a considerable proportion of individuals still experience fluctuating symptomatology.

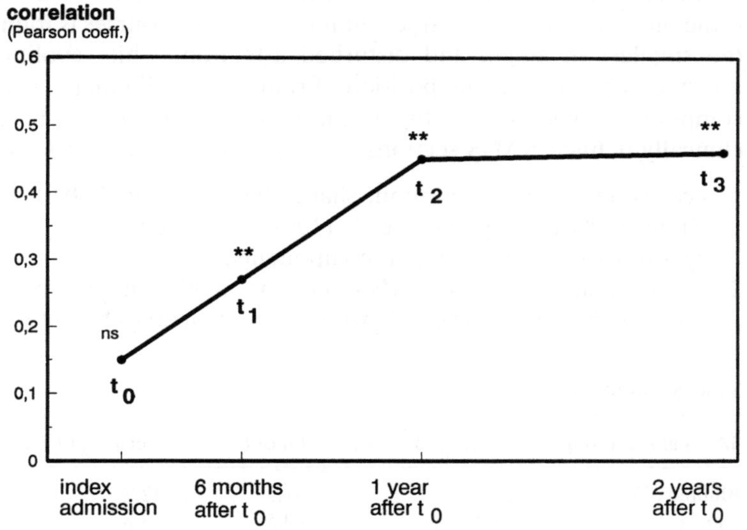

Fig. 7. Dependence or independence of positive and negative symptoms: comparison of factor-analytic PSE symptom scores and clinical (intuitive) PIRS impairment scores (n = 133). *ns*, not significant; $^{**}p \leq 0.01$

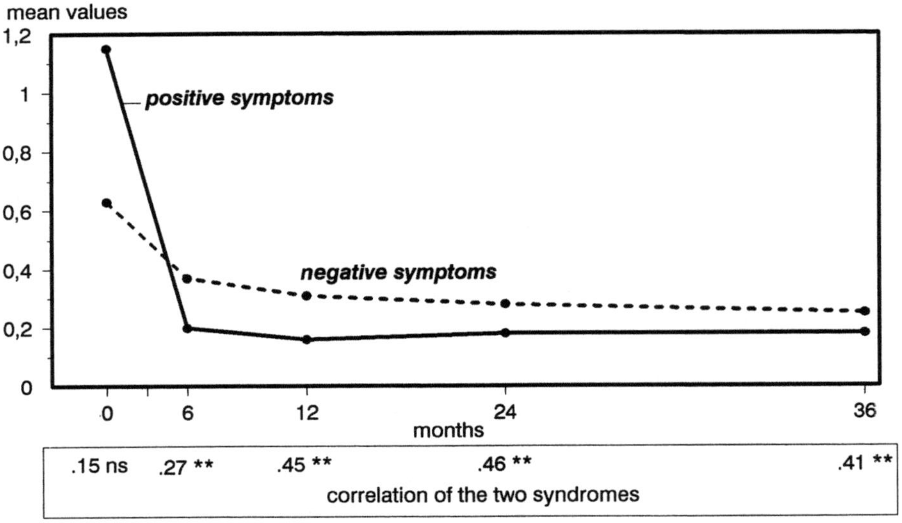

Fig. 8. Mean values of factor-analytic scores for positive and negative symptoms over 3 years – from the course sample of the Mannheim ABC schizophrenia study (n = 133). *ns*, not significant; $^{**}p < 0.01$

The negative syndrome comprises several components of presumably different etiology. With the items split up into five sections according to the SANS classification, Table 5 demonstrates that a continuous syndrome appearance in all cross-sections is the rarest phenomenon and a single appearance is the most frequent. The largest number of stable courses is found in anhedonia, the smallest in alogia and disturbed attention, while avolition and blunted affect occupy a medium position. From this preliminary analysis of various component constructs of the negative syndrome, as yet based only on the clinically defined SANS sections, we can draw two conclusions:

1. according to our assumption, shared by Kay et al. (1986), Pogue-Geile and Zubin (1988), Carpenter et al. (1991), and Rey et al. (1992), the negative syndrome comprises several components.
2. Those associated with SANS sections alogia and attentional impairment seem to be closely related, with their recurring character, to the positive

Table 5. SANS ratings over 2 years

SANS global rating	No symptom (%)	Once (%)	Recurrent (%)	Continuous (%)
Blunted affect	25.3	51.9	12.7	10.1
Alogia	53.2	40.5	3.8	2.5
Avolition	13.9	55.7	15.2	15.2
Anhedonia	19.0	40.5	13.9	26.6
Attention	50.0	38.9	4.2	6.9

symptomatology. This component of negative symptomatology is presumably above all a direct expression of the disease process. This is essentially supported by the fact that schizophrenia begins with symptoms of this cluster mostly long before the occurrence of positive symptoms.

Conversely, anhedonia, with its comparative stability, is more similar to a personality trait. It can be assumed that this symptom dimension is indeed associated with personality traits which are more often observed in persons who develop schizophrenia at a later date. This means that they are possibly – though not exclusively or specifically – associated with the genotype itself rather than with the psychotic reaction pattern of schizophrenia. The increased vulnerability for psychotic episodes of the schizophrenic type is found not only with this genotype but also with other brain dysfunctions.

Moreover, it is likely that the effects of the coping behavior of schizophrenics, e.g., the avoidance of intensive social contacts or social withdrawal in order to prevent relapse (Thurm and Häfner 1987), may usually be lumped into the total volume of negative symptoms as postulated by Carpenter et al. (1991). The discriminative analysis of the course of these components and their association with or independence of the course of other symptom dimensions were the subject of a specific study conducted by our group (Häfner and Maurer 1993b).

Social Course of Schizophrenia – When Do Social Consequences of the Disease First Occur?

With respect to the early occurrence of social and cognitive impairments, we investigated social status indicators during the prephase of schizophrenia until first admission and for 2 years there after. For this purpose we assessed the social development across three cross-sections retrospectively in the phase before and prospectively in three more cross-sections after first admission (Fig. 9). At the time of the first negative or unspecific symptom, only 10% of the schizophrenics were living on social welfare, 6% of these – as many as in the respective population at that time – were living on unemployment benefit. The remaining 4% suggest that the proportion of schizophrenics who were already considerably impaired in the premorbid stage, i.e., when prodromal symptoms appeared, was relatively small. Several years later, when the first positive symptom occurred, as many as 21% were living on social welfare and about 3 years later or 2 years after first admission 19%.

The proportion of employed patients increased until first hospitalization, as normally expected of people in adolescence and early adulthood, even though the illness process had started. But this increase reached its peak with 50%. After a clear decline to 36% 0.5 years after first admission, at the 2-year follow up 42% of the first-onset schizophrenics were in employment (see Fig. 9). The patients' mean age at that time was 32 years. At this stage one can assume that a person has normally completed his/her vocational training and the majority

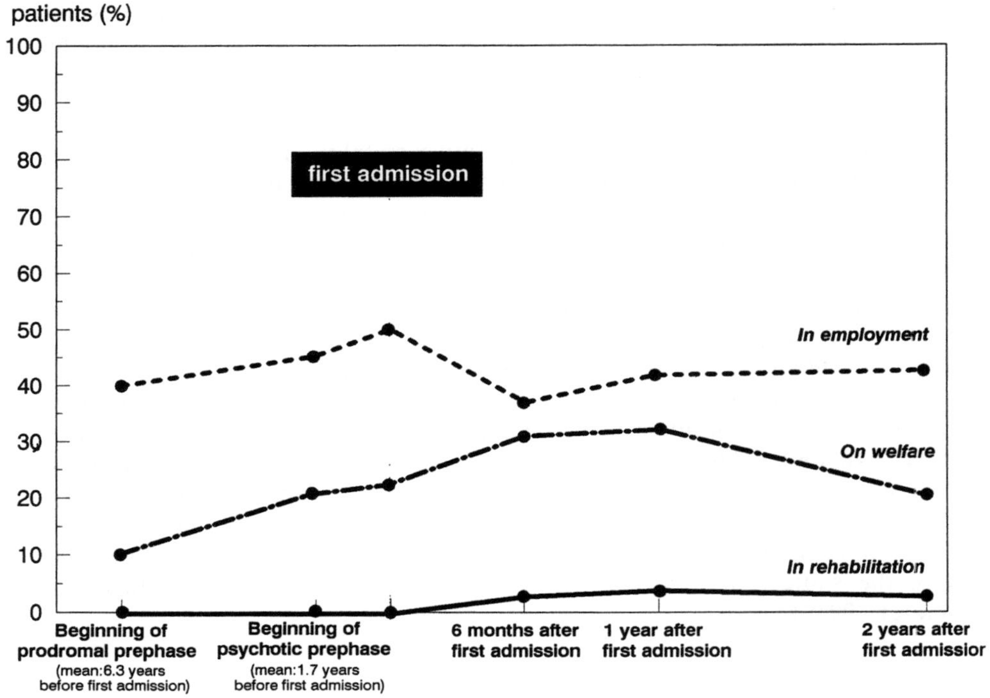

Fig. 9. Social consequences of schizophrenia from first onset until 2 years after first admission retrospectively until first admission, prospectively thereafter; $n = 133$

have already started work, live as housewives, or are registered as unemployed. In the observation period extending from the first sign via index admission until 2 years there after, there was a high individual variability of developments. Defined as proportions of the total group ($n = 133$), 27% of the patients finished their education, 16% completed their vocational training, and 23% started a job. In the same period, however, 6% had already lost existing employment, and 43% did not fulfil the social role (permanent employment) at all.

The losses in social functioning that make patients dependent upon social welfare obviously mostly occur during the prepsychotic phase before the positive symptoms, whereas rehabilitation does not start until first admission.

Long-term Course of Symptomatology and Social Indicators

Fourteen years after their first admission, an der Heiden et al. (1994) from our group made a follow up using the same assessment instruments on a smaller first admission cohort of schizophrenic patients ($n = 56$), who had first been examined prospectively in seven cross-sections over 5 years by Biehl et al.

(1986). Neither the CATEGO syndrome profiles, e.g., for delusional and hallucinatory syndromes (see Fig. 10), nor impairment (PIRS, Fig. 11) nor disability (Fig. 12) profiles had markedly changed over this long period after they had reached almost these levels on average as early as 1 year after first admission. Nevertheless, the outcome of the disease shows high individual variability. It ranges from full remission in varying phases to chronic defect states.

Early Predictors of Disability 2 Years After First Admission

Angermeyer et al. (1990) from our Institute analysed 102 studies from the international literature. Although the quality of the majority was flawed, they revealed some agreement on a more favorable course of schizophrenia in women with regard to all outcome measures (time spent in hospital, symptoms at follow up, social adaptation, etc.). In recent studies with a sound methodo-

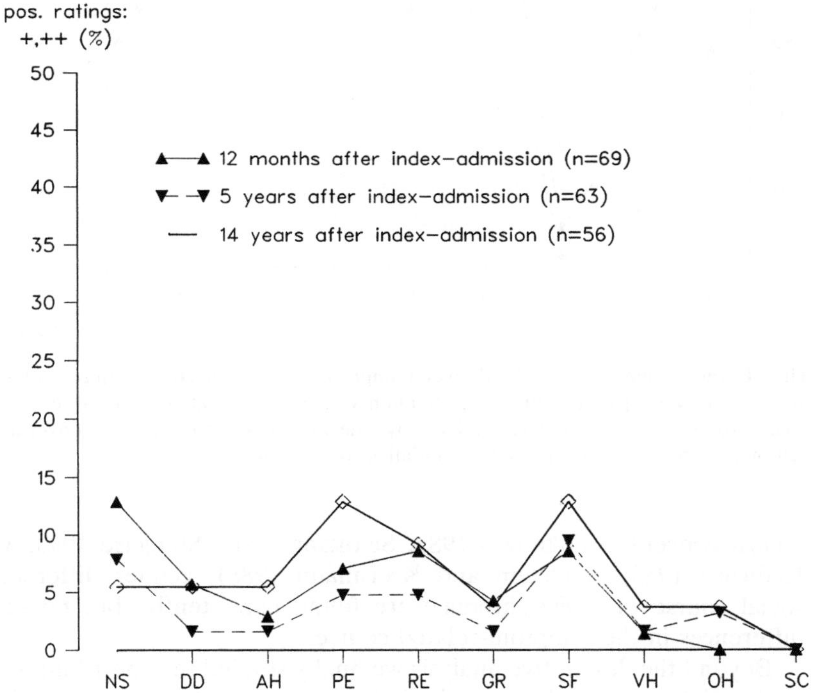

Fig. 10. PSE delusional and hallucinatory syndromes: profiles at three cross-sections over a period of 14 years. *NS*, nuclear syndrome; *DD*, depressive delusions and hallucinations; *AH*, auditory hallucinations; *PE*, delusions of persecution; *RE*, delusions of reference; *GR*, grandiose and religious delusions; *SF*, sexual and phantastic delusions; *VH*, visual hallucinations; *OH*, olfactory hallucinations; *SC*, "subcultural" delusions or hallucinations

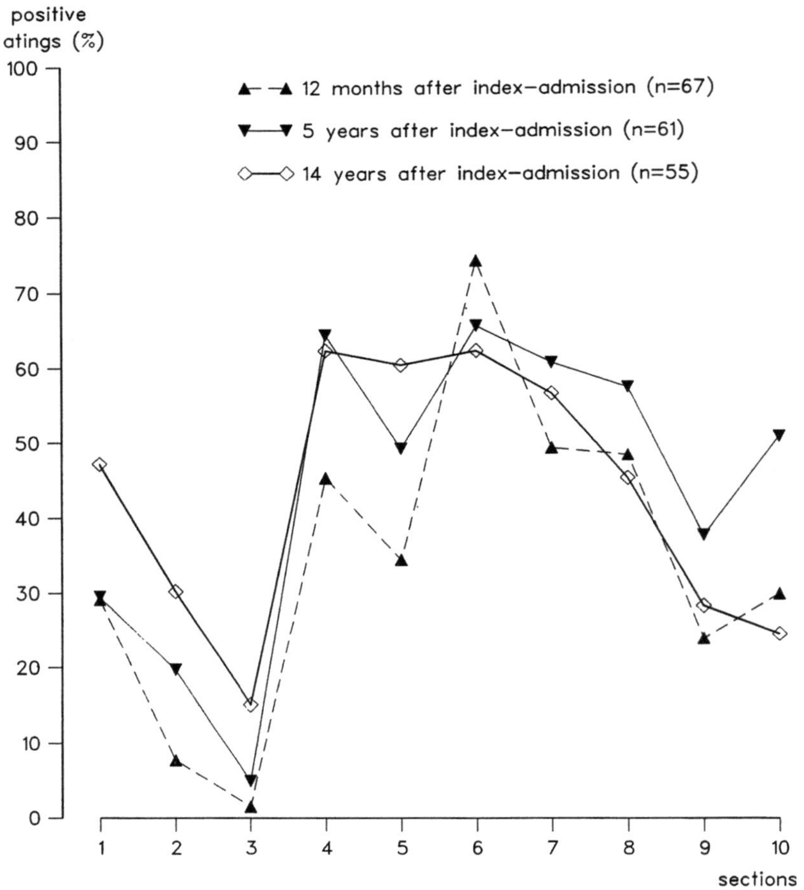

Fig. 11. Impairment profiles (PIRS: overall impressions) at three cross-sections over a period of 14 years. *1*, slowness/psychic tempo; *2*, attention withdrawal; *3*, fatiguability; *4*, communication by facial expression; *5*, communication by body language; *6*, affect display; *7*, conversation skills; *8*, self-presentation; *9*, co-operation; *10*, initiative (over- or under-)

logical concept (Salokangas 1983; Salokangas and Stengard 1990; Wattie and Kedward 1985; Goldstein and Kreismann 1988), gender differences in the social course of schizophrenia were found consistently, but no or only few differences in the symptom-related course.

Beyond the descriptive analysis we finally studied the contribution of gender, age, social status, and disease indicators at different cross-sections in the early course with a view to predicting social disability 2 years after first admission (Häfner and Nowotny 1993). The model and findings of our step-by-step regression analysis are presented in Fig. 13.

At the time of the first sign of the disease, neither a global index of social status deficits (Nowotny and Häfner 1994) nor the symptomatology contributed

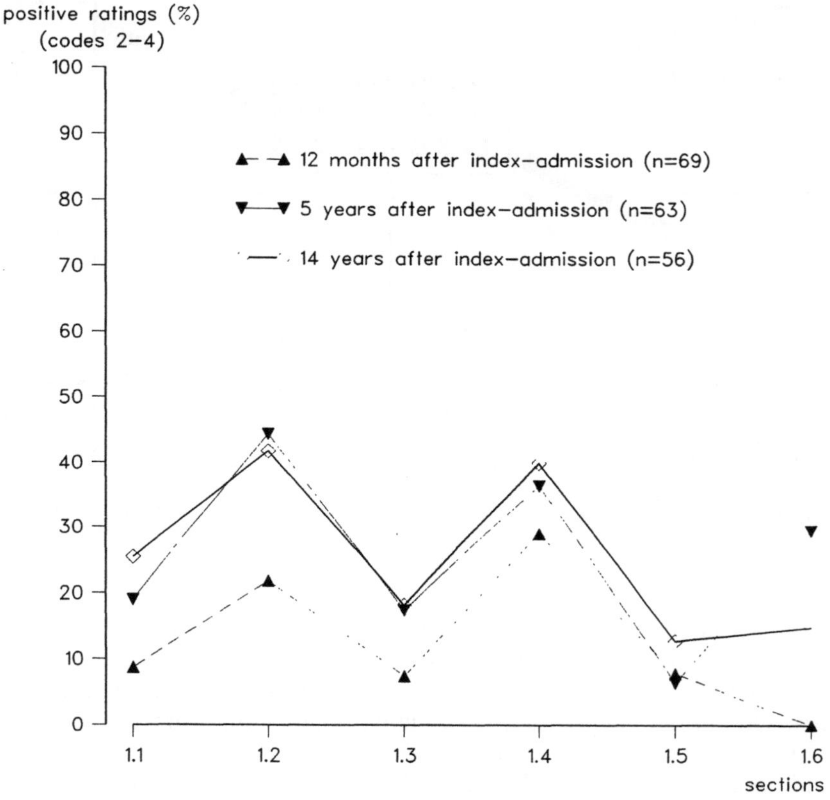

Fig. 12. Social disability: profiles at three cross-sections over a period of 14 years, Disability Assessment Schedule; section 1: general behavior. *1.1*, patient's self-care during past month; *1.2*, underactivity during past month; *1.3*, slowness; *1.4*, social withdrawal during past month; *1.5*, social contacts: friction in interpersonal relationships; *1.6*, patient's behavior in emergencies

significantly to predicting social disability, operationalized as a Disability Assessment Schedule global score ≥2 (WHO 1988; German version: Jung et al. 1989). At the beginning of the psychotic prephase, i.e., occurrence of the first psychotic symptom, the social and educational status deficits had achieved the leading predictive power for social disability (Fig. 13). As the descriptive analysis of the social course prior to first admission had demonstrated, the crucial changes determining social outcome 2 years after first admission obviously already take place during the prodromal phase. Age and gender indirectly exert their greatest influence on the early social course of schizophrenia by their different impact on the time when the illness breaks into the social development. The major impact of gender on the social course over 2 years from first admission is thus determined by the less developed social and occupational status of the earlier afflicted males at the onset of the first episode of schizophrenia.

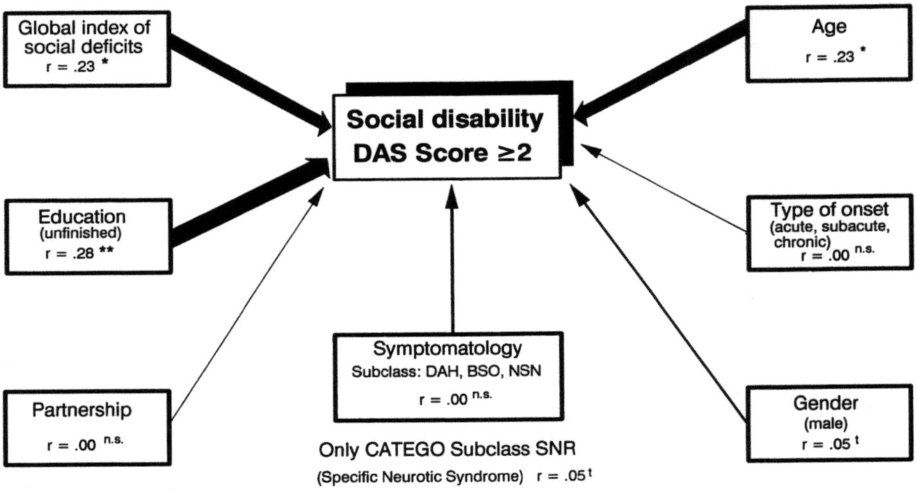

Fig. 13. Prediction model of social disability (correct overall classification 81%). *r*, partial correlation

Independently of this, however, a limited influence of age and a weak influence of gender persists on social disability. The effect of male gender is presumably due to the clearly different illness behavior of males and females assessed at first admission: male schizophrenics show a socially negative pattern of behavior, significantly more often such as self neglect, social withdrawal, and no interest in work (Häfner et al. 1993), than female schizophrenics. It can be assumed that this behavior pattern may also have bearing on a more unfavorable social course of the disease.

Conclusions

Three quarters of all schizophrenics thus present a definable prepsychotic prodromal phase with only negative and non-specific symptoms, lasting on average considerably longer than the psychotic prephase. The whole phase of disease onset culminates mostly with an exponential increase of positive, negative, and non-specific symptoms in the first psychotic episode. The decrease of the first psychotic episode also shows an almost parallel course of diminishing mean values of positive and negative symptom dimensions.

At present we do not know how often the disease comes to a standstill before the occurrence of positive symptoms. It seems very unlikely that all courses of the disorder with a given disposition proceed until positive symptoms occur, i.e., until the manifestation of psychotic episodes. This question is, however, of central importance for the genetic epidemiology of schizophrenia.

With the negative symptoms of the prodromal phase, impairments and social disabilities proceed very soon, and these seem to be already responsible for the maximum losses in social status (social drift) and for social non-starter phenomena in the total course of the psychosis. This throws a new light on the discussion of the alternative hypothesis "social drift vs. social selection" (Hollingshead and Redlich 1958; Zigler and Levine 1981; Dohrenwend 1987; Dilling and Weyerer 1987), which has so far been based on the point of first admission as defining the onset of schizophrenia. The disease indeed seems to be the major cause of the negative social gradient of schizophrenia prevalence, but a great number of its effects already occur before first admission. The social consequences of the disease affect males much harder mainly because schizophrenia begins earlier and interrupts the social development that normally progresses at a great pace in this period earlier and more often in males. An additional minor effect with unfavorable social consequences over 2 years should be considered, namely the consequences of negative illness behavior in males which had developed during the prephase of the disorder.

In the core symptomatology and the types of onset and early course, however, neither significant differences between the sexes can be found nor an impact on the social course until two years after first admission. In the long-term course of the disease, despite a great variety of individual courses, recurring psychotic episodes, and fluctuating negative symptoms, the mean levels of symptomatology and social disability seem to be fairly stable.

The traditional process model according to Kraepelin and Jaspers cannot provide an adequate picture. The course of schizophrenia is much more in accordance with the model of a self-limiting disease process with early onset characterized mainly by negative symptoms with cognitive, emotional, and social deficits. One typical feature of this process is an increased vulnerability to recurring psychotic episodes in the sense of Zubin and Spring (1977). But the nuclear symptomatology of schizophrenia in terms of a more direct relationship to the underlying brain dysfunction seems to be the negative and not the psychotic syndrome per se. Whether the detrimental courses described by earlier authors (Kraepelin 1896; Mauz 1930; Langfeldt 1956) as leading to very severe defect states in the long run or to early death in short-term catastrophic courses, were the consequences of a lack of therapy or of lasting deprivation due to unfavorable hospital conditions remains an open question.

References

an der Heiden W, Krumm B, Müller S, Weber I, Biehl H (1994) Mannheimer Langzeitstudie der Schizophrenie: erste Analysen zum Outcome 14 Jahre nach Erstbehandlung (submitted)

Andreasen NC, Olsen SA (1982) Negative v. positive schizophrenia. Arch Gen Psychiatry 39: 789–794

Andreasen NC, Flaum M, Arndt S, Alliger R, Swayze VW (1991) Positive and negative symptoms: assessment and validity. In: Marneros A, Andreasen NC, Tsuang NT (eds) Negative versus positive schizophrenia. Springer, Berlin Heidelberg New York, pp 28–51

Angermeyer MC, Kühn L, Goldstein JM (1990) Gender and the course of schizophrenia: Differences in treated outcomes. Schizophr Bull 2:293–307

Biehl H, Maurer K, Schubart C, Krumm B, Jung E (1986) Prediction of outcome and utilization of medical services in a prospective study of first onset schizophrenics – results of a prospective 5-year follow-up study. Eur Arch Psychiatry Neurol Sci 236:139–147
Bleuler E (1911) Dementia praecox oder Gruppe der Schizophrenien. In: Aschaffenburg G (ed) Handbuch der Psychiatrie. Deuticke, Leipzig, pp 1–420
Carpenter WT, Buchanan RW, Krkpatrick B et al (1991) Negative symptoms: a critique of current approaches. In: Marneros A, Andreasen NC, Tsuang MT (eds) Springer, Berlin Heidelberg New York, pp 126–133
Conrad K (1958) Die beginnende Schizophrenie. Versuch einer Gestaltanalyse des Wahns. Thieme, Stuttgart
Crow TJ (1980) Positive and negative schizophrenic symptoms and the role of dopamine. Br J Psychiatry 137:383–386
DeLisi LE, Crow TJ, Hirsch SR (1986) The third biannual winter workshop on schizophrenia. Arch Gen Psychiatry 43:706–711
Dilling H, Weyerer S (1987) Social class and mental disorders: Results from upper bavarian studies. In: Angermeyer MC (ed) From social class to social stress. Springer, Berlin Heidelberg New York, pp 46–63
Docherty JP, Van Kammen DP, Siris SG, Marder SR (1978) Stages of onset of schizophrenic psychosis. Am J Psychiatry 135:420–426
Dohrenwend BP (1987) Social class and mental disorder: the stress/selection issue. In: Angermeyer MC (ed) From social class to social stress. Springer, Berlin Heidelberg New York, pp 106–116
Goldstein JM, Kreisman D (1988) Gender, family environment and schizophrenia. Psychol Med 18:861–872
Häfner H, an der Heiden W (1986) The contribution of European case registers to research on schizophrenia. Schizophr Bull 12:26–50
Häfner H, Maurer K (1993a) Methodenprobleme der Erforschung von Krankheitsbeginn und Frühverlauf am Beispiel Schizophrenie. Fund Psychiatr 7:1–12
Häfner H, Maurer K (1993b) Epidemiology of positive and negative symptoms in schizophrenia. In: Shriqui CL, Nasrallah HA (eds) Contemporary issues in the treatment of schizophrenia. American Psychiatric Press, Washington DC
Häfner H, Nowotny B (1993) When and how does schizophrenia lead to social deficits? Paper for post-congress of 1993 WFMH World Congress, Tokyo
Häfner H, Riecher A, Maurer K, Meissner S, Schmidtke A, Fätkenheuer B, Löffler W, an der Heiden W (1990) Ein Instrument zur retrospektiven Einschätzung des Erkrankungsbeginns bei Schizophrenie (IRAOS) Entwicklung und Ergebnisse. Z Klin Psychol 19:230–255
Häfner H, Riecher A, Maurer K, Fätkenheuer B, Löffler W, an der Heiden W, Munk-Jorgensen P, Strömgren E (1991) Geschlechtsunterschiede bei schizophrenen Erkrankungen. Fortschr Neurol Psychiatr 59:343–360
Häfner H, Riecher-Rössler A, Hambrecht M, Maurer K, Meissner S, Schmidtke A, Fätkenheuer B, Löffler W, an der Heiden W (1992) IRAOS: an instrument for the assessment of onset and early course of schizophrenia. Schizophr Res 6:209–223
Häfner H, Maurer K, Löffler W, Riecher-Rössler A (1993) The influence of age and sex on the onset and early course of schizophrenia. Br J Psychiatry 162:80–86
Hamera EK, Peterson KA, Young LM, Schaumloffel MM (1992) Symptom monitoring in schizophrenia: potential for enhancing self-care. Arch Psychiatr Nursing 6:324–330
Heinrichs DW, Carpenter WT Jr (1985) Prospective study of prodromal symptoms in schizophrenic relapse. Am J Psychiatry 142:371–373
Herz MI, Melville C (1980) Relapse in schizophrenia. Am J Psychiatry 137:801–805
Herz MI, Szymanski HV, Simon JC (1982) Intermittent medication for stable schizophrenic outpatients: an alternative to maintenance medication. Am J Psychiatry 139:918–922
Hirsch SR, Jolley AG (1989) The dysphoric syndrome in schizophrenia and its implications for relapse. Br J Psychiatry 155[Suppl 5]:46–50
Hollingshead AB, Redlich FC (1958) Social class and mental illness. Wiley, New York
Huber G, Gross G (1989) The concept of basic symptoms in schizophrenia and schizoaffective psychoses. Rec Prog Med 80:646–652

Jung E, Krumm B, Biehl H, Maurer K, Bauer-Schubart C (1989) DAS-M: Mannheimer Skala zur Einschätzung Sozialer Behinderung. Beltz, Weinheim
Kay SR, Fiszbein A, Lindenmayer J-P et al (1986) Positive and negative syndromes in schizophrenia as a function of chronicity. Acta Psychiatr Scand 74:507–518
Klosterkötter J (1992) Die Entwicklung der schizophrenen Symptome ersten Ranges. Fund Psychiatr 6:81–94
Kraepelin E (1896) Psychiatrie. Ein Lehrbuch für Studierende und Aerzte. Barth, Leipzig
Kraepelin E (1909) Psychiatrie. Barth, Leipzig
Kumar S, Thara R, Rajkumar S (1989) Coping with symptoms of relapse in schizophrenia. Eur Arch Psychiatr Neurol Sci 239:213–215
Langfeldt GE (1956) Prognosis in schizophrenia. Acta Psychiatr Scand Suppl 110
Marder SR, Van Putten T, Mintz J, Lebell M, McKenzie J, Faltico G (1984) Maintenance therapy in schizophrenia: new findings. In: Kane JM (ed) Drug maintenance strategies in schizophrenia. American Psychiatric Press, Washington
Marder SR, Van Putten T, Mintz J, McKenzie J, May PRA (1987) Low and conventional-dose maintenance therapy with fluphenazine deconoate. Arch Gen Psychiatry 44:518–521
Mauz F (1930) Die Prognostik der endogenen Psychosen. Thieme, Leipzig
Murray RM, Lewis SW (1987) Is schizophrenia a neurodevelopmental disorder? Br Med J 295:681–682
Nowotny B, Häfner H (1994) Die beginnende Schizophrenie als Einbruch in die soziale Biographie – Folgen für soziale Behinderung versus soziale Wiederanpassung. Fund Psychiatr (in press)
Pogue-Geile MF, Harrow M (1985) Negative symptoms in schizophrenia: their longitudinal and prognostic importance. Schizophr Bull 11:427–439
Pogue-Geile MF, Zubin J (1988) Negative symptomatology and schizophrenia: a cenceptual and empirical review. Int J Ment Health 16:3–45
Rey ER, Beck U, Morstadt P, Oldigs J (1987) Experimentelle Untersuchungen zur Aufmerksamkeitsstörung Schizophrener. In: Gerber WD, Miltner W, Mayer K (eds) Verhaltensmedizin: Ergebnisse und Perspektiven interdisziplinärer Forschung. VCH, Weinheim
Rey E-R, Bailer J, Laubenstein D, Graβhoff U (1992) Einfluβ von prämorbider Anpassung, sozialen Netzwerkparametern und sozialer Unterstützung auf die Negativsymptomatik ersterkrankter Schizophrener. Verhaltenstherapie 2:300–308
Riecher-Rössler A, Häfner H, Maurer K, Stummbaum M, Schmidt R (1992) Schizophrenic symptomatology varies with serum estroadiol levels during menstrual cycle. Schizophr Res 6(special issue):114
Riecher-Rössler A, Häfner H, Stumbaum M, Schmidt R (1994) Can estradiol modulate schizophrenic symptomatology? Schizophr Bull 20 (in press)
Salokangas RKR (1983) Prognostic implications of the sex of schizophrenic patients. Br J Psychiatry 142:145–151
Salokangas RKR, Stengard E (1990) Gender and short-term outcome in schizophrenia. Schizophr Res 3:333–345
Statistisches Bundesamt (1987) Datenreport 3
Subotnik KL, Nuechterlein KH (1988) Prodromal signs and symptoms of schizophrenic relapse. J Abnorm Psychol 97:405–412
Tarrier N, Barrowclough C, Bamrah JS (1991) Prodromal signs of relapse in schizophrenia. Soc Psychiatry Psychiatr Epidemiol 26:157–161
Thurm I, Häfner H (1987) Perceived vulnerability, relapse risk and coping in schizophrenia. An explorative study. Eur Arch Psychiatry Neurol Sci 237:46–53
Wattie BS, Kedward HB (1985) Gender differences in living conditions found among male and female schizophrenic patients on a follow-up study. International Journal of Social Psychiatry 31(3):205–216
WHO (1988) Psychiatric disability assessment schedule. World Health Organization, Geneva
Wing JK (1978) Reasoning about madness. Oxford University Press Oxford
Zedlick D, Maurer K, Löffler W, Riecher-Rössler A, Häfner H (1993) Der Frühverlauf der Schizophrenie und seine Beziehung zu Alter, Geschlecht und Diagnose bei stationärer Erstaufnahme. Neuropsychiatrie 7:183–196

Zigler E, Levine J (1981) Premorbid competence in schizophrenia: what is being measured. J Consult Clin Psychol 90:458–467
Zubin J, Spring B (1977) Vulnerability – a new view of schizophrenia. J Abnorm Psychol 86: 103–126

Early Predictors of Onset and Course of Schizophrenia: Some Results from the Copenhagen High-Risk Study

J. PARNAS[1,2], T.D. CANNON[3], S.A. MEDNICK[1,4], and F. SCHULSINGER[1]

Background of the Study

In the late 1950s Mednick (1960) advocated a prospective longitudinal design in etiological research in schizophrenia. This idea arose from an increasing dissatisfaction with typical cross-sectional patient–control comparisons. In such comparisons, differences between schizophrenics and controls may reflect concomitants or sequelae of schizophrenia rather than suggest causal factors.

The prospective design has the following chief advantages: (a) data collected premorbidly are unbiased by knowledge of future diagnostic outcome; (b) the longitudinal aspect of the design permits temporal evaluation of events and thus facilitates etiological conjectures; (c) the study is basically epidemiological in nature and subjects are diagnosed irrespective of their contact with treatment facilities. The original hope of the prospective design was to identify risk factors which could be used in preventive interventions.

Because the morbid risk of schizophrenia in the general population is prohibitively low to draw random samples for prospective evaluation, samples are chosen on the basis of empirically elevated risk, e.g., children of schizophrenic parents.

This research paradigm allows the assessment of the morbid risks to the offspring for both schizophrenia and other disorders that may be genetically linked to schizophrenia, and the identification of early predictors of these diagnostic outcomes. Findings from the Danish Adoption Study and other evidence suggest that there is an increased risk for both schizophrenia and nonpsychotic schizophrenia spectrum conditions (e.g., schizotypal personality disorder) in the biological relatives of schizophrenic probands (Kety et al. 1978).

These findings have been interpreted by us within a spectrum–vulnerability model of schizophrenia (e.g., Meehl 1962).

This model posits that individuals at genetic risk for schizophrenia will exhibit certain fundamental symptoms (i.e., some subtle phenotypic features) summarized in the concept of "schizotypy"; among these individuals, the

[1] Institute of Preventive Medicine, Kommunehospitalet, Copenhagen, Denmark
[2] Copenhagen University, Department of Psychiatry, Hvidovre Hospital, Brøndbyøstervej 160, 2650 Hvidovre, Denmark
[3] Department of Psychology, University of Pennsylvania, Philadelphia, PA 19144, USA
[4] Social Science Research Institute, University of Southern California, Los Angeles, CA, USA

H. Häfner and W.F. Gattaz (Eds.)
Search for the Causes of Schizophrenia, Vol. III

experience of putative environmental etiological factors contributes to the expression of overt schizophrenia (e.g., Parnas and Bovet 1991). Since 1962 a group of children of schizophrenic mothers and matched control children has been followed up prospectively in Denmark (The Copenhagen High Risk Study) (Mednick and Schulsinger 1965).

In 1962, at the inception of the study, the subjects averaged 15 years of age and none was severely psychiatrically ill. The samples have been evaluated diagnostically on two occasions, first in 1972–1974 when the subjects were on average 25 years of age (Schulsinger 1976), and most recently in 1986–1989, when the subjects were on average 42 years old and thus through most of their risk period for developing schizophrenia. The diagnostic data from the project have recently been analyzed (Parnas et al. 1993) according to DSM-III-R criteria (American Psychiatric Association 1987).

The purpose of this communication is to report the life-time diagnostic DSM-III-R outcomes in this group and to report some early risk factors for the development of frank schizophrenic illness identified in this longitudinal study.

The Sample

The study began in 1962 with the selection of a group of children of schizophrenic mothers and matched controls. Social workers visited the psychiatric hospitals and departments in the Greater Copenhagen area and selected charts of hospitalized, allegedly schizophrenic women, who had children in the age range 10–18 years. The maternal hospital charts were then reviewed by a psychiatrist (Schulsinger) to assure that the diagnosis of schizophrenia fulfilled the criteria of the Danish Psychiatric Association of that time. These criteria were not operational but were anchored in the Kraepelinian-Bleulerian tradition, i.e., emphasizing chronicity and the presence of autism and formal thought disorder. Cases with uncertain diagnosis were excluded. This selection yielded 129 schizophrenic mothers who had 207 children who were selected for study (high-risk, HR, children). The maternal schizophrenia diagnoses were recently reassessed, and 91% of the mothers were found to fulfill the DSM-III-R criteria for schizophrenia. The remainder met the criteria for schizophreniform disorder and other non-affective, non-organic psychosis (Jørgensen et al. 1987). The HR children were matched to a group of control children whose parents were screened in the National Psychiatric Register and who were not found to have been by that time hospitalized for any psychotic disorder (low-risk, LR, children). The matching criteria and the resulting sociodemographic distributions in the HR and LR group are shown in Table 1. Note that the LR group cannot be considered as representative of a general population because this group was matched to the HR group on the amount of time spent in institutional rearing, which potentially biases the control sample toward more psychopathology.

None of the HR or LR children was psychotic or otherwise severely mentally disturbed at the commencement of the study as judged during a brief clinical evaluation performed by a psychiatrist using a semistructured interview.

Table 1. Characteristics of the high-risk and low-risk samples at the inception of study in 1962

	High-risk	Low-risk
Cases (n)	207	104
Boys (n)	121	59
Cirls (n)	86	45
Mean age[a] (years)	15.1	15.1
Mean social class[b]	2.2	2.3
Mean education (years)	7.0	7.3
In children's homes[c]	16.0	14.0
Mean number of years in children's	9.4	8.5
homes[c]	26.0	22.0
Rural residence[d] (%)		

[a] Age rounded to nearest whole year; range = 10–19.
[b] Based on a scale of 0 (low) to 6 (high).
[c] We only considered experience in children's homes of 5 years or greater duration in the matching procedure. Many more of the HR children had been to children's homes for brief periods while their mothers were hospitalized.
[d] Rural residence defined as living in a town with a population of 2500 persons or fewer.

Course of Project

We have conducted two major diagnostic re-evaluations of the HR and LR samples, the first in 1972–1974 (Schulsinger 1976), and most recently in 1986–1989 (Parnas et al. 1993). The sample attrition over this period appears in Table 2. As can be seen, we achieved a high rate of participation at both of the follow ups. Some additional data collections took place in this project:

1. In 1981–1982 all the children's biological fathers were examined psychiatrically in order to assess the degree of assortative mating and its potential effects on the diagnostic outcome in the offspring (Parnas 1985).
2. In the mid-1980s we performed a "lifetime" hospital chart follow up of the schizophrenic mothers in order to reassess their diagnoses as well as to enrich our previous databank on their illness. These hospital charts were coded by a psychiatrist for a variety of social, premorbid, and illness characteristics (Jørgensen et al. 1987).
3. As a part of the final diagnostic 1986–1989 assessment all HR and LR subjects who agreed to it were CT scanned with a Siemens Somatom DRG scanner with an average of 13 slices taken parallel to supraorbitomeatal line, each with a thickness of 8 mm and no overlap. Images were processed on a 256 × 256 matrix. Morphological measures were computed blindly to the risk and diagnostic status of a subject: ventricular and sulcal volume, ventricular and sulcal area, linear measurements of ventricles and cortical sulci, interhemispheric and Sylvian fissure, and measures of cerebellar structural pathology.

Table 2. Attrition in the high-risk (HR) and low-risk (LR) samples at the two adult psychiatric follow ups and on a lifetime basis

	Assessment					
	1972–1974		1986–1989		Lifetime	
	HR	LR	HR	LR	HR	LR
Diagnosed						
Interview (n)	173	91	172	94	188	101
Hospital records only (n)	3	0	1	0	4	0
	176	91	173	94	192	101
Not diagnosed						
Deceased (n)	5	0	13	1	6	0
Emigrated (n)	5	1	11	4	5	1
Refused (n)	20	10	8	3	3	0
Untraceable (n)	1	2	2	2	1	2
	31	13	34	10	15	3
Diagnosed						
Of original total[a] (%)	85.0	87.5	83.6	90.4	92.8	97.1
Of total available[b] (%)	89.3	88.3	94.5	94.9	98.0	98.1

Note: the lifetime figures account for those subjects who received diagnoses at either or both of the adult assessments.
[a] Denominators were original numbers (207 HR and 104 LR).
[b] Denominators were original numbers minus those deceased or emigrated.

Premorbid Data

In 1962 a variety of premorbid meaures were taken on HR and LR groups. The decision of what was relevant to examine was, naturally, dictated by the state of the art of that time and the investigators own interests [e.g., at that time Mednick (1960), proposed a "learning theory" of schizophrenia]. The premorbid data obtained can be grouped as follows:

Early Rearing Environment. A detailed account of each subject's upbringing was obtained, especially the amount of time spent in different rearing conditions such as being raised (a) by biological parents; (b) by a foster family; (c) in an institutional setting (for details see Parnas et al. 1985).

Obsteric History. A midwife protocol concerning delivery and perinatal conditions was obtained on each subject and scored for obstetric complications. Note that this information is "objective" and not retrospective. At that particular historical period, all deliveries in Denmark were legally required to be assisted by a midwife, who always completed a standardized form concerning the course of delivery. These midwife protocols were archived and stored on a national basis. This standardized protocol contained all relevant obstetric information (e.g., length and course of delivery, drugs used, signs of asphyxia, signs of pre-maturity, etc.) (for details see Parnas et al. 1982a).

Psychophysiological Assessment. A galvanic skin response was measured in all subjects.

Psychological Assessment. All subjects completed the Word Association Test (WISC), part of the Thematic Apperception Test, and a shortened version of the MMPI.

School Behavior Questionnaire. The schools of all children were contacted and the teacher most familiar with the subject completed a 30-question questionnaire, containing questions focused upon a variety of behavioral characteristics (e.g., "schizoid" or "acting-out" bahavior) and school achievement.

Psychiatric Assessment. All subjects were briefly (0.5–1 h) interviewed by a senior psychiatrist (Schulsinger), who completed a self-devised rating scale on a variety of characteristics after the interview and, in addition, completed an Adjective Check-list (ACL) comprising 241 adjectives which were rated as appropriate or inappropriate to the subject.

Diagnostic Assessments

The first major diagnostic assessment was performed by an experienced clinical psychologist (Dr. Hanne Schulsinger) from 1972 through 1974, when the mean age of the subjects was 25 years, blindly to the risk status of the interviewee. The interview, which lasted approximately 3 h, consisted of the Present State Examination (PSE) (Wing et al. 1974), the Current and Past Psychopathology Scales (CAPPS) (Endicott and Spitzer 1972), and other items focusing on sub-schizophrenic characteristics, e.g., formal thought disorder, perceptual aberrations, qualities of emotional rapport, and interpersonal relations (in total more than 700 items). All relevant hospital charts were obtained and scored for psychopathological and treatment information by Dr. Hanne Schulsinger blindly to the risk status of the subject. On the basis of this assessment, each subject was allocated a clinical diagnosis following the ICD-8 (WHO 1967), as well as computerized CATEGO-CAPPS and DIAGNO-PSE diagnoses.

The psychopathological data from the 1972–1974 assessment were converted into DSM-III-R diagnoses. The number and quality of the 1972–1974 psychopathological data were quite sufficient for this purpose. The conversion proceeded in two stages. First, a computer program was written which picked up the relevant symptom items for each category of DSM-III-R Axis I and II disorders, and allocated tentative diagnoses to each subject. Then, a careful review of all narrative interview summaries was performed in order to examine issues such as duration of illness and sequences of and relations between symptoms. If necessary, the computer-generated diagnoses were then modified in view of the information from the narrative summaries (blindly to the risk status of the subject). In this way each subject was allocated a set of DSM-III-R diagnoses with respect to both Axis I and II.

In 1980 we conducted a pilot project with computed cranial tomography of a subsample of the HR group (Schulsinger et al. 1984). The subsample consisted of 34 subjects who were interviewed by HS using the PSE. The psychopathological data from this assessment were used to form DSM-III-R diagnoses as described above.

From 1986 through 1989 he entire sample was reexamined by an experienced psychiatrist (Dr. Bjørn Jacobsen, who also conducted the Danish Adoption Studies). He was blind to the results of the previous psychiatric assessments and to the risk status of the interviewee. The interviews made use of the Schedule for Affective Disorders and Schizophrenia–Life Time Version (SADS-L) (Spitzer and Endicott 1987), the PSE, the PSE life-time ratings of psychotic symptomatology, the PSE Syndrome Checklist for current and lifetime psychopathology, the Scales for the Assessment of Positive and Negative Symptoms (SANS and SAPS) (Andreasen 1983, 1984), and the Personality Disorder Examination (PDE) (Loranger et al. 1985). The entire sample was also rescreened for psychiatric hospitalizations since the previous follow up, and information on psychopathology and treatment history was extracted blindly to the risk status of the subject. As before, the interviewer completed a written summary on each subject summarizing all relevant social, occupational, and psychiatric information. On the basis of this assessment, the psychiatrist allocated to each subject a set of DSM-III-R diagnoses on both Axis I and II for this last assessment.

Allocation of the Lifetime Diagnoses

Each subject was allocated a lifetime diagnosis if he or she fulfilled the criteria of a DSM-III-R disorder at any point of the course of the study. Because many subjects received multiple diagnoses during their lifetimes, a hierarchic system was constructed whereby the ultimate diagnosis was the highest in the hierarchy. Aside from the practical need to have a single summary diagnosis for each person, the hierarchic system also serves to maintain order of severity relevant to the original purpose of this high-risk study. Thus, the major psychoses were considered as the most important followed by cluster A personality disorders (with the following decreasing hierarchy: (a) schizotypal; (b) paranoid; and (c) schizoid personality disorder) as the next most important category in the hierarchy. Since such a hierarchical ordering may underestimate the frequency of other disorders among the relatives of schizophrenic probands, the aggregation of single disorders without imposing the hierarchy effect was also examined.

Diagnostic Outcome

The hierarchical and non-hierarchical frequencies of lifetime DSM-III-R diagnostic outcomes in the HR and LR groups are presented in Table 3. The rate of functional psychosis (including definite schizophrenia and other non-affective, non-organic psychoses) is significantly higher among the HR group compared to the LR group. When the diagnostic categories were considered separately, both schizophrenia and atypical psychosis were elevated in the HR group. One of the two definite schizophrenics among the LR group had a mother who was hospitalized for a paranoid psychosis after the beginning of the study in 1962. The age-corrected morbid risk for schizophrenia in the HR and LR groups (by Weinberg's abridged method, with the customary upper age

Table 3. Lifetime DSM-III-R diagnoses in the high-risk (HR) and low-risk (LR) samples

	Hierarchical				Non-hierarchical				
	HR		LR		HR		LR		
Diagnosis	(n)	(%)	(n)	(%)	(n)	(%)	(n)	(%)	X^2
Axis I functional psychosis									
Schizophrenia	31	16.1	2	2.0	31	16.1	2	2.0	13.3***
Schizoaffective disorder	1	0.5	0	0.0	1	0.5	0	0.0	0.5[a]
Schizophreniform psychosis	1	0.5	0	0.0	1	0.5	0	0.0	0.5[a]
Atypical psychosis	7	3.6	1	1.0	7	3.6	1	1.0	1.8[a]
	40	20.8	3	3.0	40	20.8	3	3.0	16.9***
Cluster A personality disorders									
Schizotypal personality	36	18.8	5	5.0	40	20.8	5	5.0	12.7***
Paranoid personality	5	2.6	0	0.0	9	4.2	0	0.0	4.9[a]
Schizoid personality	1	0.5	0	0.0	21	10.9	2	1.9	7.3**
	42	21.9	5	5.0	46	26.0	5	5.0	16.6***
Other Axis I and II disorders									
Major depression	9	4.7	8	7.9	23	12.0	9	8.9	1.2
Organic brain syndrome	5	2.6	5	5.0	13	6.8	5	5.0	0.4
Substance abuse	12	6.3	5	5.0	46	24.0	12	11.9	6.1**
Other Axis I disorder	6	3.1	7	6.9	24	12.5	10	9.9	0.4
Other Axis II disorder	10	5.2	8	7.9	43	22.4	19	18.8	0.5
	42	21.9	33	32.7	100	52.1	39	38.6	4.8*
No diagnosis (Axis I and II)	68	35.4	60	59.4	68	35.4	60	59.4	15.5***

Note: χ^2 statistics based on non-hierarchical diagnoses. Psychotics could receive Axis II diagnoses premorbidly only. Totals for non-hierarchical tabulations of Cluster A personality disorders and other Axis I and II disorders are minus those subjects with co-morbid diagnoses in the same category.

[a] One or more expected cell frequencies are less than 5; chi-square may not be a valid test.

$^{*}p < 0.05$; $^{**}p < 0.01$; $^{***}p < 0.001$.

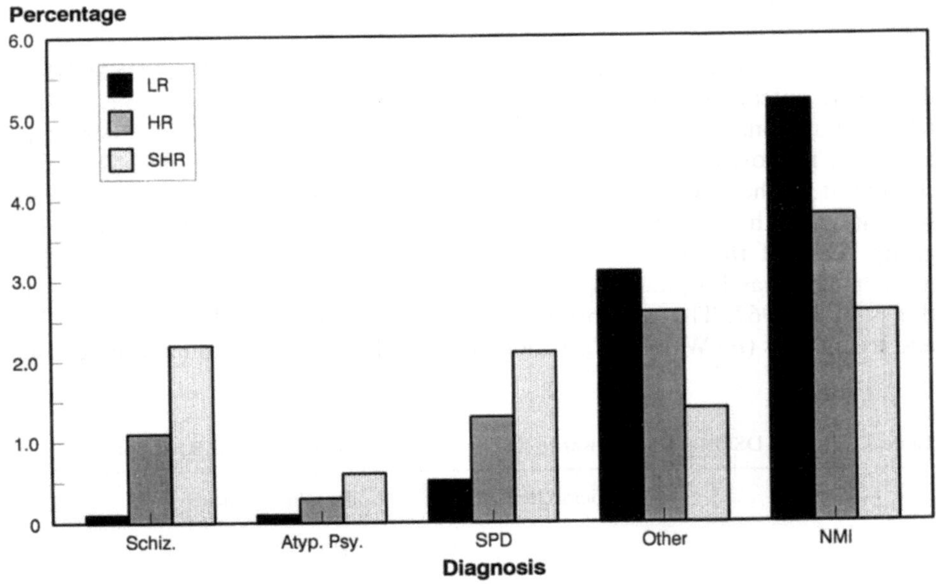

Fig. 1. Relationship between familial risk status (parental diagnosis) and offspring adult diagnosis in the Copenhagen High-Risk Project. *LR, low risk; HR*, high risk; *SHR*, super-high risk, i.e., offspring of couples with assortive mating; *Schiz*, schizophrenia; *Atyp Psy*, atypical psychosis; *SPD*, schizotypal personality disorder; *NMI*, no mental illness

limit of 40 years) is 17.1% and 2.9%, respectively. Excluding HR subjects whose mothers did not meet the DSM-III-R criteria does not change risk estimates by more than a fraction of 1%.

Cluster A personality disorders aggregate at a significantly higher rate among the offspring of schizophrenic mothers. As can be seen in Table 3, many cases of paranoid personality disorder and nearly all cases of schizoid personality disorder were "co-morbid" for schizotypal personality disorder, hence in the lierarchical comparison schizoid personality disorder does not appear to aggregate among offspring of schizophrenics.

Taken together, nearly 43% of the offspring of schizophrenic mothers received a lifetime diagnosis within the schizophrenia spectrum.

Affective disorders did not aggregate in the offspring of schizophrenic mothers, either as a primary diagnosis or as a co-morbid diagnosis among those with spectrum disorders. In addition, there were no cases of affective psychosis in either the HR or the LR group.

There was a significantly higher rate of substance abuse among the HR offspring A significantly higher proportion of the LR group did not meet criteria for any Axis I or II disorder as compared to the HR group.

It is noteworthy that 21% of the subjects with functional psychosis (finding similar to other population studies) and 51% of the subjects with Cluster A personality disorders have never experienced any psychiatric treatment in their lifetimes.

Figure 1 shows lifetime morbid risks for schizophrenia, atypical psychosis, schizotypal personality disorder, other diagnosis, and no mental illness as a function of genetic risk status of the subject (i.e., parental diagnosis). As can be seen the risk for the schizophrenia spectrum disorder in the offspring increases as a function of the degree of familial risk.

Schizophrenia: Premorbid Behavioral Characteristics

For the purpose of this communication three preliminary scales from the Adjective Checklist (ACL) completed by a psychiatrist during the 1962 premorbid assessment (see above: Premorbid Data) were constructed to examine which behavioral characteristics predicted later schizophrenia (Table 4). These scales were constructed on an a priori basis and concerned three behavioral dimensions of theoretical interest: the degree of cognitive disorganization, eccentricity–peculiarity, and the degree of withdrawn, schizoid behavior. Table 4 shows means, standard deviations and Cronbach's a for each of the scale. As can be seen, preschizophrenics as compared to the rest of the HR sample scored significantly higher on cognitive disorganization and eccentricity–peculiarity, whereas the scale measuring withdrawn-shut-in characteristics (Schizoid behavior) did not discriminate between the groups.

Table 4. Behavioral predictors of schizophrenia

	Schizophrenia		All other HR		F	*p* (Mann-Whitney)
	Mean	SD	Mean	SD		
CogDis ($\alpha = 0.60$)	1.25	1.38	0.57	0.97	11.12	0.001
Ex ($\alpha = 0.67$)	1.03	1.40	0.54	1.00	5.24	0.02
Schiz ($\alpha = 0.80$)	5.22	3.20	4.96	3.47		ns

CogDis, distracted, confused, absent-minded, split, inattentive.
Ex, peculiar, silly, clownish, bizarre, strange, eccentric.
Schiz, shy, inhibited, introvert, shut-in, nervous, reserved, withdrawn, evasive, anxious.

Schizophrenia: Early Rearing Environment

We have previously reported (Parnas et al. 1985) that schizophrenia detected by the 1972–1974 follow up was strongly predicted by institutional rearing of the HR offspring. This effect remained stable even when all relevant potential confounding effects were taken into account (statistically controlled for), such as age of onset of maternal schizophrenia, maternal and paternal absence, etc.

Some of our previous analyses indicated that early environment could be stratified in the following order of influence upon the risk of adult schizophrenia for the HR offspring: (a) rearing in a non-schizophrenic foster family exerted a protective effect; (b) rearing by a schizophrenic mother was "neutral"; (c) rearing in an institutional setting (children's homes) exerted a pathogenic influence.

We interpreted these findings by hypothesizing that early emotional deprivation may decompensate a vulnerable individual to overt schizophrenic illness. Note that institutional rearing in that particular historical period was mainly concerned with providing sufficient hygienic and nutritional conditions but was definitely insufficient in providing adequate emotional contact from stable adult figures.

Table 5 presents the data on early rearing conditions, when HR psychotics are subdivided into chronic and "non-chronic" schizophrenics. Chronic schizophrenics are characterized by the unremitting course of their psychotic symptoms and pronounced "negative" symptoms (they were diagnosed as chronic according to the DSM-III-R subtype; from the clinical point of view they were characterized by early onset, non-remitting psychotic symptoms, and elevated levels of "negative" symptoms).

It is noteworthy that chronic schizophrenics do not differ from the rest of the sample with respect to institutional rearing; it is only non-chronic schizophrenics who were exposed to prolonged institutional rearing.

Schizophrenia: Obstetric Complications

We have previously reported (Parnas et al. 1982a) that schizophrenics experienced more frequent and more severe obstetric complications that other HR subjects, especially the schizotypes. We also reported that the frequency of obstetric complications did not discriminate between HR and LR groups.

Tabel 6 shows the distribution of the total score of pregnancy and birth complications (PBC) among the HR life-time diagnostic groups. As can be seen, chronic schizophrenics experienced significantly more perinatal stress than other HR individuals, including non-chronic schizophrenics and schizotypes.

Precursors of Brain Alterations

We examined structural brain changes as a function of (a) presence or absence of obstetric (delivery) complications; and (b) the degree of genetic risk for the offspring (Cannon et al. 1993). The latter was stratified in the following way: (a) LR, i.e., control children whose parents did not have schizophrenia; (b) HR children, i.e., whose mothers were schizophrenic; (c) super-high-risk (SHR), i.e., offspring of couples with assortative mating (where the mother was schizophrenic and the father also suffered from a schizophrenia spectrum disorder, mainly the schizotypal personality disorder. The results of these analyses can be summarized as follows:

Table 5. Institutional rearing as a function of stratified psychosis diagnosis

	Non-chronic schizophrenia (n = 27)		Chronic schizophrenia (n = 12)		Cluster A (n = 42)		Other and normal (n = 108)		F	p	Contrast
	Mean	SD	Mean	SD	Mean	SD	Mean	SD			
Institutional rearing during years 1–5 (months)	13.9	20.5	3.92	11.8	3.50	7.5	3.82	9.8	5.97	0.0001	NC > C & CLA & O & N (0.01)

NC, nonchronic schizophrenia; C, chronic schizophrenia; CLA, cluster A personality disorders; O & N, other diagnosis and no mental illness.

Table 6. Pregnancy and birth complications as a function of diagnosis

	Nonchronic schizophrenia (n = 24)		Chronic schizophrenia (n = 12)		Cluster A (n = 34)		Other and normal (n = 84)		F	p	Contrast
	Mean	SD	Mean	SD	Mean	SD	Mean	SD			
PBC total score	1.13	1.7	2.38	2.5	1.29	1.7	1.35	1.9	3.57	0.06	D > ND & ClA & O & N (0.05)

PBC, pregnancy and birth complications.

1. The main effect of genetic risk for schizophrenia was significant, and the interaction of risk and brain region was not significant. This pattern indicates a generalized effect of genetic risk for schizophrenia on structural brain pathology regardless of region (Fig. 2). The scores increase in a stepwise, linear fashion with increasing level of genetic risk for schizophrenia. A contrast testing for a linear increase in the means with increasing level of risk was significant ($F[1,138] = 14.3$, $p < 0.001$).

2. There was a significant overall main effect of delivery complications, and the interaction of delivery complications with brain region was not significant. These findings indicate a generalized effect of delivery complications on enlargement of the cerebro-spinal fluid (CSF) spaces in the brain. Univariate tests indicated that the delivery complications effect was statistically significant for the ventricular measure ($F[1,138] = 10.0$, $p < 0.001$) but not for the sulcal or cerebellar measures.

3. There was a significant interaction between delivery complications, risk for schizophrenia, and brain pathology (see Fig. 2). In univariate analyses it was found that the risk by delivery complications interaction term was significant for the ventricular measure ($F[2,138] = 3.55$, $p < 0.05$) but not for the sulcal and cerebellar measures. For the ventricular measure, the degree of pathology related to delivery complications increased in proportion to the degree of genetic risk for schizophrenia (see left section of Fig. 2). Pairwise t tests indicated that delivery complications were associated with significant increases in ventricular CSF–brain ratios in the SHR and HR groups ($p <$ 0.01, 0.004, respectively) but not in the LR group ($p > 0.88$). For the sulcal

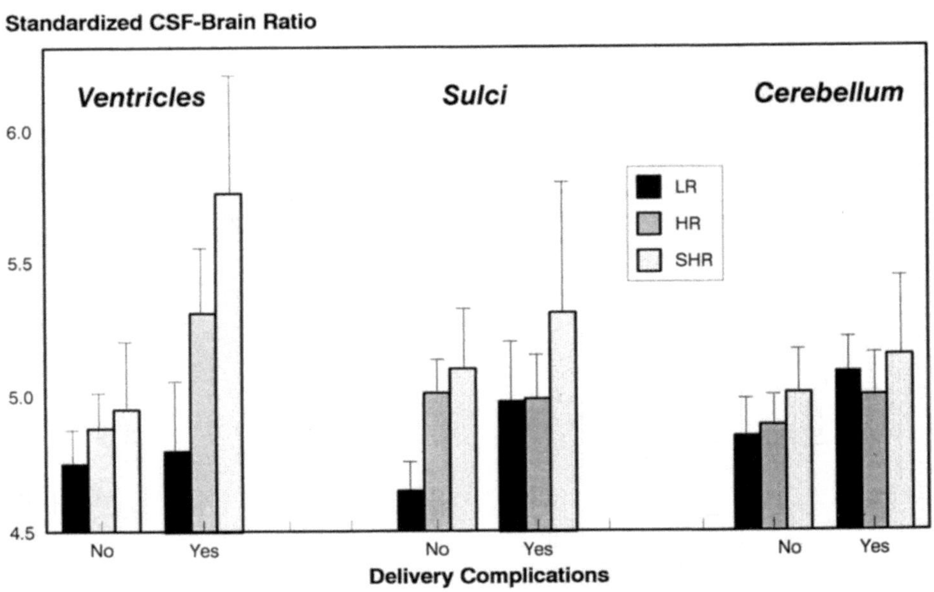

Fig. 2. Structural brain changes as a function of obstetric complications and genetic risk. *LR*, low risk; *HR*, high risk; *SHR*, super-high risk

and cerebellar measures, the effect of delivery complications did not vary as a function of risk level. Pairwise t tests indicated no significant differences in sulcal or cerebellar pathology as a function of delivery complications within any of the three genetic risk groups. ($p > 0.10$).

Structural Brain Alterations Versus Diagnosis

Figure 3 shows the distribution of structural brain abnormalities across adult diagnostic groups among HR offspring (Cannon et al. in press). Schizophrenics differ significantly from all other groups on measures of ventricular atrophy, whereas both schizophrenics and schizotypes show more pronounced cortical atrophy than other HR offspring. We interpret these findings according to a diathesis–stress model. It seems that cortical changes are more genetically determined, whereas central brain atrophy is strongly related to perinatal asphyxia (Cannon et al. 1989, 1993).

Discussion

Diagnostic Outcome

This study confirms previous evidence of significant familial aggregation of schizophrenia. The morbid risk (MR) rates in the present study are most

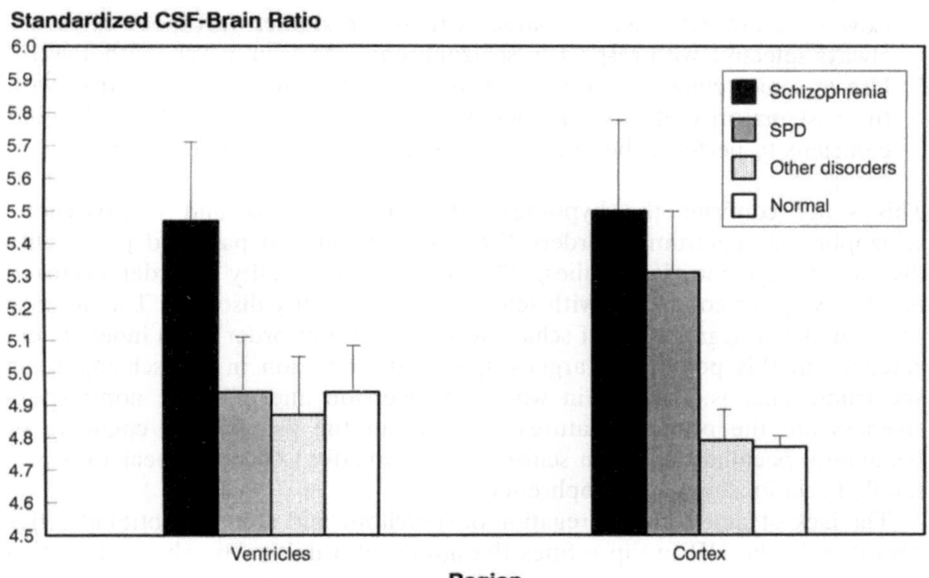

Fig. 3. Structural brain changes as a function of the diagnostic status in the HR offspring. *SPD*, schizotypal personality disorder

directly comparable to those found in other recent studies using operational criteria and blind diagnosis of relatives of probands and controls (reviewed by Kendler 1988). The rate of schizotypal and paranoid personality disorders in offspring of schizophrenics in our study is within the range found in other major recent studies (i.e., between 15% and 30%). However, the rate of schizophrenia (16.2%) in the present study is much higher than that reported in the recent studies (i.e., between 3.6% and 8.9%). There are several possible, not mutually exclusive, explanations for this difference:

1. Longitudinal design with multiple assessments (which facilitates (a) detection of 1-week active psychotic phase required in the DSM-III-R for the diagnosis of schizophrenia; and (b) evaluation of temporal stability of prodromal/residual schizophrenic symptoms).
2. It seems that our sample of schizophrenic mothers may have been more severely affected than probands in other studies. The effect of our sampling practices becomes a selective shift away from acute onset, remitting, paranoid schizophrenics and toward severely ill, chronic, schizophrenics with both positive and negative symptoms. Such selective sampling in the present study is likely to become associated with increased familial aggregation of schizophrenia. The illness severity bias of our sample of mothers is likely to amplify morbid risk for schizophrenia in the offspring both through unusually pronounced assortative mating and through unstable rearing conditions in childhood.
3. In examining offspring only, all family studies indicate that of all first-degree relatives, children have the highest morbid risk for schizophrenia (Gottesman and Shields 1982).
4. Low drop-out rates as compared with other studies (drop-out rates are always selective with respect to schizophrenia spectrum psychopathology).
5. Use of senior clinicians to perform the interviews (which is in contrast with the vast majority of modern studies which employ "reliably trained" non-clinicians to perform the interviews, sometimes even by telephone).

This study confirms the hypothesis that schizophrenia and non-psychotic schizophrenia spectrum disorders (i.e., schizotypal and paranoid personality disorders) aggregate in families. The schizoid personality disorder occurred nearly always as co-morbid with schizotypal personality disorder. The absence of a significant aggregation of schaizoid personality disorder as an independent category in this population argues against its inclusion in the schizophrenia spectrum. That is, patients in whom introversion and affective non-responsiveness are the primary features but without the presence of cognitive or behavioral peculiarities (as in schizotypal personality) do not appear to bear a familial relationship to schizophrenia.

The lack of significant aggregation of (psychotic and non-psychotic) affective disorders in the HR group refutes the notion of a unitary psychosis, i.e., that affective and schizophrenic symptomatology are different expressions of similar underlying pathology.

Behavioral Predictors of Onset

Even though schizophrenia does not appear before late adolescence, researchers and clinicians (irrespective of their theoretical commitments) share, with few exceptions, the conviction that the main etiological factors operate in the individual's very first months of life (genetic disposition and/or some very early noxious environmental factors). Premorbid traces of these etiological factors may therefore be considered as being of the utmost importance for the understanding of the development of frank schizophrenic episodes. Such premorbid phenotypic characteristics are usually designated as vulnerability traits. They may also perhaps be observed in individuals predisposed to schizophrenia who never develop a frank psychotic episode. Such traits are therefore potentially important in genetic studies of schizophrenia (Matthysse and Parnas 1992).

In our preliminary analyses we find evidence that overt schizophrenia does not arise ex nihilo, but is rather preceded by a number of behavioral anomalies which are related to Bleuler's (1911) concept of the so-called fundamental symptoms (Parnas et al. 1982b; Parnas and Jørgensen 1989; Cannon et al. 1990). The overall picture emerging from our study and other studies is that no single gross behavioral or symptomatological abnormality is identifiable in all preschizophrenics. However, disturbances are detected in several domains: aggressiveness, introversion, neophobia, difficulties in interpersonal relations, defective emotional rapport, eccentricity, flattening of affect, and formal thought disorder (Parnas and Bovet 1991; Bovet and Parnas 1993).

It is noteworthy that "pure schizoid" premorbid behavior in terms of withdrawal and introversion, is not predictive of later schizophrenia (Table 4), which is in line with our diagnostic outcome finding that pure schizoid personality disorder does not belong to the schizophrenia spectrum (Table 3). The disorganization of affect and cognition, predictive of later schizophrenia can be apprehended as indicators of a defective attunement between the individual and his or her world, or of what has been called (in the Continental phenomenological tradition) as autistic vulnerability (Minkowski 1927). In a phenomenological perspective, the term autism refers to a particular way of relating to the world which is already exhibited premorbidly and which becomes accentuated during frank schizophrenic illness. Two points have to be emphasized: (a) autism cannot be considered as a symptom of the medical model, in the way jaundice can point to some liver pathology, but it reflects a certain change of human existential pattern; (b) autism, from the phenomenological perspective, does not refer to withdrawal and rich fantasy life (as originally defined by Bleuler 1911), but points to a defective core attunement between the individual and the world. The existential pattern of a (pre) schizophrenic (the individual and his/her world) is "decentered" from intersubjectivity (i.e., from a shared network of perspectives on the world) Parnas and Bovet 1991; Bovet and Parnas 1993).

Environmental Predictors of Onset and Type of the Schizophrenia Spectrum Disorder

So far empirical studies have failed to identify environmental stressors specific for the development of schizophrenia, and our study is no exception in this respect. The overall pictures that emerges from epidemiological research is that environmental causal contribution to the schizophrenic onset is of a non-specific nature; i.e., that a variety of stressors may be operating in some sort of an apparently "random" fashion.

Our results indicate that obstetric complications contribute to the development of chronic-type schizophrenia with unremitting course of psychotic symptoms and elevated levels of "negative" symptoms. This particular schizophrenic outcome is associated with structural brain pathology, especialy ventricular enlargement. On the other hand, remitting schizophrenia and other non-affective non-organic psychosis is predicted by emotional deprivation during early childhood. HR individuals who remain at the schizotypal level seem spared from environmental insults.

These findings can be interpreted within the diathesis–stress model in which subtle schizotypal features, reflective of autistic vulnerability, are most proximate to the genetic substrate; overt schizophrenia, on the other hand, is a result of further decompensation due to other genetic influences and environmental stressors acting upon a vulnerable individual.

We find no evidence of genetic/non-genetic dichotomy corresponding to Crow's (1990) hypothesis on the origin of positive and negative schizophrenia. Similarly, our data and other recent studies (Günther et al. in press) suggest that obstetric complications cannot be considered as a substitute for a genetic liability in the etiology of schizophrenia. Rather, our results indicate that adult cerebral atrophy in schizophrenia is the result of an interaction between genetic risk and obstetric complications. In other words, the greater the genetic vulnerability, the greater the likelihood of obstetric complications contributing to neural loss. Even the original promotors of the simple negative-positive (organic–genetic) dichotomy of schizophrenia (Andreasen et al. 1992) now express some scepticism:

> The initial model that posited two discrete syndromes, a negative syndrome characterized by structural brain abnormalities and a positive syndrome characterized by reversible biochemical abnormalities, is almost certainly too simple. More complex models must be developed that continue to build our growing knowledge concerning normal brain structure and function, the compex interactions of neurocircuits connecting various brain regions, and the complexity of the psychopathology of schizophrenia.

Theoretical Considerations

The prevailing paradigm in psychiatric thinking and conventional cognitive sciences has roots in the so-called faculty psychology of mid-19th century, which divided mental functions into three domains: intellectual (cognitive), affective, and volitional (conative). The concept of a mental function was usually related to the discrete operation of an organ.

The "physicalistic" paradigm, currently prevailing in psychiatry and which is conceptually anchored in faculty psychology, postulates that each single mental content (symptom) must be somehow caused by a specific neurochemical brain process, localized in some specific brain area. The typical research paradigm in schizophrenia is therefore to search for associations between certain symptoms of the illness and corresponding dysfunctions in specific brain areas. So far not much has been gained from this approach.

This cognitive model has been sustained during the last 4 decades by a powerful community of researchers on artificial intelligence (AI). However, the limitations of AI programs based on this classical paradigm lead to the progressive development of an alternative model, variously designated as "connectionism," "parallel distributed processing," "network model," or "subsymbolic processing model." This model is now rapidly developing but is still largely confined to cognitive sciences.

The connectionist model proposes that mental capacities reflect certain global activation patterns of interconnections in very large ensembles of resonating neurons. Each neuron has multiple and changing, context-dependent responses and enters into various "cooperations" with other neurons in the production of various mental contents. A specific mental content (e.g., a delusional idea or a hallucination) is not localized, but is a function of a certain, widely distributed, pattern of activity in neuronal networks. The cardinal difference between the connectionist and the classical approach is that in the connectionist model, representational (contentual), conscious level has no "direct" single physical support. The content of consciousness is dependent on the underlying network's properties and peculiarities, as well as being rooted in the network's history. There are three interconnected consequences of this model which are relevant for research in psychopathology in general and in the domain of schizophrenia in particular:

1. A content of consciousness, e.g., a delusional idea, can neither be ascribed to a single, particular region of the brain nor can it be conceived of as a sign of a single dysfunction. In this model, the very concept of "mental function," as originally envisaged by the faculty psychology, cannot be maintained.
2. The substrate of any mental activity is inseparable from its continuous transformation and therefore rooted in the history of the organism. It implies that even two identical twins, having an identical delusional idea, will be in different "brain states" (they will have different configurations of their active neuronal networks due to their different histories).
3. The brain as a system has auto-organizing capacities which are active during its entire life span. These auto-organizing capacities are, on the one hand, inherent to the system's own peculiar properties and dynamics. They are, however, also bound to the history of the system, i.e., to the history of the system's neural interconnections. In this perspective, a mental content

(e.g., delusional idea) cannot be solely viewed as a defective product of the brain machine but also as a way of negotiating an adjustment between the organism and its world. The notion of "the organism with *its* world" (Varela et al. 1991) implies a radical departure from an objectivistic–physicalistic epistemology.

The pathogenesis of schizophrenia from a connectionist perspective involves a shift of attention from localized specific brain dysfunctions specifically responsible for abnormal symptoms to focus on malintegration of the interplay between brain functions (Parnas and Bovet, in press). Such conceptual shift should be encouraged by the "fuzziness" of research findings in schizophrenia, in that dysfunctions are being identified in almost all domains studied, but with low specificity, low sensitivity, and high context dependence with low reproducibility.

The concept of malintegration is an old idea in schizophrenia research:

> Expressions like "discordance" (Chaslin), "intrapsychic ataxia" (Stransky), "intrapsychic dysharmony" (Urstein), "loss inner unity" (Kraepelin), "dissociation" (Claude and Levy-Valensi), "schizophrenia" (Bleuler) point to the idea that it is not this or that function which is disturbed, but much more their cohesion, their harmonious interplay, in its globality. To make use of an image, the essential disorder does not alter one or many faculties, whatever their order in the hierarchy of functions be, but resides between them, in the "interstitial space." (Minkowski 1926, own translation)

Autistic vulnerability is most likely much more proximate to the underlying pathophysiology and etiology of schizophrenias than any single psychotic symptom considered as a manifestation of brain dysfunction. It could be proposed that autistic vulnerability is the manifestation of a certain configuration state of neuronal interconnections, historically shaped by interactions between specific genetic liability and environmental influences. Such a vulnerable network may become grossly dysfunctional (i.e., give rise to the phenotypic schizophrenia) due to a variety of additional, apparently non-specific stressors which change the network's functioning beyond a certain adaptive level. It has been demonstrated in simulation research that reduction of neural connectivity ("neural pruning") may leave the network well-functioning up to a certain point where dysfunctions begin to occur. Neural loss (e.g., due to obstetric complications) may lead to irreversible dysfunction, with a chronic schizophrenic picture, whereas remitting schizophrenia may reflect temporary adaptations of a vulnerable network, conditioned by its inherent characteristics and unique developmental history.

Insights from further research in the connectionist paradigm, including the nature of erratic functioning of such networks, may provide clues to the understanding of schizophrenic difficulties.

Acknowledgement. The research was supported by the NIMH grant R01 MH41469 (Dr. Parnas).

References

American Psychiatric Association (1987) Diagnostic and statistical manual of mental disorders, 3rd rev edn. Washington DC

Andreasen NC (1983) The Scale for the Assessment of Negative Symtoms (SANS). University of Iowa, Iowa City

Andreasen NC (1984) The Scale for the Assessment of Positive Symptoms (SAPS). University of Iowa, Iowa City

Andreasen NC, Flaum M, Arndt S, Alliger R, Miller D, O'Leary D (1992) Positive and negative symptoms and syndromes: Assessment and biological correlates. In: Ferrero FP, Haynal AE, Sartorius N (eds) Schizophrenia and affective psychosis. Nosology in contemporary psychiatry. Libbey, Rome, pp 89–96

Bleuler E (1911) Dementia Praecox oder Gruppe der Schizophrenien. In: Aschaffenburg G (ed) Handbuch der Psychiatrie, part 4, 1 Deuticke, Leipzig (English translation: Dementia praecox or the group of schizophrenias. International University Press, New York)

Bovet P, Parnas J (1993) Schizophrenic delusions: a phenomenological approach. Schizophr Bull 19:579–597

Cannon TD, Mednick SA, Parnas J (1989) Genetic and perinatal determinants of structural brain deficits in schizophrenia. Arch Gen Psychiatry 46:883–889

Cannon TD, Mednick SA, Parnas J (1990) Antecedents of predominantly negative and predominantly positive-symptom schizophrenia in a highrisk population. Arch Gen Psychiatry 47:622–632

Cannon TD, Mednick SA, Parnas J, Schulsinger F, Præstholm J, Vestergaard Å (1993) Developmental brain abnormalities in the offspring of schizophrenic mothers: 1. Contributions of genetic and perinatal factors. Arch Gen Psychiatry 50:551–564

Cannon TD, Mednick SA, Parnas J, Schulsinger F, Præstholm J, Vestergaard Å. Developmental brain abnormalities in the offspring of schizophrenic mothers: 2. Structural characteristics of schizophrenia and schizotypal personality disorder. Arch Gen Psychiatry (in press)

Crow TJ (1990) The continuum of psychosis and its genetic origins. Br J Psychiatry 156:788–797

Endicott J, Spitzer R (1972) Current and Past Psychopathology Scales (CAPPS). Arch Gen Psychiatry 27:678–687

Gottesman II, Schields J (1982) Schizophrenia. The epigenetic puzzle. Cambridge University Press, Cambridge

Günther F, Bovet P, Hohlfeld P. Obstetric complications and schizophrenia: a case-control study. Br J Psychiatry (in press)

Jørgensen A, Teasdale TW, Parnas J, Mednick SA, Schulsinger F (1987) The Copenhagen high risk project: the diagnosis of maternal schizophrenia and its relation to offspring diagnosis. Br J Psychiatry 151:753–757

Kendler KS (1988) Familial aggregation of schizophrenia and schizophrenia spectrum disorders. Evaluation of conflicting results. Arch Gen Psychiatry 45:377–383

Kety SS, Rosenthal D, Wender PH, Schulsinger F, Jacobsen B (1978) The biologic and adoptive families of adopted individuals who became schizophrenic: prevalence of mental illness and other characteristics. In: Wynne LC, Cromwell R, Matthysse S (eds) The nature of schizophrenia. Wiley, New York, pp 28–37

Loranger AW, Sussman VL, Oldham JM, Russakoff LM (1985) Personality disorder examination. A structured interview for making diagnosis of DSM-IIIR personality disorders. Cornell Medical College, White Plains, NY

Matthysse S, Parnas J (1992) Extending the phenotype of schizophrenia: implications for linkage analysis. J Psychiatr Res 26:329–344

Mednick SA (1960) The early and advanced schizophrenic. In: Mednick SA, Higgins J (eds) Current research in schizophrenia. Edwards, Ann Arbor, pp 69–78

Mednick SA, Schulsinger F (1965) A longitudinal study of children with a high risk for schizophrenia: a preliminary report. In: Vandenberg S (ed) Methods and goals in human behavior genetics. Academic, New York, pp 255–296

Meehl PE (1962) Schizophrenia, schizotypy and schizotaxia. Am Psychologist 17:827–838
Minkowski E (1926) La notion de perte de contact vital avec la réalité et ses applications en psychopathologie. Jouve and Cie, Paris
Minkowski E (1927) La schizophrénie. Psychopathologie des schizoïdes et des schizophrènes. Payot, Paris
Parnas J (1985) Mates of schizophrenic mothers: a study of assortative mating from the American-Danish high risk project. Br J Psychiatry 146:490–497
Parnas J, Bovet P (1991) Autism in schizophrenia revisited. Compr Psychiatry 32:1–15
Parnas J, Bovet P (1994) Negative/positive symptoms of schizophrenia: clinical and conceptual issues. Nordic J Psychiatry (in press)
Parnas J, Jørgensen A (1989) Premorbid psychopathology in schizophrenia spectrum. Br J Psychiatry 155:623–627
Parnas J, Schulsinger F, Teasdale TW, Schulsinger H, Feldman PN, Mednick SA (1982a) Perinatal complications and clinical outcome within the schizophrenoa spectrum. Br J Psychiatry 140: 416–420
Parnas J, Schulsinger F, Schulsinger H, Teasdale TW, Mednick SA (1982b) Behavioral precursors of the schizophrenia spectrum. Arch Gen Psychiatry 39:658–664
Parnas J, Teasdale TW, Schulsinger H (1985) Institutional rearing and diagnostic outcome in children of schizophrenic mothers. A prospective high-risk study. Arch Gen Psychiatry 42:762–769
Parnas J, Cannon T, Jacobsen B, Schulsinger H, Schulsinger F, Mednick SA (1993) Life-time DSM-IIIR diagnostic oucomes in offspring of schizophrenic mothers: the results from the Copenhagen High Risk Study. Arch Gen Psychiatry 50:707–714
Schulsinger H (1976) A ten year follow-up of children of schizophrenic mothers. A clinical assessment. Acta Psychiatr Scand 53:371–386
Schulsinger F, Parnas J, Petersen ET, Schulsinger H, Teasdale TW, Mednick SA, Møller L, Silverton L (1984) Cerebral ventricular size in the offspring of schizophrenic mothers: a preliminary study. Arch Gen Psychiatry 4:602–606
Spitzer RL, Endicott J (1977) Schedule for affective disorders and schizophrenia-lifetime version, 3rd edn. New York State Psychiatric Institute, New York
Varela FJ, Thompson E, Rosch E (1991) The embodied mind. Cognitive science and human experience. MIT Press, Cambridge
Wing JK, Cooper JE, Sartorius N (1974) The measurement and classification of psychiatric syndromes. Cambridge University Press, Cambridge
World Health Organization (1967) Manual of the international classification of diseases, injuries and causes of death, 8th edn. WHO, Geneva

Neurobehavioral Studies of Schizophrenic Children: A Developmental Perspective on Schizophrenic Disorders

R.F. Asarnow, R. Caplan, and J.R. Asarnow

Introduction

Attempts to understand the nature of the CNS dysfunction in schizophrenia have to contend with two important facts: (a) there is no unitary brain lesion found in all schizophrenic patients; and (b) the brain lesions found in some schizophrenic patients are not unique or specific to schizophrenia "because other disorders that are associated with pathology in similar brain areas usually do not present as schizophrenia" (Weinberger 1987). In addition, there often appears to be a significant delay (more than a decade) between the time wher the brain lesions first occur and the first appearance of positive symptoms of schizophrenia. Post mortem studies suggest that brain lesions occur very early in life (in many instances during foetal development) and are static, non-progressive (Weinberger 1987).

Weinberger (1987) has proposed a "neurodevelopmental model" of schizophrenia to account for the delay between the creation of the brain lesion and the first onset of psychotic symptoms. It is suggested that schizophrenia is a "neurodevelopmental disorder in which a fixed brain lesion from early in life interacts with certain normal maturational events that occur much later" (Weinberger 1987). The fixed brain lesion involves congenital, static pathology of the limbic system and prefrontal cortex. What is "pathophysiologically" distinct about schizophrenia, is neither the location nor the cause off the "lesion", but is, instead, the interactions between this "lesion" and the normal course of maturation of the neural system affected by it (Weinberger 1987). The lesion remains relatively silent until early adulthood when the normal maturation of certain brain structures, notably dorsolateral prefrontal cortex occurs. Animal studies suggest that the dorsolateral prefrontal cortex is relatively uncommitted or plastic until adulthood, and as a consequence is relatively silent.

The neurodevelopmental model provides a useful framework to begin to understand the complex pathways from CNS dysfunction to the symptoms of schizophrenia. In this paper we present neurobehavioral data which can inform neurodevelopment models of schizophrenia. These data are products of our

Department of Psychiatry, University of California–Los Angeles, 48-256 NPI, 760 Westwood Plaza, Los Angeles, CA 90024, USA

H. Häfner and W.F. Gattaz (Eds.)
Search for the Causes of Schizophrenia, Vol. III

ongoing family genetic study of schizophrenic children. Our research project's working assumption is that schizophrenia is a disorder involving impaired CNS functioning. For many patients we believe there may be a genetic basis for the CNS disturbance. We are studying schizophrenic children because of prior research (for reviews see Fish 1977; Asarnow 1994) suggesting that they might represent a more homogeneous and more severe form of schizophrenia in which the CNS dysfunction is more readily discernable than in adult onset schizophrenia.

In this paper we will summarize two complimentary sets of data which help elucidate the nature of the neurobehavioral impairment in schizophrenia. The first set of data are the results of studies which retrospectively characterize neurobehavioral development in schizophrenic children. These studies reveal that prior to the onset of psychotic symptoms there are certain neurobehavioral impairments in schizophrenic children. The second set of data are a series of cross-sectional studies examining cognitive/neuropsychological functioning and thought and language disorder in schizophrenic children. These studies attempt to delimit the cognitive processes which are impaired in schizophrenic children.

Subjects

The children included in the studies summarized below were ascertained and diagnosed by the University of California–Los Angeles (UCLA) Childhood Schizophrenia Research Program. A case registry maintained by the Los Angeles County Department of Mental Health containing information (e.g., demographic and diagnostic data, the name and location of the agency providing treatment, and recent history of treatment contacts) for the more than 40 000 children and adolescents who receive services funded by the Los Angeles County Department of Mental Health Agencies each year was used to identify the agencies which were most likely to treat schizophrenic children. We then identified geographically contiguous clusters of agencies so that we could recruit subjects from those agencies in an efficient manner given the geographical expanse of Los Angeles County.

To be included in the study, the children had to have an onset of schizophrenic psychosis prior to 13 years of age. Children were interviewed by an experienced clinician (either a child clinical psychologist or child psychiatrist) using a semi-structured interview, the Kiddie-SADS-E (Orvaschel and Puig-Antich 1987) supplemented by the psychosis and mood disorders sections of the K-SADS-P to provide more detailed coverage of the severity of psychotic and mood symptoms. The best-informed parent was first interviewed about the child's history of psychiatric symptoms; then the child was interviewed. In addition, hospital and school records were collected. The clinician interviewing the parent and child reviewed the diagnostic interviews as well as ancillary information with the second clinician to arrive at a consensus diagnosis. Only children who obtained consensus diagnoses of DSM-III-R schizophrenia were included in the UCLA Childhood Schizophrenia Research Program. Children were excluded if they were retarded (Full Scale I.Q. < 70), had a history of central nervous system diseases which could produce psychotic-like symptoms (e.g., temporal lobe epilepsy), or had taken drugs which could produce psychotic symptoms.

Normal children were drawn from local school districts that have children who are demographically comparable to the schizophrenic children.

Different subsets of our cohort of schizophrenic subjects are included in the various studies summarized below. To indicate the demographic characteristics of this group of children, we have summarized the demographic characteristics of the entire cohort in Table 1.

Precursors of Psychotic Symptoms in Schizophrenic Children

To identify the precursors of psychotic symptoms in schizophrenic children we employed a follow-back design, similar to that used in studies of adult schizophrenia (Watt and Lubensky 1976) in which school hospital and other early records of a group of children who subsequently developed schizophrenia were examined and objectively rated [see Watkins et al. (1988) for details of this study]. This provided a means of rating the presence of behavior and symptoms at different age intervals using multiple, contemporaneous reports of the children at each interval.

Two rating scales were used to characterize the children: a DSM-III Symptom Rating Scale and the Achenbach Child Behavior Checklist (CBCL) (Achenbach and Edelbrock 1983). The DSM-III Symptom Rating Scale was used to rate each symptom of schizophrenia, schizotypal personality disorder, childhood onset pervasive developmental disorder and infantile autism included in DSM-III. The CBCL provided broad coverage of behavioral problems associated with a wide variety of child psychiatric disturbances that were not covered by the Symptom Rating Scale. Symptoms were rated at each of four age ranges:

Table 1. Characteristics of schizophrenic children

	Mean	SD
Age (years) (n = 47)	10.5	1.9
Peabody picture vocabulary test – Revised (n = 47) (full-scale IQ)	93.9	14.7
Gender (n = 47)		
Males (n)	34	
Females (n)	13	
Handedness (n = 45)		
Right (n)	40	
Left (n)	5	

0–30 months, 31 months to 5 years 11 months, 6 years to 8 years 11 months, and 9 years to 11 years 11 months.

A symptom rating database was developed by obtaining copies of records from all hospitals, clinics, schools, physicians and other relevant professionals seen by the children prior to referral to UCLA. The database consisted of: (a) copies of all school records including cumulative folders, teacher reports, and school psychologist reports from each school, including preschool, that the child had attended; (b) copies of case records from clinics, hospitals, and agencies (e.g., regional centers, infant care programs) that had evaluated or treated the child since birth, including information from the family physician or pediatrician; and (c) all medical records for each child describing diagnostic and treatment procedures. This symptom rating database thus provided contemporaneous accounts of each child's behavior, medical status, and school performance in each of the four age ranges.

The majority of schizophrenic children had significant developmental delays beginning in infancy. Gross deficits in language development or no language prior to 30 months were reported in 72% of the schizophrenic children. Motor development problems were also seen in schizophrenic children, including delayed milestones and poor coordination in 72% of the children and hypotonia in 28% of the children. Comparison across the four age ranges showed clearly that the frequency of language deficits in the schizophrenic children gradually decreased.

There appeared to be two somewhat different patterns of symptom development prior to the onset of psychotic symptoms in these children. The children with the most severe language problems had a number of other autistic-like symptoms including peculiar speech, pervasive lack of responsiveness and self-mutilation. The balance of the children tended to have less severe language problems and no autistic-like symptoms.

The children with the more severe language problems and other autistic-like symptoms generally manifested schizophrenic psychotic symptoms one age range earlier (during the 6 year to 8 years 11 months) than the balance of the children. During this age range, the schizophrenic symptoms consisted primarily of incoherence, loosening of associations, other symptoms of formal thought disorder and flat or inappropriate affect in 71% of the children, with only two showing diagnostically significant delusions or hallucinations. By the 9 years to 11 years 11 months age range; however, 71% of these children developed diagnostically significant delusions and hallucinations in addition to formal thought disorder and flat or inappropriate affect. Differences in symptoms between children with and without autistic-like symptoms faded with age as rates of hallucinations and delusions increased for both groups in the 9 years to 11 years 11 months age range.

The other group, schizophrenic children without autistic-like symptoms, showed a relatively chronic gradual worsening of symptoms. While less disturbed during infancy than children with autistic-like symptoms, the majority of these children also had significant developmental delays beginning in infancy.

Most of the children *without* early histories of autistic like symptoms first manifested schizophrenic psychotic symptoms during the 9 years to 11 years 11 months age range. During the age range immediately preceding the onset of psychotic symptoms; however, 82% of these children were rated as *both* socially impaired and presenting with at least one of the following potential prodromata of schizophrenia identified in prior research: excessive anxiety and panic, constricted or inappropriate affect, magical thinking, suspiciousness, and undue social anxiety and hypersensitivity to criticism.

Interestingly, given other data suggesting that depressive symptoms are associated with more positive outcomes, schizophrenic children without autistic-like symptoms showed higher rates of depressive symptoms on the CBCL at the time of the onset of the schizophrenic episode, when compared to schizophrenic children *with* autistic-like symptoms. Prior to the first onset of psychotic symptoms, symptoms associated with the hyperactive factor of the CBCL were frequently observed in both groups of children including: speech problems, clumsy, acts too young for age, poor school work, daydreams, cannot concentrate, impulsive, hyperactive and confused were seen in over 50% of each group.

In general, there was a gradual developmental unfolding of a broad spectrum of symptoms affecting social, cognitive, sensory and motor functioning in these children, beginning many years before the appearance of schizophrenic symptoms-usually in early infancy. Overall, there were far more severe symptoms and far more social impairment prior to the onset of schizophrenia in these children than is seen in the childhood histories of schizophrenic adults.

Additional support for the view that impairments can be identified in schizophrenic children prior to the onset of the schizophrenic episode is provided by follow-back analyses conducted by Asarnow and Ben-Meir (1988) and Asarnow et al. (1994). Using a sample composed of schizophrenic children who had been psychiatric inpatients [a small subset of whom were also in the Watkins et al. (1988) sample], Asarnow et al. (1994) found that schizophrenic children were most likely to present with insidious as opposed to acute onset of disorder with only one of 21 schizophrenic children (4.76% of the sample) presenting with an acute onset. Schizophrenic children also showed impairments prior to the onset of their schizophrenic episodes. When compared to a group of child psychiatric inpatients with major depression, schizophrenic children showed significantly poorer levels of overall premorbid adjustment, as well as significantly poorer premorbid functioning in the area of peer relationships, scholastic performance, school adaptation, and interests.

Schizophrenic children did show variability in the level of premorbid adjustment, with the subgroup of schizophrenic children with depressive symptoms tending to show higher levels of premorbid adjustment.

The sex differences found in our cohort of schizophrenic children parallel those reported in the early histories of adult schizophrenics, where males are more likely to show early signs of deviant development than females (Goldstein 1980; Mednick et al. 1978).

High-Risk Studies

It is interesting to note that early delays and/or impairments similar to those described during the premorbid periods of childhood-onset schizophrenics, have also been observed among children considered at-risk for schizophrenia based on schizophrenia in their biological parents. In considering this literature, however, several points merit note. First, because most high-risk studies have defined risk on the basis of schizophrenic parentage, results may not generalize to the larger population of schizophrenics whose parents do not also suffer from schizophrenia. Second, variations in design, sampling, and assessment characteristics of high-risk studies, have made rigorous analyses of replicability very difficult. Third, it is important to remember that most of the children of schizophrenic parents will not develop schizophrenia. Despite these limitations prospective high-risk studies are particularly powerful for distinguishing between early impairments that pre-date the onset of the full-blown schizophrenic disorder and the consequences of having had the illness.

Several studies have reported findings of neurobehavioral impairments among children of schizophrenic parents. As elegantly reviewed by Fish et al. (1992), strong evidence is emerging across several studies pointing to neurointegrative deficits in a subgroup of children at-risk for schizophrenia. Although the cross-study evidence is not entirely consistent, and comparisons across diverse assessment procedures and samples are problematic, there is currently strong and replicated evidence that a subgroup of children at-risk for schizophrenia show early patterns of "pandysmaturation." As defined by Fish et al. (1992), pandysmaturation is characterized by (a) a transient delay in gross motor and/or visual motor development; (b) an abnormal profile of function on a developmental evaluation in which a child fails simple items while passing more complex items; and (c) retardation in skeletal growth. Evidence is also emerging that suggests that pandysmaturation during infancy is associated with higher levels of subsequent impairment, including schizophrenic and schizotypal symptoms.

Neurobehavioral impairments have also been observed among high-risk children during middle childhood. For example, in a recent report from the Jerusalem Infant Development Study, Marcus et al. (1993) noted that during middle childhood 44% of children of schizophrenic parents showed neurobehavioral dysfunctioning on perceptual-cognitive and/or motor tasks, as compared to 24% of children of parents with other forms of mental illness and 15% of children of parents with no mental illness. Interestingly, a subgroup composed of 40% of the children of schizophrenic parents showed neurobehavioral dysfunction *both* during infancy and middle childhood. This pattern of relatively persistent impairment was not observed in children whose parents were not schizophrenic.

Several studies have also reported impairments in attention and information processing among children at high risk for schizophrenia. These difficulties are most apparent on relatively complex tasks that tax children's information processing capacities (for review see Asarnow 1988). Reports have suggested that impaired attention and information processing performance is associated with children's adjustment levels, although evidence indicating that child or adolescent performance on attention and information processing measures predicts the onset of schizophrenia is lacking (e.g., Asarnow 1983; Cornblatt and Erlenmeyer-Kimling 1985). The promise of measures of attention and information processing in understanding the pathogenesis of schizophrenia is suggested, however, by other research documenting that attention and information processing impairments (a) remain relatively constant across changes in clinical state and characterize schizophrenic patients during periods of remission as well as acute disturbance; (b) show some degree of specificity to patients with schizophrenia; and (c) are under some degree of genetic control (for review see Asarnow et al. 1989).

Comment

Our retrospective studies of neurobehavioral development in schizophrenic children and prospective studies of children at risk for schizophrenia reveal the presence of certain neurobehavioral impairments well in advance of the first onset of psychotic symptoms. The earliest neurobehavioral impairments are observed during infancy and early childhood. The impairments involve delays in the acquisition of gross motor skills and language. Somewhat later, delays in the acquisition of fine motor coordination are also observed. One clear implication of these findings for neurodevelopmental models of schizophrenia is that the static brain lesions, which may occur very early in the development of some individuals who subsequently manifest schizophrenic psychotic symptoms, are silent only in the sense that they do not give rise to psychotic symptoms until adolescence. The neurobehavioral impairments observed early in the development of many schizophrenic children and children at risk for schizophrenia may represent the earliest manifestations of the underlying brain pathology in many schizophrenic patients.

Neuropsychological Functioning in Schizophrenic Children

Survey of Neuropsychological Functions

This section will summarize a series of studies conducted in R. Asarnow's laboratory which survey a range of neuropsychological functions using performance tasks shown to be sensitive to focal brain damage in patients with neurological disease. The goal of this survey was to identify the task demands and/or experimental conditions which elicit the least and most impairment in the performance of schizophrenic children. This is a critical first step in delimiting the cognitive processes which are impaired in schizophrenic children. The studies used either well validated clinical neuropsychological tasks or experimental versions of those tasks.

The presence of a differential deficit (i.e., impaired performance in certain functions and relatively normal performance in others) in schizophrenic children is vividly illustrated by a comparison (Asarnow et al. 1987) of the performance of non-retarded autistic and schizophrenic children on the WISC-R (Wechsler 1974), a standardized measure of intellectual functioning. The children's performance on the WISC-R was summarized by three independent factors which have been consistently found to account for the variance on the WISC-R. These factors (Sattler 1974; Kauffman 1979) are called "verbal comprehension," "perceptual organization," and "distractibility." The mean Full Scale IQs of the schizophrenic (mean = 87.0, SD = 9.6) and autistic mean = 91.4, SD = 16.1) children were similar to each other and were within the low normal range. The groups differed dramatically, however, in the pattern of factor scores.

Figure 1 summarizes the performance of the two groups across individual subtest scores (expressed as standard scores) grouped according to their loading on the three factors. The autistic and schizophrenic children did not significantly differ on the verbal comprehension and perceptual organization factors, but the schizophrenic children had significantly lower scores on the "distractibility" factor than the autistic children. The scores of the schizophrenic children on

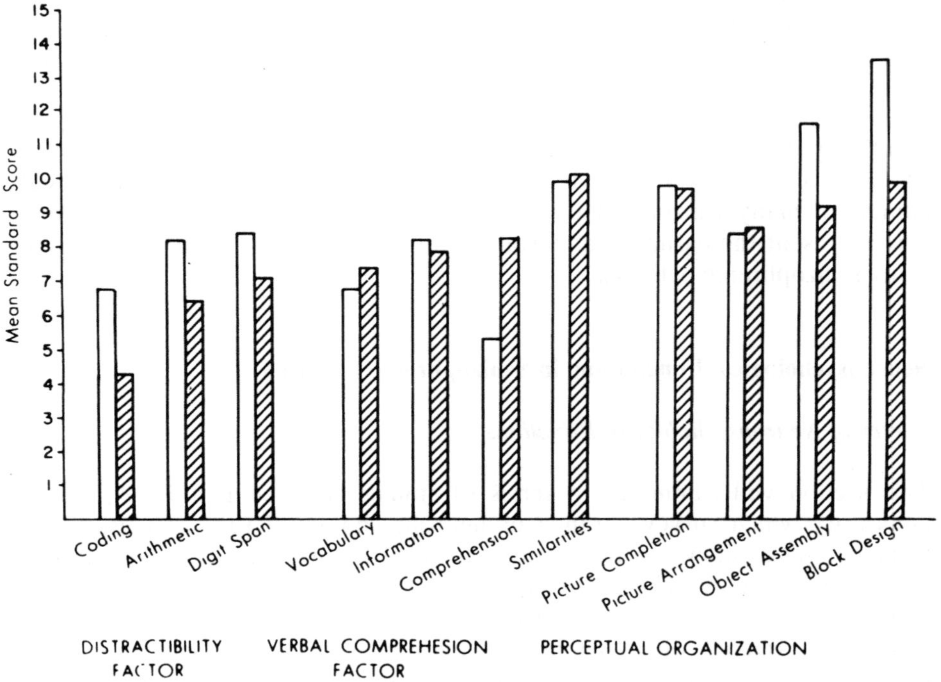

Fig. 1. Mean WISC-R subtest standard scores for non-retarded autistic (*open columns*) and schizophrenic (*hatched columns*) children. (From Asarnow et al. 1987)

the distractibility factor were more than one standard deviation below the mean of the standardization sample, and most importantly, significantly lower than the scores the schizophrenic children obtained on the verbal comprehension and perceptual organization factors. While there is dispute over the precise meaning of the distractibility factor, the subtests included in this factor (arithmetic, coding, and digit span) make extensive demands on controlled attentional processes because of their requirements for working memory, attention and speed of responding (Sattler 1974). This impairment appears to be somewhat specific since the schizophrenic children performed within the normal range on tasks tapping overlearned abilities (such as those tapped by the verbal factor).

Language Functioning

As noted above, the schizophrenic children in these studies scored within the normal range on the verbal comprehension factor of the WISC-R, which includes the information, similarities, vocabulary, and comprehension subtests. In addition, the schizophrenic children score within the normal range on the Peabody Picture Vocabulary Test–Revised (Dunn and Dunn 1981) a measure of receptive vocabulary (Schneider and Asarnow 1987).

The schizophrenic children do show impairments, however, on two other tasks which require the processing of auditory stimuli. Schizophrenic and normal children were administered the Token Test from the Multilingual Aphasia Examination (Benton and Hamsher 1989). The Token Test screens for receptive aphasia by requiring subjects to follow oral instructions of increasing complexity. For example, a child is instructed to pick up a green square, or to touch a large red square and then a small yellow circle. Inspection of Fig. 3 reveals that the schizophrenic children made significantly more errors than the normal controls on this test. The schizophrenic children had particular difficulty understanding complex grammatical transpositions or instructions that involved carrying out sequences of behavior involving multiple steps (Watkins et al., to be published).

The schizophrenic children's pattern of performance on the Seashore Rhythm and Token Tests does not indicate central impairments in the primary processing of auditory stimuli. Rather, it appears that the errors made by schizophrenic children on these two tasks (one using linguistic and the other non-linguistic stimuli) may reflect impairments in attention and/or memory since both tasks require subjects to sustain attention and to remember sequences of verbal instructions (on the Token Test), and to attend to and remember and reproduce sequences of non-verbal auditory stimuli (on the Seashore Rhythm Test).

Fine Motor Functioning

Visual motor coordination and fine motor speed were examined (Watkins et al., to be published) in schizophrenic children using the Purdue Pegboard

(Tiffin 1968). The Purdue Pegboard requires subjects to place small metal pegs into a line of small holes as rapidly as possible. Schizophrenic children performed significantly worse than normal control children and the standardization sample on the Purdue Pegboard using the dominant hand, non-dominant hand, and both hands. Impairments in visual motor coordination have also been observed in the children of schizophrenic parents (Fish 1977; Fish et al. 1992; Marcus et al. 1985; Walker and Emory 1983) as well as schizophrenic adults (Manschreck 1983). That schizophrenic children are impaired bilaterally on the task is *not* consistent with models that schizophrenic patients suffer from dysfunctions confined to only one hemisphere.

Perceptual Functioning

As noted above, schizophrenic children obtain normal scores on the perceptual organization factor of the WISC-R. The subtests included in the perceptual organization factor (picture arrangement, picture completion, object assembly, and block design) require subjects to arrange pictures similar to comic strip panels to make a story, to identify the missing part in a picture of a familiar object, to arrange the pieces of a puzzle to create a familiar object, and to arrange blocks to reproduce a picture of a design.

In addition, schizophrenic children perform within normal limits (Schneider and Asarnow 1987) on the Benton Judgement of Line Orientation Task (Benton et al. 1975), a task that requires subjects to discriminate among lines which differ only in angular orientation.

We recently examined in some detail a more circumscribed aspect of visual perception, form discrimination abilities, using two tests developed by Benton (Watkins et al., to be published). One task, the Visual Retention Test (Benton 1974) assesses form discrimination abilities by requiring subjects to draw designs presented to them in a booklet. A second task, the Visual Form Discrimination Test (Benton et al. 1977), assesses form discrimination abilities using a multiple choice format which makes much fewer demands on graphomotor functions than the Visual Retention Test. On the Visual Form Discrimination Test, subjects are required to pick one of four comparison designs which match a sample design placed above the comparison stimuli. Figure 2 summarizes the results of this study. Consistent with the findings, schizophrenic and mental age matched normal children did not differ in their ability to discriminate between simultaneously presented designs (Visual Form Discrimination Test) or to reproduce a design by copying (Visual Retention Test).

When memory demands were added to both tasks, however, schizophrenic children showed impairments relative to normal control children (see Fig. 2). A 15-s delay was interposed between the presentation of the design and when the subject was allowed to begin drawing the design on the Visual Retention Test. On the Visual Form Discrimination Test, after a 10-s presentation, the sample was removed so that the subject matched the comparison stimuli to their memory of the standard stimulus. The schizophrenic children performed significantly worse than the matched normal controls on both tests in which

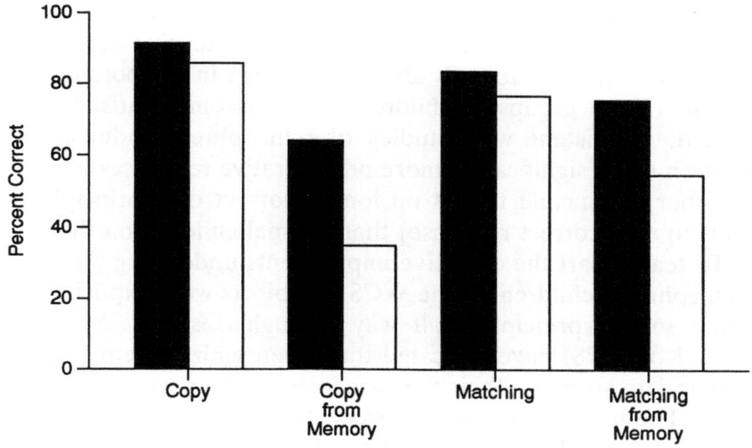

Fig. 2. Visual Retention Test and Visual From Discrimination Test: mean scores for schizophrenic (*open columns*) and normal control (*solid columns*) children when subjects copy or discriminate between designs in conditions with and without memory demands. (From Asarnow et al., in press)

subjects were required to hold the sample stimulus in short-term memory before responding.

The two Benton tests are sensitive measures of the kinds of visual–spatial impairments found in patients with focal lesions involving the right posterior cortex (Lezak 1983). The normal performance of the schizophrenic children on the visual retention and visual form discrimination tests indicates that they do not suffer from the kinds of visual–spatial impairments typically found in patients with focal lesions confined to the right posterior lobe. In contrast, adding a memory demand to these tests elicited impaired performance in schizophrenic children regardless of whether they responded by simple pointing or by copying a design.

Executive Functions

One interesting executive function is guided visual search. Rey's Tangled Lines Test (Rey 1964) measures guided visual search by requiring subjects to visually track a series of tangled lines that cross and overlap at multiple points. Schizophrenic children showed a tendency (which did not attain statistical significance) to perform worse (to perform more slowly and make more errors) on the Rey's Tangled Lines Test than normal children matched to them in mental and chronological age (Schneider and Asarnow 1987).

To begin to characterize the impairments in executive functions found in schizophrenic children we also tested subjects (Schneider and Asarnow 1987) on the Wisconsin Card Sorting Test (WCST) (Grant and Berg 1981). In numerous studies schizophrenic adults (e.g., Goldberg et al. 1987; van der Does and van den Bosch 1992) show deficits on the WCST. The WCST requires subjects to sort cards containing figures that differ in number, form and color into categories according to a single sorting principle that must be discovered by the subject. This task makes extensive demands both on verbal mediation and the subject's ability to engage in temporally extended problem solving. Three groups of children (schizophrenic, autistic, and normal) were studied. Consistent with studies of schizophrenic adults, the schizophrenic children made significantly more perseverative responses (sorts made according to either a principle that is no longer correct or a principle that consistently leads to an incorrect response) than normal children on the WCST.

To tease apart the cognitive impairments underlying the performance of the schizophrenic children on the WCST, subjects were explicitly taught the appropriate sorting principles half way through this task. Some previous studies (e.g., Koh 1978) have indicated that when schizophrenic adults are taught the appropriate strategy, their performance can be normalized on certain memory tasks. The major difference between the schizophrenic and autistic children in this study was that the schizophrenic children *increased* the number of non-perseverative errors (random responding) from the first to the second half of the WCST while autistic and normal children showed no change (see Fig. 3). Providing a sorting strategy *did not* facilitate the performance of the schizophrenic children on the WCST. Rather, their performance significantly deteriorated after they were taught the relevant sorting principles. Stuss et al. (1983) also found that a similar manipulation had a deleterious effect on the WCST performance of chronic schizophrenics with prefrontal leukotomies. In addition, K.M. Puschel (1980, Neuropsychological assessment and "hypothesis testing" in schizophrenic and brain damaged patients, unpublished manuscript) found that providing schizophrenic adults verbal directions impaired their performance on a hypothesis testing task. These findings are reminiscent of Rodnick and Shakow's (1940) and Steffy's (1978) reaction time studies where they found

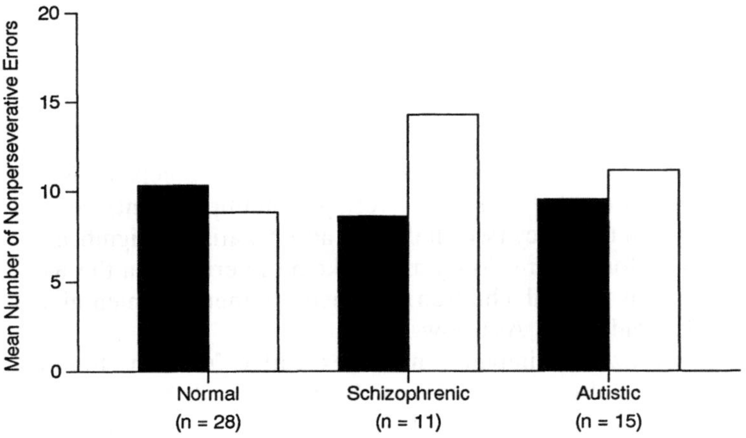

Fig. 3. Wisconsin Card Sorting Test: mean number of non-perseverative errors made by schizophrenic, autistic and normal children before (*solid columns*) and after (*open columns*) children were taught the relevant dimensions. (From Asarnow et al., in press)

that providing relevant information to schizophrenic adults sometimes impaired their performance. The relevant information provided during the second half of WCST performance requires the momentary integration of information from a variety of sources to direct an ongoing activity. Providing task relevant information to the schizophrenic child, whose processing capacity may already be taxed, resulted in the seemingly paradoxical effect of further impairing their performance.

Conclusions

It appears that rote language skills and simple perceptual functions are not impaired in schizophrenic children. In contrast, schizophrenic children perform poorly on a number of tasks which tap executive functions, fine motor speed and/or which place extensive demands on attention and/or short-term memory.

Tasks like the WCST tap multiple cognitive processes. Which of those processes is responsible for the impaired performance shown by schizophrenic children? In the next section, we will review a series of studies examining visual attention which attempt to isolate the cognitive processes which are impaired in schizophrenic children.

Studies of Visual Information Processing

This section summarizes a series of studies (for details see Asarnow and Sherman 1984; Asarnow et al., in press) which were designed to delimit the cognitive processes which are impaired in schizophrenic patients. We attempted to accomplish this by determining which of the multiple cognitive processes tapped by a forced choice, span of apprehension task is impaired in schizophrenic children. The span of apprehension task was originally developed by Estes and Taylor (1964) to provide an index of the rate of visual information processing. This task detects dysfunction not only in actively psychotic schizophrenic patients but also in individuals vulnerable to schizophrenia, including schizophrenic adults in clinical remission and the unaffected first degree relatives of schizophrenic probands (for a review see Asarnow et al. 1991).

Our attempts to isolate the cognitive processes impaired by schizophrenia are guided by information processing models. Information processing models describe "the structures and processes by which individuals register, encode, select, maintain, transform, store and retrieve information" (Nuechterlein and Asarnow 1989). Early information processing models emphasized the role of structural limitations on cognitive functions. Modern information processing models, in contrast, acknowledge the role of processing resources (capacity) in setting limits on the information processing that can be carried out in a fixed unit of time. Processing resources are broadly conceptualized as the limited fuels, processes and skills that are available at a given moment to enable performance of cognitive tasks (Hirst and Kalmar 1987).

In experiment I of Asarnow and Sherman (1984), schizophrenic, mental age-matched normal children, and a group of younger normal children were administered the same span of apprehension task used in studies of adult schizophrenics and their relatives. Subjects were told that either a **T** or **F** would be flashed briefly (for 50 ms) on a backward projection screen along with other letters. They were instructed to report after each trial which of the two target letters had been presented. The target stimulus was embedded in arrays containing either 0, 2, 4, or 9 *non-target* letters. Subjects received four scores, which were the number of correct detections of the target stimuli for each array size. Schizophrenic children showed impaired performance relative to mental age-matched normals on the five- and ten-letter arrays of the span task but not on the one- and three-letter arrays.

Next, we (Asarnow et al., to be published) determined the diagnostic specificity of these impairments by comparing the schizophrenic children to another group of children with attentional impairments, children with Attention Deficit Disorder. The slope of the function relating the probability of detecting the target to array size was steeper in the schizophrenic than attention-deficit disorder (ADD) children suggesting that schizophrenic children have a slower rate of visual information processing than ADD children.

The results from experiment I suggest that all groups of subjects were engaged in serial search when performing on the partial report span task. Serial processing demands focal attention. It is characterized by directing attention "serially to different locations, to integrate the features within the same spatio-temporal 'spotlight' into a unitary percept" (Treisman and Gelade 1980). One of the defining characteristics of a serial mode of processing is that there is an incremental cost (increased reaction time or errors) when subjects are required to detect targets in displays with increasing numbers of distractors. This is exactly what happened on the Estes and Taylor span task in experiment I. As the number of distractors increased from 0–2 to 4–9, the target detection rates for all groups decreased. The fact that the schizophrenic children and the younger normals showed a greater "cost" with increased number of distractors than mental age-matched normals suggests that their serial search is either initiated more slowly or employed less efficiently than that of the older normal group.

A convergent result emerged from experiment II of Asarnow and Sherman (1984). Visual scan paths were studied by modifying the stimulus matrix so that each target stimulus appeared eight times in each quadrant. When the target is in the upper quadrants, as opposed to the lower quadrants, both the schizophrenics and the mental age-matched normals showed a significantly greater probability of correctly detecting the target. This suggests both the schizophrenic and older normal children consistently began their serial search in the upper quadrants, and that their iconic image of the stimulus display faded before they could adequately process the lower quadrants.

The results of experiments I and II suggest that both schizophrenic and normal children use serial search to detect the target on partial report span of apprehension tasks. Schizophrenic children appeared to be delayed in the initiation of serial search and/or carry out serial search more slowly than normal children (see Sherman and Asarnow 1985). Serial search makes extensive demands on processing resources.

Taken collectively, these experiments lead to the working hypothesis that schizophrenic children show impairments on tasks which make extensive demands on processing resources and perform relatively normally on tasks which make less extensive demands on processing resources.

Developmental Studies of Formal Thought Disorder

Another set of data which can help delimit the cognitive processes which are impaired in schizophrenic children are detailed studies of key cognitive symptoms of schizophrenia. These symptoms may represent the interface between the clinical manifestations of schizophrenia and the impaired neurobiological processes which underlie this disorder. In this section we will first summarize a series of studies conducted in R. Caplan's laboratory which describe the clinical manifestations of Formal Thought Disorder (FTD) in childhood onset schizophrenia. We will then briefly review another set of Caplan's studies which suggest that different signs of FTD are linked to different cognitive impairments.

Formal thought disorder is a composite of clinical signs that reflect the aberrant way a patient sometimes presents his or her thoughts to a listener. FTD has been regarded a hallmark of schizophrenia since the early descriptions of this illness (Kraepelin 1896). Included in the clinical signs of FTD are loose associations and incoherence (DSM-III-R) (American Psychiatric Association 1987), illogical thinking, loose associations, incoherence, and poverty of content of speech (DSM-III) (American Psychiatric Association 1978), and derailment, digressive speech, tangentiality, circumstantiality, echolalia, and perseveration (Andreasen 1979). Although widely studied in schizophrenic adults (Andreasen and Grove 1986; Holzman et al. 1986; Harrow et al. 1989; Oltmanns et al. 1985) FTD has not been extensively studied in schizophrenic children.

Any examination of FTD in children must be informed by current information on the normal development of communicative abilities in children (Caplan 1994). A child's ability to present his/her thoughts to a listener in a coherent manner begins to develop between ages 2 and 3 years (Shatz 1982). It undergoes a major advance during middle childhood and reaches mature adult levels during adolescence (McTear 1985; Shatz 1982; Tunmer and Herriman 1984). For this process to evolve adequately, a child must develop cognitive, linguistic, and pragmatic competence. FTD is likely to involve impairments in one or more of these abilities.

The assessment of FTD during middle childhood confronts two major methodological problems (Caplan et al. 1989). First, young normal children are often unaware of an adult listener's needs and assume that the adult makes logical (Piaget 1959) and linguistic (Maratsos 1976) connections for them. Without developmentally appropriate operational definitions and norms for formal thought disorder in middle childhood, clinicians could, therefore, overdiagnose or underdiagnose such immature communication skills as instances of illogical thinking and loose associations.

A second problem in measuring formal thought disorder in youngsters is the relative paucity of spontaneous or expansive speech in middle childhood. Whereas adults talk spontaneously in "paragraphs" – thus providing the listener with speech samples long enough for rating formal thought disorder – young children use only one to two utterances to express themselves. In addition, children under the age of 9 years may resist answering interview questions, particularly those that probe for psychotic symptomatology.

Caplan et al. (1989) developed the Story Game (Caplan et al. 1989), an indirect interviewing technique, to elicit adequate speech samples for the rating of FTD from children of different ages. A trained clinician administers the three parts of the Story Game to the child. In the first and last part the child hears an audiotaped story, retells the tale, and answers standardized questions on each story. In the second part of the Story Game, the child makes up a story chosen from four topics.

Caplan et al. (1989) operationalized the four DSM-III signs of FTD for use in children and developed the Kiddie FTD Rating Scale (K-FTDS), an instrument based on Andreasen's Thought, Language, and Communication Scale (Andreasen 1979). An initial study (Caplan et al. 1989) demonstrated that two of the four DSM-III signs, illogical thinking and loose associations, could be reliably rated in schizophrenic and schizotypal children. The two remaining DSM-III signs of FTD, incoherence and poverty of content of speech, could not be reliably measured in children. Younger (under 7 years of age) schizophrenic patients showed significantly more illogical thinking and loose associations than schizophrenic patients over 7 years of age (Caplan et al. 1989, 1990a) (Fig. 4). Eighty percent of the schizophrenic children with a mental age under 7 years had illogical thinking compared to 58% and 36% of the schizophrenic children above ages 7 and 9.6 years, respectively. Loose associations were found in 90%, 75%, and 50% of the schizophrenic children with a mental age below 7 years, between 7 and 9.6 years, and above 9.6 years. Incoherence and poverty of content of speech occurred infrequently in both schizophrenic and normal children.

As expected, schizophrenic (Fig. 4) children showed more illogical thinking and loose associations scores than normal children with whom they were yoked by chronological and mental age (Caplan et al. 1989, 1990a). The age of onset, not the duration of illness correlated significantly with both the severity of the schizophrenic children's illogical thinking and loose associations scores (Caplan et al. 1990a).

Illogical thinking and loose associations were also found in normal children under age 7 years (Caplan et al. 1989, 1990a). Normal children older than age 7 years had minimal illogical thinking, but no loose associations (Caplan et al., in press).

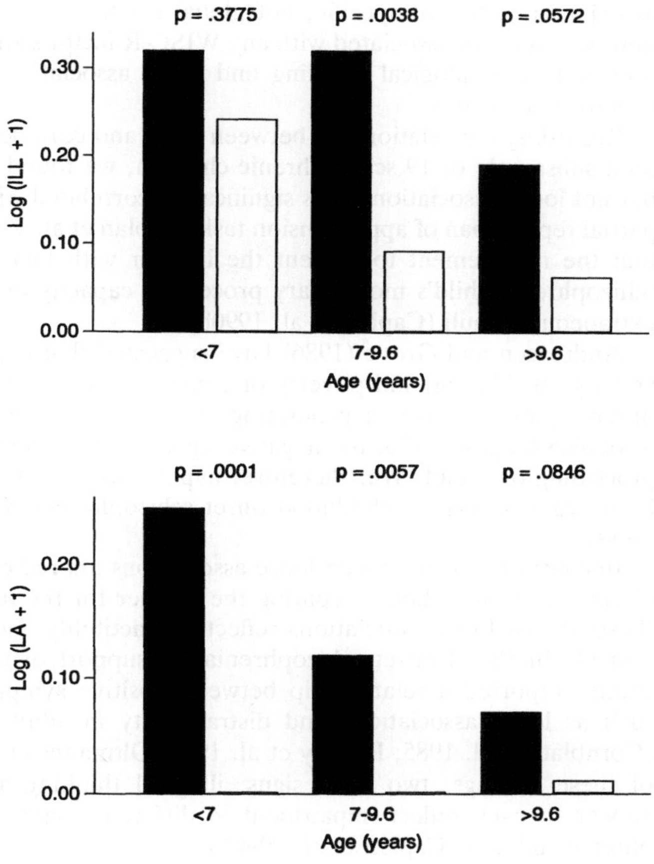

Fig. 4. Illogical thinking and loose associations in schizophrenic (*solid columns*) and normal (*open columns*) children. (From Asarnow et al., in press)

Cognitive Correlates of FTD

The next set of studies examining the cognitive correlates of FTD in schizophrenic children addressed three questions. First, is FTD associated with a global cognitive deficit in schizophrenic children? Second, what is the relationship between FTD and the performance of schizophrenic children on a partial report span of apprehension task? Third, does FTD reflect impaired cognitive development?

Regarding the relationship between FTD and global deficits, in schizophrenic children illogical thinking and loose associations were not correlated with fullscale I.Q. scores (Caplan et al. 1990a, in press). Loose associations scores were, however, significantly correlated with the WISC-R distractibility and verbal comprehension factor scores (Caplan et al. 1990b) suggesting that loose associations reflected specific, not global cognitive deficits. Illogical thinking, however, was not associated with any WISC-R factor score raising the intriguing possibility that illogical thinking and loose associations could have different cognitive correlates.

Regarding the relationship between FTD and core deficits of schizophrenia, on a subsample of 19 schizophrenic children, we found that illogical thinking, but not loose associations, was significantly correlated with performance on the partial report span of apprehension task (Caplan et al. 1990b). We hypothesized that the requirement to present the listener with logical reasoning taxes the schizophrenic child's momentary processing capacity and ability to screen out extraneous stimuli (Caplan et al. 1990b).

Andreasen and Grove (1986) have suggested that negative FTD signs, such as illogical thinking and poverty of content, reflect the difficulty adult schizophrenic patients have in generating conversation. Neuchterlein et al. (1986) have also suggested that the negative signs of adult schizophrenics could reflect processing overload. We, therefore, hypothesized that illogical thinking might be a negative sign of childhood onset schizophrenia (Caplan et al. 1990b, in press).

In contrast, the child with loose associations unpredictably changes the topic of conversation without preparing the listener for the topic change. We hypothesized that loose associations reflect distractibility and that this is a positive sign of childhood onset schizophrenia. In support of this hypothesis, several studies reported a relationship between positive symptoms of schizophrenia, such as loose associations and distractibility in adult schizophrenic patients (Cornblatt et al. 1985; Harvey et al. 1988; Oltmanns et al. 1985). On the basis of these findings, two FTD signs, illogical thinking and loose associations, appear to each reflect impairment in different cognitive functions in schizophrenic children (Caplan et al. 1990b).

Finally, to answer the third question on the relationship between FTD and cognitive development, we examined the relationship between K-FTDS scores and Piagetian measures of conservation using the Goldschmid-Bentler Concept Assessment Kit-Conservation (Goldschmid and Bentler 1968). With this instrument, a child successfully performs a conservation task if he/she understands that changes in the form or shape of matter do not involve changes in the quantity of matter (Piaget 1959). Schizophrenic and schizotypal children performed more poorly than normal children in the conservation of continuous and discontinuous matter (Caplan et al. 1990c). To conserve continuous matter, the child needs to recognize that there is no change in the amount of water transferred from a narrow and tall container to a flat and shallow container. In the discontinuous matter task a glass full of popcorn is divided between six smaller glasses.

The performance of the schizophrenic and schizotypal children on these two tasks was associated with their illogical thinking scores (Caplan et al. 1990c). Their ability to conserve matter on tasks which did not involve a concomitant change in the matter, as well as in the container, was associated with the presence of both illogical thinking and loose associations (Caplan et al. 1990c).

These findings provided additional evidence that illogical thinking and loose associations reflect different cognitive deficits.

From an information processing perspective, the acquisition of conservation skills in normal children involves the ability to attend to multiple cues simultaneously (O'Bryan and Boersma 1971; Flavell 1985), to discriminate salient cues (O'Bryan and Boersma 1971; Flavell 1985), and to respond to perceptual stimuli using socially-appropriate reality-based reasoning (Acredelo 1982; Flavell 1985; Mehler and Bever 1967). Caplan et al. (1990c) concluded that to conserve continuous and discontinuous matter successfully, the child needs to screen out the extraneous information derived from the containers and to focus his/her attention on the matter. In the other conservation tasks, however, the child was only exposed to the changes in matter and there is less of a demand to screen out extraneous stimuli. The findings of the conservation and span of apprehension studies, therefore, are convergent and support the association between information processing deficits and illogical thinking (Caplan, in press).

Discourse Skills

To present the listener with coherent information, the speaker needs to use such linguistic devices, as cohesion and reference patterns, to tie together his/her clauses (sentences) (Halliday and Hassan 1976). These linguistic devices enable the listener to follow who and what the speaker is referring to and how what he/she says relates to what the speaker or the listener previously said.

Adult schizophrenics with and without FTD underutilize some of these discourse devices and overutilize others (Harvey 1983; Harvey et al. 1986; Rochester and Martin 1979). Using Halliday and Hassan's (1976) analysis of cohesion on transcriptions of the videotaped Story Game (Caplan et al. 1989), we found that, like schizophrenic adults, schizophrenic children spoke less and did not provide the listener with enough links (cohesive ties) to previous utterances or with enough reference (referential cohesion) to people, objects, or events mentioned in earlier utterances compared to normal mental age matched subjects (Caplan et al. 1992). Schizophrenic children also broke the flow of conversation by referring to people, objects, or events in their immediate surroundings and by not focussing on the ongoing conversation (exophoria) (Caplan et al. 1992). Like thought-disordered schizophrenic adults (Harvey 1983; Harvey et al. 1986; Rochester and Martin 1979), schizophrenic children with loose associations confused the listener by the unclear and ambiguous way they referred to people, objects, and events (unclear/ambiguous reference) (Caplan et al. 1992).

Schizophrenic children, however, had three discourse deficits which distinguished their speech from adult schizophrenic patients (Caplan et al. 1992). They provided the listener with fewer connectives between contiguous clauses (conjunction) and with less repetition of words or word roots (lexical cohesion) than normal children. Finally, they also omitted part of a previous clause on the presumption that the listener retained enough information from this clause (ellipsis). By using fewer conjunctions, less lexical cohesion, and more ellipsis than normal children, the conversing schizophrenic child made it difficult for the listener to piece together the links in his/her speech.

As was the case with the cognitive correlates of FTD, illogical thinking and loose associations had different discourse correlates. Illogical thinking was associated with exophora, while loose associations correlated with referential cohesion, lexical cohesion, and ellipsis (Caplan et al., in press).

Model of the Communication Deficits of Schizophrenic Children

A principal components analysis was conducted to test the working hypothesis that the communication impairment of schizophrenic children is heterogenous and reflects two communication factors: a loose associations/distractibility/discourse factor and an illogical thinking/conservation/span of apprehension factor.

This analysis confirmed that the communication deficits of schizophrenic children are not homogenous, but have three separate principal components. The first component includes discourse measures other than exophora. The second factor included loose associations, the WISC-R distractibility factor score, and the WISC-R verbal factor score. The third factor consisted primarily of illogical thinking and exophora.

In this study span of apprehension and conservation scores were not included in the principal components analysis because they were available for only some of the children. A Pearson pairwise correlation revealed no statistically significant correlation between the span of apprehension and discourse scores of 19 schizophrenic children. An examination of the relationship between conservation and the linguistic/pragmatic measures, however, revealed that schizophrenic and schizotypal children with better conservation competence used more conjunctions, were more verbally productive (words per clause), and had higher verbal comprehension WISC-R factor scores than those with poor conservation competence. An analysis of covariance adjusted for the FTD measures revealed a main effect of loose associations on the relationship of conservation competence with the scores for conjunction and verbal productivity (i.e., words per clause).

The use of conjunctions, verbal productivity, and loose associations appears, therefore, to be related to conservation competence. To join the statements presented in two contiguous clauses with a conjunction, the child needs to perceive the contiguity between these two clauses. Similarly, to converve competently, the child needs to perceive the invariance of matter despite the change in its shape. The child with loose associations does not present the listener with the conversational links that maintain the topic of conversation. This constellation of findings suggests that in schizophrenic children difficulty using conjunctions, conserving, and maintaining the topic of conversation represent diverse manifestations of the same cognitive impairment (Caplan et al., in press).

The independence of the second and third communication components confirms the findings of previous studies that loose associations and illogical thinking are separate communication deficits (Caplan et al. 1990a,b,c). As noted above, illogical thinking is associated with low scores on the partial report span of apprehension task and with conservation deficits on the Bentler and Goldschmid (1968) task.

Cognitive Processes Impaired in Schizophrenic Children

Our survey of a broad range of neuropsychological functions and detailed studies of visual attention generally yielded converging results. The tasks which elicit impaired performance in schizophrenic children tap a wide diversity of computational functions across sensory modalities. A key characteristic that differentiates the tasks which elicit impaired performance in schizophrenic children from those that do not is that they make more extensive demands on processing resources. This is consistent with the hypothesis developed in earlier reviews of cognitive/neuropsychological studies of schizophrenic children (Sherman and Asarnow 1985; Asarnow et al. 1986, in press) and adults (Nuechterlein and Dawson 1984) that schizophrenic individuals have subtle limitations in processing capacity.

The studies of the cognitive and discourse correlates of FTD provide a link between the cognitive impairments detected by certain cognitive/neuropsychological tasks and the clinical manifestations of schizophrenia. As noted above, the schizophrenic child with the greatest deficit in processing resources (as indexed by performance on a partial report span of apprehension task) tended to show illogical thinking, have conservation deficits on the Bentler and Goldschmid (1968) task and break the flow of conversation to refer to the immediate environment. What do these three measures have in common? Instead of screening out extraneous stimuli from the testing environment and focusing on the linguistic context of the conversation, schizophrenic children with exophora break the flow of conversation to refer to stimuli in the testing situation or environment. The requirement to present the listener with logical reasoning, to focus on the linguistic context of a conversation, and to detect the invariance of matter despite the change in the container tap cognitive processes which make extensive demands on processing resources (Caplan, in press). The correlations between these three measures may reflect this common demand.

Psychophysiological studies of the processes involved in the mobilization and allocation of processing resources are important complements to the behavioral studies reviewed in this paper. Strandburg et al. (1994) summarized a series of studies of event related potentials in schizophrenic children. Some of the behavioral tasks used in our laboratory (e.g., partial report span of apprehension tasks) were administered to schizophrenic children with the major dependent variables in these studies being the event related potentials elicited by the span stimuli.

Across a number of studies, schizophrenic children produced significantly less endogenous negative activity between 100 and 300 msec after the onset of span stimuli than age matched normal children. Endogenous negative activity reflects attention effort associated with serial search and stimulus identification. The reduced endogenous negative activity of schizophrenic children reflects an inability to allocate adequate processing resources for the processing of span stimuli. The results of the studies provide converging evidence that schizophrenic individuals have limited processing resources.

Implications for Neurodevelopmental Model

Our retrospective studies of the neurobehavioral antecedents of psychotic symptoms in schizophrenic children, cross-sectional studies of neurobehavioral functioning, as well as FTD and communication impairments in schizophrenic children provides some intriguing findings which can inform neurodevelopmental models of schizophrenia. Our retrospective studies of neurobehavioral development in schizophrenic children indicate that well in advance of the first onset of psychotic symptoms, schizophrenic children manifest certain neurobehavioral impairments. During infancy and early childhood their language acquisition is slow (particularly for expressive language) and gross motor functioning is impaired. Somewhat later there are impairments in fine motor coordination. These neurobehavioral impairments may be early manifestations of the brain lesions posited by neurodevelopmental models. As noted above, in the neurodevelopmental model these brain lesions are thought to be silent. The results of our studies, as well as studies of children at risk for schizophrenia, suggest that the lesions are silent only in the sense that they do not immediately produce active psychotic symptoms. It is worth noting that the expressive language and motor skills are subserved by structures in the frontal lobes. It is these areas where many of the early brain lesions are also found (Weinberger 1987).

If subtle neurobehavioral impairments reflects early damage to the frontal and temporal lobes an interesting question arises. Why is there a delay in the production of psychotic symptoms? Is there some level of functional connectivity required between the frontal lobes and other parts of the brain, particularly the occipital and temporal lobes, required to generate hallucinations? That degree of functional connectivity is not required presumably to support simple motor and language functions.

There is an interesting disparity between the results of our detailed cross-sectional neurobehavioral assessments and the results of the retrospective studies of neurobehavioral development prior to the onset of psychotic symptoms. As noted above, a great majority of schizophrenic children were slow to acquire expressive language. For example, the children were delayed in speaking in two and three words phrases. Curiously, the results of the neuropsychological evaluations conducted when the children were 9 years of age revealed that simple expressive and receptive language skills were among the *least* impaired functions in these children. It appears that the early impairments in language function were delays, not static neuropsychological deficits. On the other hand, when schizophrenic children had to process complex verbal and non-verbal information (e.g., by holding it in short-term memory while they executed a sequence of actions) their performance was impaired. A parallel situation obtains in regard to motor functioning. There were delays in motor functioning involving gross motor coordination. At age 9 schizophrenic children did not show impairments in gross motor functioning. They did, however, show impairments on the Purdue Pegboard, a measure of speeded motor performance which places extensive demands on fine motor speed and motor coordination. Placing cognitive systems under time pressure (e.g., increasing the amount of information that must be processed in a fixed unit of time) makes greater demands on processing resources (Kahneman 1973). This constellation of findings indicating a delay in the acquisition of certain skills, yet residual problems when the skills have to be performed under conditions which involve more complex processing may offer some important insights into the nature of neurobehavioral impairments in schizophrenia.

What mechanisms could produce this constellation of delayed acquisition of skills, which even once acquired are not necessarily deployed as efficiently? Impairments still can be revealed by putting system under time pressure, for example. The developmental delays might be an analogue of a delay of automization of skills. Research examining the ontogeny of skilled performance may provide a framework for beginning to elucidate the mechanisms the developmental delays observed in schizophrenic children.

An informative parallel to the concept of developmental delays can be found in the literature examining the acquisition of skilled motor performance and perceptual discrimination learning. It appears to be the case that schizophrenic children are most likely to be delayed in acquiring those skills (e.g., speaking in two- and three-word utterances) that are at the cusp of development (that is emerging skills that are just coming on-line developmentally). It is not the case that the schizophrenic children are unable to speak in two and three word utterances, rather they are delayed in the acquisition of the skill since some months later they can speak in two and three word utterances.

Developmental delays may be analogous to the delay in learning a skill. Why should schizophrenic children show delays in learning skills? An intriguing answer to this problem emerges from our earlier hypothesis that schizophrenic children have a reduced processing capacity. Controlled processes are used when an individual is first attempting to learn a skill. For example, on tasks like the span of apprehension controlled processes modify long-term memory and are used to search the arrays and decide responses. As practice proceeds these responses become automated, i.e., they make less demands on processing resources. In this way the limited controlled processing system lays down the "stepping stones" of automatic processing (Schneider et al. 1984). Reduced processing resources can lead to deficient development of automatic operations. In particular, it can lead to a delay in the time taken to make a task automated (i.e., make it resource free). Perhaps the developmental delays we have observed in schizophrenic children represent in a broader time scale the greater time it takes schizophrenic individuals to automate certain information processing tasks in the laboratory.

There is a reciprocal relationship between automatic and controlled processes. Reduced availability of processing resources leads to greater limitations on controlled processes, which in turn leads to slow development of automatic processes. Subsequently, reduced utilization of automatic operations requires that resources demanding controlled processes be used for more components of the task, which further reduces the amounts of available resources. "It is this cascading system of interactions between automatic and controlled processes, which allows a limited capacity central processor to carry out very complex cognitive tasks in normal persons, that appears to be somehow disrupted in patients with schizophrenia" (Granholm et al. 1991).

References

Achenbach TM (1966) The classification of children's psychiatric symptoms: a factor analytic study. Psychol Monogr 80(7):615

Achenbach TM, Edelbrock CS (1983) Manual for the child behavior checklist and the revised behavior profile. Department of Psychiatry, University of Vermont, Burlington

Acredelo C (1982) Conservation/nonconservation: alternative explanations. In: Brainerd CJ (ed) Progress in cognitive development, vol 1. Springer, Berlin Heidelberg New York

American Psychiatric Association (1978) Diagnostic and statistical manual of mental disorders, 3rd rev edn. Association, Washington DC

American Psychiatric Association (1987) Diagnostic and statistical manual of mental disorders, 3rd rev edn. Association, Washington DC, pp 335, 340–342

Andreasen NC (1979) Thought, language, and communication disorders: I. Clinical assessment, definition of terms, and evaluation of their reliability. Arch Gen Psychiatry 36:1315–1323

Andreasen NC, Grove WM (1986) Thought, language, and communication in schizophrenia: diagnosis and prognosis. Schizophr Bull 12:348–359

Asarnow RF (1983) Schizophrenia. In: Tarter RE (ed) The child at psychiatric risk. Oxford University Press, New York

Asarnow JR (1988) Children at risk for schizophrenia. Schizophr Bull 14:613–632

Asarnow JR (1994) Childhood-onset schizophrenia. J Child Psychol Psychiatry (in press)

Asarnow JR, Ben-Meir S (1988) Children with schizophrenia spectrum and depressive disorders: a comparative study of premorbid adjustment, onset pattern and severity of impairment. J Child Psychol Psychiatry 29:477–488

Asarnow RF, Sherman T (1984) Studies of visual information processing in schizophrenic children. Child Dev 55:249–261

Asarnow RF, Sherman T, Strandburg RJ (1986) The search for the psychobiological substrate of childhood onset schizophrenia. J Am Acad Child Psychiatry 26:601–614

Asarnow RF, Tanguay PE, Bott L, Freeman BJ (1987) Patterns of intellectual functioning in non-retarded autistic and schizophrenic children. J Child Psychol Psychiatry 28:273–280

Asarnow RF, Asarnow JR, Strandberg R (1989) Schizophrenia: a developmental perspective. In: Cicchetti D (ed) The emergence of a discipline: Rochester Symposium of Developmental Psychopathology. Erlbaum, Hillsdale, New Jersey, pp 189–219

Asarnow RF, Granholm E, Sherman T (1991) Span of apprehension in schizophrenia. In: Steinhauer SR, Gruzelier JH, Zubin J (eds) Neuropsychology, psychophysiology, and information processing. Elsevier, Amsterdam, pp 335–370 (Handbook of schizophrenia, vol 5)

Asarnow JR, Tompson MC, Goldstein MJ (1994) Childhood-onset schizophrenia: a follow-up study. Schizophr Bull (in press)

Benton AL (1974) Revised visual retention test, 4th edn. Psychological Corporation, New York
Benton AL, Hamsher K deS (1989) Multilingual aphasia examination, 2nd edn. Benton Laboratory of Neuropsychology, Division of Behavioral Neurology, Department of Neurology, University Hospitals, Iowa City, IA
Benton AL, Hannay HJ, Varney NR (1975) Visual perception of line direction in patients with unilateral brain disease. Neurology 25:907–910
Benton AL, Hamsher K deS, Stone B (1977) Visual retention test: visual form discrimination. Division of Behavioral Neurology, Department of Neurology, University Hospitals, Iowa City, IA
Cannon TD, Mednick SA, Parnas J, Schulsinger F, Praestholm J, Vestergaard A (1993) Developmental brain abnormalities in the offspring of schizophrenic mothers: I. Contributions of genetic and perinatal factors. Arch Gen Psychiatry 50:551–564
Caplan R (1994) Communication deficits in childhood schizophrenia spectrum disorders. Schizophr Bull (in press)
Caplan R, Guthrie D (1992) Communication deficits in childhood schizotypal personality disorder. J Am Acad Child Psychiatry 35:961–967
Caplan R, Guthrie D, Fish B, Tanguay PE, David-Lando G (1989) The Kiddie Formal Thought Disorder Rating Scale (K-FTDS) Clinical assessment, reliability, and validity. J Am Acad Child Adoles Psychiatry 28:208–216
Caplan R, Perdue S, Tanguay PE, Fish B (1990a) Formal thought disorder in childhood onset schizophrenia and schizotypal personality disorder. J Child Psychol Psychiatry 31:1103–1114
Caplan R, Foy JG, Asarnow RF, Sherman TL (1990b) Information processing deficits of schizophrenic children with formal thought disorder. J Psychiatr Res 31:169–177
Caplan R, Foy JG, Sigman M, Perdue S (1990c) Conservation and formal thought disorder in schizophrenic and schizotypal children. Dev Psychopathol 2:183–190
Caplan R, Guthrie D, Foy JG (1992) Communication deficits and formal thought disorder and in schizophrenic children. J Am Acrd Child Adoles Psychiatry 31:151–159
Cornblatt BA, Erlenmeyer-Kimling L (1985) Global attentional deviance as a marker of risk for schizophrenia: specificity and predictive validity. J Abnorm Psychol 94:470–486
Cornblatt BA, Lenzenweger MF, Dworkin RH, Erlenmeyer-Kimling L (1985) Positive and negative schizophrenic symptoms, attention, and information processing. Schizophr Bull 11:397–408
Crow TJ (1985) The two-syndrome concept: origins and current status. Schizophr Bull 11:471–486
Dunn LM, Dunn LM (1981) Peabody picture vocabulary test-revised. American Guidance Service, Circle Pines, Minnesota
Estes WK, Taylor HA (1964) A detection method and probabilistic models for assessing information processing from brief visual displays. Proc Natl Acad Sci USA 52:446–454
Fish B (1977) Neurobiologic antecedents of schizophrenia in children. Arch Gen Psychiatry 34:1297–1313
Fish B, Hagin R (1973) Visual-motor disorders in infants at risk for schizophrenia. Arch Gen Psychiatry 28:900–904
Fish B, Marcus J, Hans SL, Auerbach JG, Perdue S (1992) Infants at risk for schizophrenia: sequelae of a genetic neurointegrative defect: a review and replication analysis of pnadysmaturation in the Jerusalem Infant Development Study. Arch Gen Psychiatry 49:221–235
Flavell JH (1963) Developmental psychology of Jean Piaget. Van Nostrand, New York
Flavell JH (1985) Cognitive development. Prentice-Hall, Englewood Cliffs, NJ
Goldstein MJ (1980) The course of schizophrenic psychosis. In: Brim OG, Kagan J (eds) Constancy and change in human development. Harvard University Press, Cambridge, pp 325–358
Goldschmid ML, Bentler PM (1968) Manual: concept assessment kit – conservation. Educational and Industrial Testing Service, San Diego, CA
Goldberg TE, Weinberger DR, Berman KF, Pliskin NH, Podd MH (1987) Further evidence for dementia of the prefrontal type in schizophrenia? A controlled study of teaching the Wisconsin Card Sorting Test. Arch Gen Psychiatry 44:1008–1014
Granholm E, Asarnow RF, Marder SR (1991) Controlled information processing resources and the development of automatic detection responses in schizophrenia. J Abnorm Psychol 100:22–30
Grant DA, Berg EA (1981) Wisconsin card sorting test. Psychological Assessment Resources, Odessa, FL

Halliday MAK, Hassan R (1976) Cohesion in spoken and written English. Longmans, London
Harrow M, Lanin Kettering I, Miller JG (1989) Impaired perspectives and thought pathology in schizophrenic and psychotic disorders. Schizophr Bull 15:605–623
Harvey PD (1983) Speech competence in manic and schizophrenic psychoses: the association between clinically rated thought disorder and performance. J Abnorm Psychol 92:368–377
Harvey PD, Earle-Boyer EA, Levinson JC (1988) Cognitive deficits and thought disorder: a retest study. Schizophr Bull 14:57–66
Hirst W, Kalmar D (1987) Characterizing attentional resources. J Exp Psychol Gen 116:68–81
Holzman PS, Shenton ME, Solovay MR (1986) Quality of thought disorder in differential diagnosis. Schizophr Bull 12:360–372
Kahneman D (1973) Attention and effort. Prentice-Hall, Englewood Cliffs, NJ
Kauffman AS (1979) Intelligent testing with the WISC-R. Wiley, New York
Koh SD (1978) Remembering of verbal materials by schizophrenic young adults. In: Schwartz S (ed) Language and cognition in schizophrenia. Erlbaum, Hillsdale, NJ, pp 55–99
Kraepelin E (1896) Psychiatrie. Ein Lehrbuch für Studierende und Ärzte. Barth, Leipzig
Lezack M (1983) Neuropsychological assessment. Oxford University Press, New York
MacCrimmon DJ, Cleghorn JM, Asarnow RF, Steffy RA (1980) The clinical and behavioral characteristics of foster children at risk for schizophrenia. Arch Gen Psychiatry 37:671–674
McTear MF (1985) Children's conversation. Basil Blackwell, Oxford
Manschreck TC (1983) Psychopathology of motor behavior in schizophrenia. In: Maher BA (ed) Progress in experimental personality research, vol 12. Academic, New York, pp 53–99
Maratsos MP (1976) The use of definite and indefinite reference in young children. Cambridge University Press, Cambridge
Marcus J, Hans SL, Mednick S, Schulsinger P, Michelsen N (1985) Neurological dysfunctioning in offspring of schizophrenics in Israel and Denmark: a replication analysis. Arch Gen Psychiatry 42:753–761
Marcus J, Hans SL, Auerbach JG, Auerbach AG (1993) Children at risk for schizophrenia: the Jerusalem Infant Development Study: II. Neurobehavioral deficits at school age. Arch Gen Psychiatry 50:797–809
Mednick SA, Schulsinger F, Teasdale TW, Schulsinger H, Venables PH, Roch DR (1978) Schizophrenia in high risk children: sex differences in predisposing factors. In: Serban G (ed) *Cognitive defects and the development of mental illness*. Brunner/Mazel, New York, pp 169–197
Mehler J, Bever TG (1967) Cognitive capacities of young children. Science 158:141–142
Nuechterlein KH, Asarnow RF (1989) Perception and cognition. In: Kaplan HI, Sadock BJ (eds) Comprehensive textbook of psychiatry V. Williams and Wilkins, Baltimore
Nuechterlein KH, Dawson ME (1984) Information processing and attentional functioning in the developmental course of schizophrenic disorders. Schizophr Bull 10:160–203
Neuchterlein KH, Edell WS, Norris M, Dawson ME (1986) Attentional vulnerability indicators, thought disorder, and negative symptoms. Schizophr Bull 12:408–426
O'Bryan KG, Boersma FJ (1971) Eye-movements, perceptual activity, and conservation development. J Exp Child Psychol 12:157–169
Oltmanns TF, Murphy R, Berenbau H, Dunlop SR (1985) Rating verbal communication impairment in schizophrenia and affective disorders. Schizophr Bull 11:292–299
Orvaschel H, Puig-Antich J (1987) Schedule for affective-disorder and schizophrenia for school age children: epidemiologic version. Medical College of Pennsylvania, Eastern Pennsylvania Psychiatry Institute
Parnas J, Cannon TD, Jacobsen B, Schulsinger H, Schulsinger F, Mednick SA (1993) Lifetime DSM-III-R diagnostic outcomes in the offspring of schizophrenic mothers: results from the Copenhagen High-Risk Study. Arch Gen Psychiatry 50:707–714
Piaget J (1959) The language and thought of the child. Humanities Press, New York
Rey A (1964) L'Examen clinique en psycologie. (The clinical exam in psychology.) Presses Universitaires de France, Paris
Rochester SR, Martin JR (1979) Crazy talk: a study of the discourse of schizophrenic speakers. Plenum, New York

Rodnick EH, Shakow D (1940) Set in the schizophrenic as measured by a composite reaction time index. Am J Psychol 97:214–225

Sattler JM (1974) Assessment of children's intelligence. Saunders, Philadelphia

Schneider W, Dumais ST, Shiffrin RA (1984) Automatic and controlled processing and attention. In: Parasuraman R, Beatty J, Davies J (eds) Varieties of attention. Academic, San Diego, pp 1–27

Schneider SG, Asarnow RF (1987) A comparison between the cognitive/neuropsychological impairments of non-retarded autistic and schizophrenic children. J Abnorm Child Psychol 15:29–46

Shatz M (1982) Communication. In: Musson P (ed) Carmichael's manual of child psychology. Wiley, New York, pp 841–889

Sherman T, Asarnow RF (1985) The cognitive disabilities of the schizophrenic child. In: Sigman M (ed) Children with emotional disorders and developmental disabilities: assessment and treatment. Grune and Stratton, New York

Steffy RA (1978) An early cue sometimes impairs process schizophrenic performance. In: Wynne LC, Cromwell RL, Matthysse S (eds) The nature of schizophrenia: approaches to research and treatment. Wiley, New York, pp 225–232

Strandburg R, Marsh JT, Brown WS, Asarnow RF, Guthrie D, Higa J (1994) Continuity of information processing deficits across childhood and adult onset schizophrenia: ERP correlates. Schizophrenia Bulletin (in press)

Stuss DT, Benson DF, Kaplan EF, Weir WS, Naesser MA, Liberman I, Ferrill D (1983) The involvement of orbitofrontal cerebrum in cognitive tasks. Neuopsychologia 21:235–248

Tiffin J (1968) Purdue pegboard. Science Research, Chicago

Treisman AM, Gelade G (1980) Afeature-integration theory of attention. Cogn Psychol 12:97–136

Tunmer WE, Herriman ML (1984) The development of metalin-guistic awareness: a conceptual overview. In: Tunmer WE, Pratt C, Herriman ML (eds) Metalinguistic awareness in children. Springer, Berlin Heidelberg New York, pp 12–35

Van der Does AJ, van den Bosch RJ (1992) What determines Wisconsin Card Sorting performance in schizophrenia? Clin Psychol Rev 12:567–583

Walker E, Emory E (1983) Infants at risk for schizophrenia: offspring of schizophrenic parents. Child Dev 54:1269–1285

Watkins JM, Asarnow RF, Tanguay PE (1988) Symptom development in childhood onset schizophrenia. J Child Psychol 29:865–878

Watkins J, Asarnow RF, Williams M (submitted) Qualitative variation in neuropsychological test performance in schizophrenic children

Watt NF, Lubensky AW (1976) Childhood roots of schizophrenia. J Consult Clin Psychol 44:363–375

Wechsler D (1974) Wechsler intelligence scale for children-revised manual. Psychological Corporation, New York

Weinberger DR (1987) Implications of normal brain development for the pathogenesis of schizophrenia. Arch Gen Psychiatry 44:660–669

Antecedents, Onset, and Early Course of Schizophrenia: Discussion

L.E. DeLisi

Introduction

This symposium and its proceedings have been assembled, now for the third time, because we are, as the title says, deeply committed to searching for the cause or causes of schizophrenia. Meanwhile, time is passing, progress has been slow, and we are now hoping that the new scientific technologies such as molecular genetics and brain imaging will provide answers. However, we still need to use clues from the past and what we presently know about the clinical characteristics of schizophrenia. The quality of the early signs of illness and age at onset of schizophrenia is the single most important aspect of the illness that could yield clues to its underlying biology. Based on present knowledge, the onset of schizophrenia appears to be a genetically programmed process, slowly developing until it reaches some threshold of clinical detection. Detecting the occurrence of its pathology, will then depend on the sophistication of our knowledge about its consequences and the ability to make use of new diagnostic tools. Diabetes mellitus is an analogous illness of a chronic nature. If one depended on the clinical presentation of a patient in a diabetic coma or crisis for diagnosis the age of onset might be stated as middle-aged adulthood. However, the process has been building up over several years microscopically in blood capillaries, and in the pancreas. Diagnosis might be made many years earlier through characteristic ocular signs, elevated post-prandial blood sugars, other physiological problems, and eventually genetic markers, only to be clinically considered a "disease" at a much later time.

Thus, any discussion of the onset for schizophrenia should focus on the following central questions: (a) what is the definition of "onset"; that is, what symptoms or signs characterize the onset; (b) what causes the onset – or is it a clue to the cause, and can it be modified by genetic or nongenetic factors; and (c) what are the differences in age of onset among ill individuals, and do these imply biological differences, causal differences, or indirectly normal brain developmental differences? The onset represents a process, and all of the above are factors affected by this process. Actual age of onset may also represent severity of this process, and thus outcome. The final implication is that future treatments may be developed to prevent onset, if the cause can be

Department of Psychiatry, HSC, T-10, SUNY Stony Brook, Stony Brook, NY 11794, USA

H. Häfner and W.F. Gattaz (Eds.)
Search for the Causes of Schizophrenia, Vol. III

determined, and what the earliest detectable onset of the disorder is. Once the antecedents of schizophrenia, as described in this session, are present, one must ask if the process can still be reversible. Will new neuroleptics be able to prevent further deterioration and development of the psychotic state?

Consistent Trends Across Studies

What is age of onset? Most studies in the literature, particularly those that are retrospective use age at first hospitalization as the age of onset, while others use age at first appearance of "psychotic symptoms" and others still use age at appearance of first behavioral change, the latter two being less reliable retrospectively.

The first trend that runs through all the studies in this session is that schizophrenia is not an illness will an abrupt onset in an individual in early adulthood who has never had any indications of premorbid pathology. The implication is that subtle behavioral and cognitive skills are lacking in many preschizophrenic subjects that lead to difficulties in interpersonal communication. It is unclear from these studies what percentage of these children are deviant, or are they all deviant from what could have been their normal potential?

The first signs of illness in all these studies are very nonspecific signs of nervousness, anxiety, decreased concentration, lack of energy, and depressive symptoms. These could result from any number of causes. No cluster or syndromes have emerged that could distinguish the preschizophrenic child from others. The percentage of children with these problems that go on to become schizophrenic is likely to be small. One has to be cautious and perhaps only use these as warning signs in those individuals who also have a strong family history for schizophrenia.

None of the studies has begun to examine the underlying cause to these symptoms, although the implication is that it is genetically controlled and brain developmental in nature.

The clearest trend across previous literature and the presentations in this symposium are that sex differences exist in age of onset from the earliest indications of disease to the age at first hospitalization. Not only is the appearance of subtle communication problems delayed until later in childhood, but that some childhood representations of early signs tend to be different among males and females – males more externalizing and females more internalizing. However, in other ways the presentation is similar, but just delayed in time. Are these differences genetic or epigenetic? There is indication that the gender differences are primarily a delay in onset and that age at onset determines everything after. There is also literature to suggest a strong genetic influence on age of onset (reviewed in Crow and Done 1986).

Specific Issues Raised by Each Presenter

Crow et al.

The study reported by Crow et al. makes use of a United Kingdom national birth cohort surveying all newborns born 1 week in March of 1958. All admissions to psychiatric institutions from 1974 to 1986 were then examined by record review and in relation to childhood follow-up developmental data. It was noted that the mothers of the schizophrenic children had poorer prenatal care, and other nonspecific potentially damaging risks to their pregnancies, although there was no evidence of greater exposure to influenza or an excess of complications with the births of the preschizophrenic child. They note that at age 7, preschizophrenic boys are more anxious and hostile towards adults and peers, and have poor concentration. By age 11 they are additionally depressed, while females do not have detectable disturbances appearing until the age 11 follow-up contact. Academic performance was in general more impaired in the preschizophrenics of both sexes than their nonschizophrenic pairs at ages 7, 11, and 16 when follow-up data were collected. It is interesting also to note, that while there were sex differences in behavioral signs, there were no sex differences in language and mathematical scores, schizophrenics uniformly doing poorer than other diagnostic groups. It is suggested that the speech and language deficits and overall impairment of intellectual abilities probably have an impact on ability to communicate. However, one thing we do not know is, if children with these behaviors could be singled out, what proportion of them would become schizophrenic? It is also not mentioned whether there was a sex difference in early language acquisition or in physiological signs, such as bedwetting, etc. What is clear is that there is a difference in the time sequence of development of illness among the sexes. Crow et al. say that "disturbances in behavior cannot be accounted for by intellectual deficits or vice versa. It seems to be a disorder of both intellect and personality or perhaps the relationship between the two." Crow et al. point out further that there are sex differences in personality traits, such as cooperativeness and sociability. They discuss the complexities of social interaction, and the ability to take account of the dispositions and intentions of other individuals, certainly an aspect unique to the evolution of man. Deficits of this ability are at the heart of many schizophrenic syndromes. It follows that if one can understand how genetic variation in intelligence, language and social ability arise, one will have insight into how the psychoses arise.

Crow et al. say that a single genetic factor could account for the premorbid deviations, and the sex differences in onset of psychosis. But what is the critical rate-limiting factor they describe? Could it be the rate of regional brain growth? Crow et al. propose that "perhaps various combinations of structure size and asymmetry are associated with different cognitive and personality styles, and this diversity is maintained by the selective factors that have increased brain size and retarded its maturation. The extremes of this variation may be expressed as the cognitive and behavioral deviations that we recognize as psychosis." If I take this a little further: if a gene or group of genes determines the rate of regional brain growth, one region relative to another, then schizophrenia is a disregulation of this growth (see Fig. 1). Thus, if there is a variation among individuals in the rate of relative growth, or the amount of production of these genes, when a certain threshold is reached, signs of psychosis will begin (i.e., age of onset). It is more likely to be "milder" or slower in females (emergence of symptoms later), either because of a relationship to normal brain sexual dimorphisms, modifying hormonal factors, or inheritance of a defective gene on the X chromosome with normal gene products produced by the second X chromosome of females (provided the gene is not subject to inactivation). Data to support this latter hypothesis have previously been reported (Crow et al. 1989; DeLisi and Crow 1989). One recent study relevant to this issue (Reiss et al. 1993) describes functional and structural brain differences among monozygotic twins discordant for Turner's syndrome (X chromosome monosomy), probably as a result of mitotic nondysjunction. The twin with Turner's syndrome had larger CSF to brain ratios particularly in prefrontal, parietal occipital and parietal–perisylvian regions, with the largest grey matter discrepancy in the left parietal-perisylvian region and right prefrontal. Thus this study would strongly suggest, given identical genes in the rest of the genome for this pair, that brain growth in the regions relevant to schizophrenia could be influenced by a gene or genes on the X chromosome.

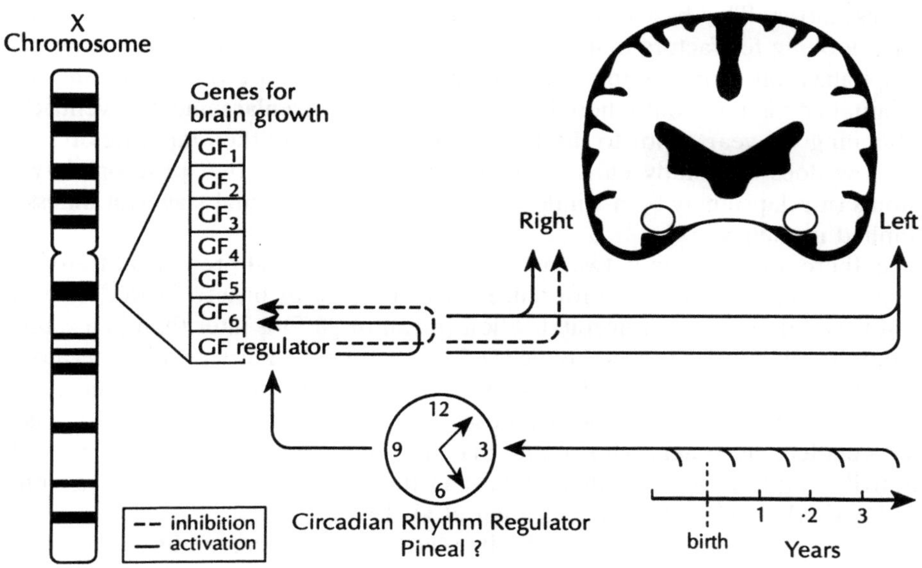

Fig. 1. Regulation of regional brain growth

Walker et al.

Walker et al. also studied the behavior of children at similar ages – mean of 9.7 years and then 1 year later. Externalized behavior problems were increased in males; at the second follow up these began to increase in females. Females at high risk, on the other hand, showed the greatest internalizing behavior at time two, despite males showing it at both times one and two. Externalized behavior was increased in male preschizophrenics, in all but the birth to 4-year period. What is most impressing is the consistent pattern across study designs and an increase in behavioral problems with age. What they add is specific studies of facial expression and motor deficits detected by observations of home movies. No sex differences in neuromotor deviations were noted.

These investigators emphasize the dopamine hypothesis and that estrogen reduces the behavioral effects of dopamine in rats. This effect may be mediated by a reduction in dopamine D-2 receptor sensitivity and that estrogens having antidopaminergic properties. They say that sex differences are due to hormonal and neuroanatomical factors, with hormonal influence predominating between pubescence and menopause. However, they fail to speculate about what causes the neuroanatomical sexual dimorphisms. And these may be the key to schizophrenia.

Häfner et al.

Häfner et al. find that a mean period of 7.6 years passes from the occurrence of first symptoms until the first episode in 232 cases of a first hospital admission for psychosis. This has serious implications for biological studies, such as my own, looking for factors that are present at the "onset of psychosis" using first hospitalization. For example, large cerebral ventricles are present by the first hospital admission; could there have been progressive enlargement of ventricles beginning 7.5 years prior to the first hospitalization and then tapering off?

How does one study early signs of illness? The idea that prodromal symptoms of relapse may be a window to antecedent prodromal stages of illness is implied in their work.

Is there a distinction between early signs of illness and premorbid traits?

Are there geographical differences in actual age of onset? In the study by Häfner et al. age at first hospitalization (mean of 30.5) is later than in a similar Stony Brook, United States sample (DeLisi et al. 1992). Perhaps this is because a broad definition of schizophrenia is used in the Häfner et al. cohort that includes atypical nonaffective psychoses. This is probably correct to do, based on Kendler and others who report that the same genetic component is likely contributing to a broader phenotype than the classical Kraepelinian syndrome or DSM-III-R schizophrenia (Kendler et al. 1993).

Age at onset then appears to be determined by an additive model of accumulation of nonspecific symptoms over time. The correlation of positive and negative symptoms, to these investigators, imply one biological process is present. However, they have not ruled out the presence of two biological processes that are dependent on each other.

Upon analyzing their data, Häfner et al. became aware of, and thus emphasize, the importance of longitudinal studies versus horizontal cross-sectional views of patients at different stages of illness.

Finally, they note that the majority of the social effects occur prior to the first admission. Subsequently, there is increased vulnerability to psychotic episodes. In Kraepelin's day, however, the course after the first hospitalization was more severe, perhaps, they end by saying, due to a lack of therapy of presence of unfavorable hospital conditions.

Parnas et al.

It is now approximately 30 years since the 1962 start of this high-risk study. The 207 children were all 10–18 years old in 1962 and thus now are all over age 40. Evaluations were performed 10 and 25 years later with only approximately 10% attrition. Schizophrenia and functional psychoses, as well as related personality disorders, were higher in the high risk children compared with low risk. There did not appear to be any aggregation of affective disorders. These authors emphasize that this implies that there is no unitary continuum of illness from schizophrenia through affective disorders. However, there are three potential caveats to this theory: (a) these individuals may not be through the age of risk for development of affective disorder, which is generally later than schizophrenia; (b) it has been noted in the past that affective disorder in parents frequently leads to schizophrenia in offspring, but the reverse is rare (reviewed in Crow 1986) – the present study only focuses on offspring of schizophrenic mothers, and as such would then not be expected to find an excess of affective disorder in the high risk children; (c) lastly, the number of female offspring studied is small – since affective disorder, particularly unipolar, is more prevalent in females, it is possible that this could account for its scarcity.

Other findings of interest from this study are the higher rate of substance abuse among the high-risk group and that institutional rearing predicted a schizophrenic outcome. They suggest that early emotional deprivation may decompensate the vulnerable individual toward overt schizophrenia. Birth complications were higher among the schizophrenics although this did not distinguish high- and low-risk groups.

As in the other studies, this high-risk cohort displayed nonspecific behavioral antecedents to illness; however, there was no typical "schizoid" predisposition as has been described in the past.

These researchers favor a diathesis stress model for schizophrenia characterized by subtle schizotypal features (the reflection of autistic vulnerability, they say would be most proximate to a genetic substrate) and that the onset of overt schizophrenia would be a result of further decompensation. However, they fall short of suggesting how this would come about.

Asarnow et al.

Asarnow et al. have contributed studies of children with early onset of overt schizophrenia (prior to age 13). These children have neurodevelopmental anomalies beginning in infancy and also similarly show delays in language acquisition, as will as motor functioning. The delays in language acquisition are particularly consistent to findings in the Crow et al. studies and also are similar to the findings from the retrospective analyses from my own studies of new onset schizophrenics, showing later mean ages for beginning to talk and read for both male and female schizophrenic patients (DeLisi et al. 1991).

Significance of These Findings to the Cause of Schizophrenia

Thus, taken together, these five separate studies all suggest that deviance in normal brain growth over time leads to schizophrenia. An understanding of the genetic control of this process and its progression over time could come from knowledge of the evolution of brain structures that provide the ability for complex social interactions characteristic of human beings. How the brain normally develops in modern homo sapiens, the developmental differences between the sexes, and what brain development is like in the preschizophrenic child, is a result of thousands of years of evolutionary expansion of specific brain regions.

Crow et al. note that "psychosis provides important clues to the evolution of man. . . ." Perhaps it is the reverse – that the evolution of man (i.e., brain evolution) proides important clues to the origin of psychoses.

It is clear that there is a long evolutionary history in the primate brain for specific growth and adaption of left hemispheric structures for thought and oral communication. These left hemispheric structures may have subsumed such processes because of an earlier specialization for sequential, analytic, time dependent functions, such as in thinking how to make tools. Alternatively, these functions may have originated from prior normal communicative skills controlled by the left hemisphere. It is not clear whether all these processes are genetically prespecified, or are instead themselves subject to, or perhaps largely determined by, environmental influences during development. It is also unclear whether left hemispheric specialization precedes or is a consequence of a prior right hemisphere specialization for emotional and/or spacial functions. While there must be evolutionary precedents for human language, nevertheless there is a qualitative uniqueness that is a result of complex social and genetic interactions (Bradshaw 1991). The relatively late appearance of Broca's area, not at all recognizable in apes, and Brodmann's areas 39 and 40, and only barely, if at all, detectable in orangutans and chimpanzee, suddenly appears in our ancestor, homo habilis with fuller development occuring up the evolutionary ladder towards homo sapiens (Eccles 1989).

The superior posterior plane of the temporal lobe (planum temporale), the perisylvian cortical region in the center of these language processing areas (reviewed in Wada et al. 1975), has recently received a lot of attention in schizophrenia, some studies reported of reduced lateralization of the brains of adult schizophrenics in this region (Falkai et al. 1992; Rossi et al. 1992; DeLisi et al., in press). The majority of normal adults have larger left than right planums and this asymmetry is measurable by the 29th week of gestation. It is generally thought, however, despite the already present morphologic and physiologic asymmetry, that the critical time period for the ascribing of speech and language to one hemisphere develops slowly from the 2nd to 12th years of life. There also appears to be a greater asymmetry in adult than infant brains implying that the left planum continues to grow larger with age into adulthood. The planum is larger bilaterally in male than female infants and only on the left in male adults, thus females are more likely to have less asymmetry than males in adulthood, although these data on sex differences are only preliminary. Asymmetries of this specific plane have not been found in the brains of rhesus monkeys and baboons, thus suggesting a specific significance to the evolution of the human brain. Wada et al. suggest that "the ultimate pattern of cerebral speech and language representation . . . may depend on the dgree of predetermined morphological asymmetry and the availability of uncommitted contralateral cortex . . ." particularly in reference to insults to the developing brain.

Brain lateralization must have a direct relationship to the development of the corpus callosum, a midline structure which contains bundles of fibers connecting both cerebral hemispheres. The total corpus callosum area has been found to be larger in women than men and the splenium particularly, is described as larger and more bulbous in women, although not all studies show this (DeLacoste-Utamsing and Holloway 1982; Witelson and Kigar 1988). A sex difference in the corpus callosum is seen as early as the 26th prenatal week, providing evidence that sex differences in brain structure occur early during brain growth. Consistent with the corpus callosum differences are reports that the degree of brain lateralization is greater in males compared with females, assuming the size of the corpus callosum is inversely related to the degree of lateralization. This structure is also reported to be different in shape and size in schizophrenia, with a loss of the normal sexual dimorphism, although the results from several studies of corpus callosum measurements are inconsistent (reviewed in Hauser et al. 1989). It is of interest that the corpus callosum enlarges slowly with age in humans and with different time sequences in males and females. The female corpus callosum does not obtain its maximum width until age 41–50, while the male corpus callosum reaches a peak width at 20 years and then declines in size. These changes are not linear with time and complex (Cowell et al. 1992). If the process of brain functional lateralization was related to the development and expression of schizophrenia, the sex differences in corpus callosum growth may underlie some of the sex differences in illness onset. An evolutionary examination of the corpus callosum may also yield some insight into schizophrenia.

Sex differences in other brain structures (reviewed in DeLisi et al. 1989) have been found, i.e., the preoptic area of the rat, hamster, and rhesus monkey, and in the hypothalamic arcuate nucleus and medial amygdala of the rat. Similarly, in humans, the superchiasmatic nucleus of the preoptic hypothalamus is two to three times larger in men compared with women. The functional significance of this nucleus is unknown, as is the timing for development of a sex difference in its structure.

Other Aspects of Normal Brain Growth with Possible Relationship to Schizophrenia

Brain lateral and third ventricular size is significantly enlarged to varying degrees in chronic schizophrenia. A deviation in the normal maturation of the relation of ventricular to brain tissues may be occurring in schizophrenia. In one study (Lodin 1968) the widths of the anterior horns of the lateral ventricles increased up to age 2, after which they remained stable over time in relation to total brain size. The left horn was found to be generally wider than the right. The width of the third ventricle was somewhat smaller during the 1st year of life, but stable afterward. Thus, if enlargement of the ventricles in schizophrenia is not a result of atrophy, it is likely to be related to a developmental process taking place prior to age 2.

Decreases in temporal, frontal, and limbic cortical tissue have also been described in chronic schizophrenia, including a small overall decrease in brain size, although these findings are all controversial. An understanding of the timing of normal brain growth in males and females might yield further clues to the development of these differences in schizophrenia. Determination of the normal growth spurt in the brain as a whole is complicated by variations in growth rates in different brain regions. An increase in the cranial volume may occur into adulthood, especially in the superior prefrontal region. This small increase is consistent with the fact that the cranial sutures do not completely fuse until after 30 years of age. Nevertheless, cortical volume decreases with age and is not related to the cranial increases. Changes are localized to the superior cortical regions (frontal and parietal convexities) and are reductions in grey matter with widening of the CSF spaces. Volume reductions were also seen in subcortical structures, such as the caudate, and correlated with an increase in ventricular volume. A sex effect is present in most regions except the superior anterior cortical regions (Jernigan et al. 1991).

The hippocampus is a structure that appears to be increasing in size in the course of human evolution, as seen by studies of nonhuman primate brain showing progressive increase in size from gorilla to chimpanzee to man (reviewed in Kretschmann et al. 1986). The hippocampus has been a focus also in schizophrenia because of its functional relationship to memory and emotion, and has been found to be reduced in size bilaterally in chronic schizophrenic adults. While far too few studies carefully delineate the normal growth of this structure, and are limited to male brains, a large volumetric increase occurs over the second half of gestation, and an even larger spurt peaks at approximately the 2nd postnatal month in normal development; myelination slowly increases over the first 2–3 years of life. No left/right differences are present.

Studies of neuronal density, glial cell counts, and nucleic acid amounts have been focused on prenatal and postnatal periods only up to approximately 3 years of age. By age 3, neuronal and glial counts appear fixed and myelination is approximately 90% of adult values. However, myelination of the cerebral commissures is not complete until 9 years of age and myelination of intracortical white matter continues throughout life (Thatcher et al. 1987). Although overall growth can be described by continuous trajectories, there are also growth spurts in specific regions that are observable. The human brain growth spurt – a transient period of rapid brain growth, begins when the adult neuronal number has been largely reached, beginning about midgestation. The enormous multiplication of cells, which occupies its early part is predominantly glial. The later phases of the growth spurt is myelination which follows the establishment of the glial cell population (Dobbing and Sands 1973). The peak in the growth spurt in humans occurs by about 3–4 years of age. "The tremendous increase in dendritic complexity accompanied by the establishment of synaptic connectivity must inevitably be later, and almost certainly coincides and is coextensive with it. . . ." Thatcher and Giudice conclude that ". . . the divergent electrophysiological patterns that are present in brain growth studies suggest an ontogenetic hypothesis of human cortical development in which there is a genetically programmed unfolding of specific corticocortical connections at relatively specific postnatal ages." Thus, interference with this latter process, which unfortunately is difficult to measure, may be one explanation for the development of major psychiatric disease.

Neurochemical and neuroendocrine relationships to brain growth are being studied, although the integration of developmental neurochemical studies and gross morphological observations has yet to be done. The serotonergic system plays a major role in the maturation of the brain (Azmitia and Whitaker-Azmitia 1991; Whitaker-Azmitia 1991). 5HT1A receptors on neurons and astrocytes are most dense during these early developmental periods. In astrocytes activation of these receptors stimulate the release of the growth factor S-100. In the adult brain, however, the developmental role of 5HT is dormant. However, other neurochemical systems are also likely to have roles in development and the regulation and timing of the interaction between them must be crucial.

Most investigators are quick to attribute sex differences to hormonal differences from prenatal periods on through adulthood. However, there are some inconsistencies with this simplified view. The timing of puberty had no effect on the distinction between those who develop psychosis and those who do not according to Crow and colleagues, age of onset does not appear to be correlated with age of puberty (DeLisi et al., in press), and hormonal therapy has not proven useful in the treatment of psychotic symptoms (reviewed in DeLisi et al. 1989). Nevertheless, the antidopaminergic effect of estrogens is well known and thought to provide a protective measure against psychotic symptoms. Thus, the observations that postpartum psychoses and a rise of psychosis in females after the age of 40 is generally attributed to the lowering of estrogen levels. In summary, the data describing sex hormone dysregulation in schizophrenia is sparse and based on old not well controlled studies; males and females have not been studied separately. It is also unknown whether the onset of hormonal deficiencies Noted in some patients appears early in life, or only after the onset of illness as an indirect consequence. Further studies are need before the importance of sex hormonal imbalance to the direct development of schizophrenia can be determined. They may be peripherally related, however, based on their effects on brain growth, lateralization and structural size.

Crow et al. focus on an important point in relation to sex differences and that is when the onset of illness is taken into account, sex, no longer significantly predicts outcome (Eaton et al. 1992). This suggests that the key variable for gender differences in schizophrenia is age of onset, not sex per se. Crow et al. conclude that "A single variable concerned with timing an aspect of central nervous system development is a major determinant of the course of the disease." But what is this variable concerned with cerebral development? "Precursors could provide clues to the identity of the functions that become deviant in psychotic disturbance." It is probable that the growth of sexually dimorphic brain structures and why these structures are different among the sexes is the final key to the cause of schizophrenia.

References

Azmitia EC, Whitaker-Azmitia PM (1991) Awakening the sleeping giant: anatomy and plasticity of the brain serotonergic system. J Clin Psychiatry 52: 12[Suppl]:4–16

Bradshaw JL (1991) Animal asymmetry and human heredity: dextrality, tool use and language in evolution – 10 years after Walker (1980). Br J Psychol 82:39–59

Cowell PE, Allen LS, Zalatimo NS, Denenberg VH (1992) A developmental study of sex and age interactions in the human corpus callosum. Dev Brain Res 66:187–192

Crow TJ (1986) The continuum of the psychoses and its implication for the structure of the gene. Br J Psychiatry 149:419–429

Crow TJ, Done DJ (1986) Age of onset of schizophrenia (1986): a test of the contagion hypothesis. Psychiatry Res 18:107–117

Crow TJ, DeLisi LE, Johnstone EC (1989) Concordance by sex in sibling pairs with schizophrenia is paternally inherited: evidence for a pseudoautosomal locus. Br J Psychiatry 155:92–97

DeLacoste-Utamsing C, Holloway RL (1982) Sexual dimorphism in the human corpus callosum. Science 216:1431–1433

DeLisi LE (in press) Editor's comment: puberty and schizophrenia. Schizophrenia Research

DeLisi LE, Crow TJ (1989) Sex chromosomes and psychosis. Schizophr Bull 15:431–440

DeLisi LE, Dauphinais ID, Hauser P (1989) Gender differences in the brain: are they relevant to the pathogenesis of schizophrenia? Compr Psychiatry 30:197–208

DeLisi LE, Boccio AM, Riordan H, Hoff AL, Dorfman A, McClelland J, Kushner M, Van Eyl O, Oden N (1991) Familial thyroid disease and delayed language development in first admission patients with schizophrenia. Psychiatry 38:39–50

DeLisi LE, Stritzke P, Riordan H, Holan V, Boccio A, Kushner M, McClelland J, Van Eyl O, Anand A (1992) The timing of brain morphological changes in schizophrenia and their relationship to clinical outcome. Biol Psychiatry 31:241–254

DeLisi LE, Hoff AL, Neale C, Kushner M (in press) A study of asymmetries in the temporal cortex of male and female first-episode patients with schizophrenia. Schizophr Res

Dobbing J, Sands J (1973) Quantitative growth and development of human brain. Arch Dis Childhood 48:757–767

Eaton WW, Mortenson PB, Herrman H et al (1992) Long-term course of hospitalization for schizophrenia: I. Risk for hospitalization. Schizophr Bull 18:217–228

Eccles JC (1989) Evolution of the brain: creation of the self. Routledge, London, pp 81–92

Falkai P, Bogerts B, Greve B, Pfeiffer U, Machus B, Folsch-Reetz B, Majtenyi C, Ovary I (1992) Loss of sylvian fissure asymmetry in schizophrenia: a quantitative postmortem study. Schizophr Res 7:23–32

Hauser P, Dauphinais ID, Berrettini W, DeLisi LE, Post R (1989) Corpus callosum dimensions measured by magnetic resonance imaging in bipolar affective disorder and schizophrenia. Biol Psychiatry 26:659–668

Jernigan TL, Trauner DA, Hesselink JR, Tallal PA (1991) Maturation of human cerebrum observed in vivo during adolescence. Brain 114:2037–2049

Kendler KS, McGuire M, Gruenberg AM, O'Hare A, Spellman M, Walsh D (1993) The Roscommon family study: I. Methods, diagnosis of probands, and risk of schizophrenia in relatives. Arch Gen Psychiatry 50:527–540

Kretschmann HJ, Kammradt G, Krauthausen I, Sauer B (1986) Growth of the hippocampal formation in man. Bibl Anat 28:27–52

Lodin H (1968) Size and development of the cerebral ventricular system in childhood. ACTA Radiol Diagn 7:385–392

Reiss AL, Freund L, Plotnick L, Baumgardner T, Green K, Sozer AC, Reader M, Boehm C, Denckla MB (1993) The effects of X monosomy on brain development: monozygotic twins discordant for Turner's syndrome. Ann Neurol 34:95–107

Rossi A, Stratta P, Matlei P, Cupillari M, Bozzoo A, Gallucci M, Lasacchia M (1992) Planum temporale in schizophrenia: a magnetic resonance study. Schizophr Res 7:19–22

Thatcher RW, Walker RA, Giudice S (1987) Human cerebral hemispheres develop at different rates and ages. Science 23:1110–1113

Wada JA, Clarke R, Hamm A (1975) Cerebral hemispheric asymmetry in humans. Arch Neurol 32:239–246

Witelson SF, Kigar DL (1988) Anatomical development of the corpus callosum in humans: a review with reference to sex and cognition. In: Molfese DL, Segalowitz SJ (eds) Brain lateralization in children: developmental implications. Guilford, New York, pp 35–57

Whitaker-Azmitia PM (1991) IV. Role of serotonin and other neurotransmitter receptors in brain development: basis for developmental pharmacology. Pharmacol Rev 43:553–561

Part II
Schizophrenia: Disease or Syndrome?

Brain Imaging, Psychopathology and Neuropathology

E.C. JOHNSTONE

Historical Introduction

The view that psychiatric illnesses were the result of disorders of the brain gained support during the 18th century (Hunter and MacAlpine 1963) and was later developed by Griesinger, Professor of Psychiatry and Neurology in Berlin. He considered that normal mental processes are dependent upon the integrity of the brain and that lesions in some parts of the brain are more likely than those of others to be associated with mental disorder (Griesinger 1857). This idea that the brain was the seat of insanity led to the view that psychiatric illnesses could be investigated in a similar way to other illnesses, and in Germany there were extensive and sometimes successful attempts by Alzheimer, Wernicke and others to correlate psychiatric conditions with neuropathology. Although neuropathological techniques were limited, replicable findings could be made in relation to both gross post-mortem change and histology. It was on the basis of the work of these neuropathologists that Kraepelin stated in 1907, "The fact is decisive that the morbid anatomy has disclosed not simple inadequacy of the nervous system but destructive morbid processes as the background of the clinical picture".

Development of Controlled Studies

In 1924, however, Dunlap described a study in which three independent observers made cell counts of the cortical layers where changes had been observed by Alzheimer and found no difference between schizophrenics and controls. Cellular changes similar to those which had been described in schizophrenia were found in both groups. This investigation cast a shadow of scepticism over the histological study of schizophrenia in later reviews (David 1957). Although at the time of the first neuropathological studies of schizophrenia there was no means of examining the structure and function of the brain until after death, since the 1920s the issue of structural change in the brain in schizophrenia has been examined by imaging studies as well as by

---

Department of Psychiatry, University of Edinburgh, The Kennedy Tower, Royal Edinburgh Hospital, Morningside Park, Edinburgh, EH10 5HF, Scotland, UK

H. Häfner and W.F. Gattaz (Eds.)
Search for the Causes of Schizophrenia, Vol. III

pathological means. In 1929 the American Roentgen Ray Society declared that it was unethical to use normal controls in pneumoencephalographic studies, thus limiting the methodology of many of those conducted subsequently. The introduction of the non-invasive technique of computer assisted tomography by Hounsfield (1973) overcame this difficulty, and a series of studies comparing the structure of the brain in schizophrenics began in 1976 (Johnstone et al. 1976). Most, but not all, studies showed significant differences between schizophrenics and controls, and it may fairly be concluded that an appreciable percentage of schizophrenic patients have subtle CT abnormalities, including ventricular and possibly sulcal enlargement (Zec and Weinberger 1986). Clinical studies of the brain using structural magnetic resonance imaging began about 1980 and since the study of Smith et al. in 1984, numerous investigations of the structure of the brain in schizophrenia using MRI have been published. The better resolution and grey/white matter contrast and the ease of axial, sagittal and coronal imaging have ensured that MRI has clearly taken over from CT as the technique of choice in investigating intracerebral structure. The findings of structural imaging indicating that schizophrenic patients have less brain tissue than normal controls have stimulated a revival of interest in histopathological studies of the brain in schizophrenia (Stevens 1982; Kovelman and Schiebel 1984; Jakob and Beckman 1986), but the generality of these studies have been retrospective, depending upon case note diagnosis and such methodology is unlikely to be successful in addressing relationships between psychopathology and neuropathology.

Relationship Between Psychopathology and the Findings of Imaging Studies

The findings of studies examining the relationship between psychopathology and structural and indeed functional imaging studies have in general been somewhat disappointing. It has been satisfactorily demonstrated that the most robust finding – that of lateral ventricular enlargement – is a feature of the illness and is not due to treatment (Owens et al. 1985; Zigun and Weinberger 1992). Various clinical features have been correlated with increased ventricular size in patients with schizophrenia. These include neuropsychological impairment, poorer premorbid adjustment, worse outcome, more suicides, worse negative symptoms, poorer response to neuroleptics, more marked movement disorder, more EEG abnormalities, more neurological soft signs, increased disordered smooth-eye pursuit and decreased neurotransmitter activity (Zigun and Weinberger 1992). Although each of these correlations has been shown positive by some workers, they have not been replicated by others, and from the generality of studies of structural imaging, specific conclusions about the clinical correlates of ventricular enlargement cannot really be drawn. It could, however, legitimately be concluded that in a general sense ventricular enlargement is a feature of more severe or "chronic" illnesses. This is indeed rather disappointing after 15 years of research and more than 50 controlled studies (Owens 1992), but it may be that recent technical advances will allow relationships that have previously been hidden by imprecision of measurement to be revealed. The recent work of Shenton et al. (1992), showing a positive relationship between thought disorder and the degree of the reduction in volume of the left posterior superior temporal gyrus, may be an indication of the findings that the most recent generation of scanners may provide.

Functional imaging, using positron emission tomography (PET) and single photon emission tomography (SPET) is limited by the expensive technology required especially for PET, by the problems of quantification and the restriction in serial examinations imposed by the use of radioactive tracers. Functional MRI has been introduced in the past year (Belliveau et al. 1991). Its usefulness in studies of psychiatric patients is yet to be established, but it may have considerable potential. Functional imaging studies using PET and SPET in schizophrenics have tended to concentrate on the issues of hypofrontality, laterality and the effects of treatment (Reveley and Taylor 1990), and so far results are not entirely consistent. Some recent studies have related PET findings to psychopathological variables. In a comparison of normal controls, chronically hallucinated schizophrenics and schizophrenics without hallucinations, Cleghorn et al. (1990) found no significant differences between schizophrenics as a whole and normal controls, but there was a pattern of correlations between regions that was unique to hallucinating subjects and anterior cingulate metabolism correlated significantly with hallucination scores. Liddle et al. (1992) using PET found that each of three previously identified syndromes of persistent schizophrenic symptoms (psychomotor poverty, disorganisation and reality distortion) was associated with a distinct pattern of cerebral blood flow. These issues will require further study and it may be that a way forward lies in linking studies of the structural, functional and neuropathological correlates of psychopathological variables more closely than has been done in the past.

Northwick Park Prospective Clinicopathological Study of Schizophrenic Patients: Introduction

As noted above, neuropathological studies in schizophrenia have in general been retrospective and the difficulties of considering psychopathological correlates of neuropathological findings in such studies are very considerable. While I worked at Northwick Park, I was involved in a prospective clinico-neuropathological study of schizophrenic patients. The neuropathological findings have already been reported (Bruton et al. 1990) but detailed analysis of the clinical aspects has just been completed and will be reported here.

Brains from 56 schizophrenic patients were included (Bruton et al. 1990). Thirty-eight of them had been included in the original large study of schizophrenic inpatients in Shenley Hospital (Owens and Johnstone 1980) and some of them survived to be included in the later studies (Ferrier et al. 1983; Owens et al. 1982, 1985; Johnstone et al. 1987). Historical information was recorded for the initial study, but the methods used for the mental state assessments were the same for all of the investigations and the ratings used for this study were those conducted closest to the patient's death. All 38 patients fulfilled the St. Louis criteria for schizophrenia (Feighner et al. 1972). Details of the post-mortem procedure and histological assessments have already been described (Bruton et al. 1990). Each brain was coded and later examined without knowledge of the clinical history. At the end of the study, when the neuropathological code numbers were compared with the relevant case histories, it was found that five of the Shenley schizophrenic patients had superadded neurological complications. Three of these related to previous treatment (leucotomy) and one patient had suffered a terminal cerebrovascular accident. The remaining patient (while working as a ballerina) had developed a psychotic illness in the absence of detectable neurological abnormality, and this illness fulfilled the St. Louis criteria for schizophrenia (Feighner et al. 1972). She later developed motor and cerebellar signs and post-mortem confirmed the diagnosis of multiple sclerosis. Thirty-three brains therefore remained from patients who fulfilled the St. Louis criteria for schizophrenia, had been documented and systematically examined in life, and who were found not to have had superadded neurological complications. A large amount of data had been collected on the patients in Shenley at the time of the original survey and it would have been inappropriate to examine this very large number of variables in relation to the relatively small number of brains available for post-mortem. Full data were not available in every case because of some incomplete early histories, but details of the historical information used are shown in Table 1.

These patients had also been examined in life, although one was too unwell (and indeed near death) when she was first seen to cooperate with mental state examination or cognitive testing, and thus was only assessed from the point of view of movement (hyperkinetic) disorder.

The assessments were positive features (Krawiecka et al. 1977) – available in 32 cases; negative features (Krawiecka et al. 1977) – available in 32 cases; cognitive functioning (Withers and Hinton 1971) – available in 26 cases; and hyperkinetic disorder assessed as present or absent – available in 33 cases. The scale by Krawiecka et al. (1977) and the Withers and Hinton (1971) test battery had been used in the repeated assessments of these patients and could therefore be used for this study, but the assessments of movement disorder were changed (Owens and Johnstone 1980; Owens et al. 1982) and the simple assessment of present or absent was the most accurate available to us.

Statistical Analysis

In the prospective clinico-neuropathological comparisons there were five variables that distinguished schizophrenics from controls (Bruton et al. 1990): brain weight, brain length, ventricle size, focal damage and gliosis. Gliosis and focal damage were related and are independent of the other more general brain changes. The intention of this study was to see whether any of these neuropathological changes relate to clinical features (either historical or examination variables). This was done using stepwise regression. For example,

Table 1. Historical information related to post-mortem findings

Specification of variable	Patients in whom variable available (n)	Source of information
Sex, age, age at first admission, duration of illness	33	Original Shenley survey (Owens and Johnstone 1980)
Premorbid academic functioning (university or equivalent; average attainment; remedial teaching or equivalent)	32	Original Shenley survey (Owens and Johnstone 1980)
Premorbid global functioning (occupational and social attainments as recorded for survey) (best sustained occupational and social attainments as recorded for survey)	32	Re-examination of survey (Owens and Johnstone 1980) to derive composite score, done blindly before PMs
Family history of mental disorder	33	Original Shenley survey (Owens and Johnstone 1980)
Physical treatment (ECT; insulin coma; neuroleptics; cardiazol; TAB vaccine; dexedrine)	33	Original Shenley survey (Owens and Johnstone 1980) + updating done for (Owens et al. 1982)
Other generalised illness (e.g., diabetes, DU surgery for hypernephroma)	33	Original Shenley survey (Owens and Johnstone 1980) + review of notes at death done blindly before PMs
Peculiar circumstances (any unusual circumstances which could be of relevance, e.g., inmate of DP camp; family history of major neurological anomaly)	33	Original Shenley survey (Owens and Johnstone 1980) + review of information obtained for same
Birth trauma/head injury	19	Information recorded from patients' mothers at time of first admission; not recorded in remaining 14 cases

the schizophrenic patients were divided into two groups on the basis of whether or not they had focal damage. Logistic stepwise regression was then used to see whether the membership of these two groups could be predicted from clinical variables (BMDP LR). The clinical variables used were subclassified as follows:

1. Sex, age, age at first admission, duration of illness
2. Positive features, negative features, premorbid functioning (academic and global)
3. Withers and Hinton (cognitive functioning)
4. Birth trauma, head injury
5. Movement disorder
6. Additional variables: past physical illness, family history, "peculiar circumstances", etc.

Because of incomplete data (Table 1) these variables were examined in successively larger subsets, i.e. set 1, set 1 + 2, set 1 + 2 + 3, set 1 + 4, set 1 + 5, set 1 + 6. In linear regression the model was forced to apply age and sex before any of the other variables, since they are known to be strong determinants of measures such as brain weight and brain length. Movement disorder was also examined.

Results

Relationships Between Brain Weight and Clinical Variables

As would be expected, there were significant relationships between brain weight and age and sex, such that females had lower brain weights than males ($p < 0.01$) and greater age was associated with lower brain weight ($p < 0.01$). In addition, lower brain weight was significantly associated with poorer premorbid global function ($p < 0.05$) and with poorer cognitive function. Lower brain weight was significantly related to both poorer premorbid academic record ($p < 0.01$) and to poorer performance on the Withers and Hinton (1971) test battery ($p < 0.01$) when examined for the survey (Owens and Johnstone 1980) or later. The effects of these two variables were independent. In other words, the Withers and Hinton scores predicted brain weight even when premorbid academic level was taken into account. This does not necessarily indicate a role for cognitive decline. Premorbid academic level was defined by only three categories: (a) university or equivalent; (b) average; (c) poor.

Relationships Between Brain Length and Clinical Variables

Women's brains were significantly shorter than those of men ($p < 0.01$). In addition there were significant associations ($p < 0.05$) between shorter brains and poorer premorbid global function and between shorter brains and more severe negative symptoms ($p < 0.02$).

Relationships Between Ventricle Size and Clinical Variables

As would be expected, there was a significant relationship ($p < 0.01$) between increasing age and greater ventricle size. Otherwise there were no relationships between clinical variables and ventricular size.

Relationships Between Focal Damage and Clinical Variables and Between Gliosis and Clinical Variables

When the total samples of schizophrenic patients and controls were examined some degree of localised cerebral pathology (focal damage) was found in 44% of the schizophrenic patients and 21% of controls ($p < 0.01$) (Bruton et al. 1990). The nature and localisation of such pathology was diverse and included such apparently unrelated entities as degeneration of the substantia nigra, calcification of the hippocampus, a ganglioglioma of the pons, focal softening of the cerebellum and infarction of the striatum. In the group of 33 schizophrenic patients in the present study, focal damage was significantly related only to sex, being less likely to be found in men ($p < 0.02$). In the total sample (Bruton et al. 1990), there was a strong positive relationship ($p < 0.01$) between both cortical and white matter gliosis and the presence of focal damage. In the 33 cases considered here, there was no relationship between periventricular gliosis and any clinical variables, and the only relationship with cortical gliosis was a marginal relationship with age ($p < 0.09$). The statistical analysis of the historical information involved grouping together variables which were not necessarily related and the individual details in patients with and without focal damage were therefore scrutinised (Table 2), but apart from some differences in exposure to unusual treatments, the two groups were closely similar.

Other Relationships

Movement disorder was not related to any pathological finding, nor were birth trauma, head injury, any form of physical treatment including insulin coma, the presence of generalised illness or "peculiar circumstances".

Conclusions from These Findings

The clinicopathological comparison between the brains of schizophrenics and those of controls showed that the schizophrenic brains are shorter, lighter and more often show focal pathological change. The present study was concerned to examine the possibility of relating the neuropathological findings in the brains of schizophrenic patients to clinical features of their condition. It is well known that the clinical picture, course and outcome of schizophrenia are by no means homogeneous (Johnstone 1992).

Although these variations are less in this population, which consisted of patients with illnesses so severe that they had spent many years up until their deaths as long-stay inpatients, there is a range of both historical and examination variables in this sample which allows us to examine relationships to pathological features. There was no evidence that birth trauma, head injury, physical illness, unusual circumstances, exposure to ECT, insulin coma treatment or neuroleptics, related either to reduction in brain size or to focal damage. The exposure to unusual treatments of the patients who had focal damage was not the same as those who showed no such pathological change but it seems implausible that TAB vaccine and cardiazol could be of serious relevance to this issue.

Lower brain weight and shorter brain length were associated with impairments in terms of global function, cognitive function and negative symptoms. The relationship between low brain weight and poor educational attainment

Table 2. Comparison of items of historical information between those with and without focal damage

	Focal damage (4 males : 8 females)	No focal damage (14 males : 7 females)
Birth trauma	7 no trauma 5 no information	1 forceps 1 hospitalised prematurity 7 no trauma 12 no information
Head injury	3 no information 2 injury in childhood 1 fractured skull – psychotic behaviour	12 no information 6 no injury 2 injury in childhood 1 later injury
Other generalised illness	7 no illness 1 diabetes 1 duodenal ulcer 1 surgery to nose 1 nephrectomy and splenectomy 1 unsustained positive test for syphilis on admission	16 no illness 1 diabetes 1 hysterectomy 1 pneumonia and peritonitis 1 bilirubinaemia ? cause 1 no information
"Peculiar circumstances"	12 no history of this	1 inmate of camp for displaced persons 1 2 children died (1 spina bifida, 1 meningitis)
History of treatment with ECT at time of last assessment before death	4 had ECT 8 no ECT	8 had ECT 13 no ECT
History of neuroleptic ingestion at time of last assessment defore death	5 no neuroleptics 7 standard doses of various neuroleptics	10 no neuroleptics 11 standard doses of various neuroleptics
Unusual physical treatment	6 none 4 TAB vaccine 2 cardiazol	17 none 2 dexedrine 1 ephedrine 1 modified narcosis
Insulin coma treatment	10 none 2 some	20 none 1 some
Family history of schizophrenia	2 no information 4 no family history 2 family history of nervous illness 1 family history of permanent hospitalisation for nervous illness 3 family history of schizophrenia	10 no information 3 no family history 3 family history of schizophrenia (in all cases multiple history)

observed in this group of schizophrenic patients need not necessarily reflect some process specific to this disorder. The additional and independent relationship between low brain weight and poor recent cognitive levels (the Withers and Hinton test that was administered in the last few years before death) is more difficult to interpret. While it might reflect a neurodegenerative process in some patients, this seems unlikely since other studies, including one of the post-mortem material described here (Bruton et al. 1990) have found little evidence for such degeneration. It is possible that the Withers and Hinton test is simply a more sensitive measure than educational attainment of a long-standing intellectual deficit. There was no relationship between focal damage and many historical or examination variables, the only significant association being with gender (females more focal damage than males $p < 0.02$). In considering these results the two main pathological findings – (a) reduction in brain weight and length; and (b) the presence of focal damage – require separate discussion.

In the present study brain weight and length were related to poor premorbid academic and global function, relationships that might be explained in terms of an early failure of development.

By contrast the presence of focal damage was unrelated to aspects of the clinical state (although more frequently seen in females).

Our original interpretation (Bruton et al. 1990) was that schizophrenia is a developmental abnormality of cerebral structure in which the structurally abnormal brain is excessively vulnerable to acquired damage in later life. Another, perhaps at least at first glance more straightforward explanation, is that schizophrenia is a final common pathway along which both developmental and acquired brain damage may become clinically expressed. The differing clinical correlates of this reduction in brain size and the focal damage strengthen the case for the latter hypothesis. These findings could be considered consistent with the view that there is a neurodevelopmental basis for schizophrenia in some cases but not others, and that the neurodevelopmental form is move common in men (Castle and Murray 1991). In support of this view it is argued that male patients have an earlier onset, a poorer outcome, are more likely to exhibit poor premorbid adjustment and to have structural abnormalities of the brain demonstrated by imaging techniques. Females, on the other hand, have a later onset (Lewine 1981; Häfner et al. 1989) and sometimes a more favourable outcome, at least in social terms (Johnstone et al. 1981).

The association of focal damage with female sex would support the idea that when schizophrenia occurs in women it is sometimes on the basis of pathology, separate from or additional to developmental factors. Such an interpretation is in ways appealing. It seems to make sense of various apparently unrelated pieces of information. It is, however, important to consider what acceptance of this interpretation would imply. It would mean that various diverse focal pathologies which can and do occur in persons who are evidently well are causative of schizophrenia in some people. Either this must mean that these focal pathologies which look the same are in fact different, and that those which are associated with schizophrenia are affecting neural structures which are not involved in other cases, or it must mean that susceptibility to the effects of the focal damage varies in the population. This view has much in common with the idea that schizophrenia is associated with a structurally abnormal brain which is excessively vulnerable to acquired damage in later life (Bruton et al. 1990). The study did not offer us any solid evidence about the cause of the reduced size of the schizophrenic brains. We are left with speculations about genetic factors and early insults, but in this investigation at least, no clear evidence of either.

Combined Study of Clinical Variables and Structural and Functional Imaging Guided by Neuropathological Findings

A further study currently in progress combines the study of clinical and psychological variables with linked structural and functional scans and a second phase of the investigation will address issues arising out of neuropathological findings with ligand binding techniques.

The study concerns 40 (20 male : 20 female) schizophrenic patients of the Royal Edinburgh Hospital, 20 being treatment responsive according to the criteria of May et al. (1988) and 20 being treatment resistant. These groups were matched for age, sex and duration of illness and compared in terms of clinical, social and psychological variables. Linked SPET and MRI scans were conducted and results of these examinations will be presented. Marked differences between the groups in terms of the Rivermead Behavioural Memory test were demonstrated, the treatment-resistant cases performing very poorly.

Single photon emission tomography investigations within the Department (Ebmeier et al. 1993) have shown a reduction in ^{99m}technetium exametazime uptake in anterior cingulate cortex in schizophrenic patients (Ebmeier et al. 1993) and this is also found in some (but not all) of the study sample (Lawrie et al. 1993). The cingulate has been implicated in schizophrenia by the work of Cleghorn et al. (1990) as described. Furthermore, Benes et al. (1991) have described a reduction in the number of gabaergic basket cell interneurons in both the prefrontal and anterior cingulate regions of chronically psychotic patients. It was hypothesised that this loss of gabaergic interneurons would be accompanied by a compensatory up-regulation of $GABA_A$ receptors in schizophrenic patients, and this finding was actually observed at post-mortem in a recent study (Benes et al. 1992). ^{123}Iomazenil is one of the first specific ligands for SPET and has high affinity and specificity for the GABA/BDZ receptor. Patients from the study with relatively high and relatively low uptake of exametazime will be examined with iomazenil and compared with age-matched controls, to test the hypothesis that in schizophrenic patients cell loss in the anterior cingulate cortex is associated with receptor supersensitivity for $GABA_A$/BDZ receptors. If the hypothesis is supported, the other detailed assessments in these patients will allow us to determine the clinical, psychological and structural correlates of the cell loss and associated receptor supersensitivity.

These are our present aspirations for this study, but the limited conclusions that can be drawn from the many investigations conducted over the past 15 years warn against allowing natural optimism to triumph too confidently over experience.

References

Belliveau JW, Kennedy DN, McKinsery RC, Buchbinder BR, Weisskoff RM, Cohen MS, Vevea JM, Brady TJ, Rosen BR (1991) Functional mapping of the human visual cortex by magnetic resonance imaging. Science 254:716–719

Benes FM, McSporran J, Bond ED, Vincent SL, San Giovanni LP (1991) Deficits in small interneurons in schizophrenic cortex. Arch Gen Psychiatry 48:996–1001

Benes FM, Vincent SL, Alsterberg G, Bird ED, San Giovanni JP (1992) Increased $Gaba_A$ receptor binding in superficial layers of cingulate cortex in schizophrenics. J Neurosci 12:924–929

Bruton CJ, Crow TJ, Frith CD, Johnstone EC, Owens DGC, Roberts GW (1990) Schizophrenia and the brain: a prospective cliniconeuropathological study. Psychol Med 20:285–304

Castle DJ, Murray RM (1991) The neurodevelopmental basis of sex differences in schizophrenia. Psychol Med 21:565–575

Cleghorn JM, Garrett ES, Nahmius C, Brown GM, Kaplan RD, Szechtman H, Szechtman B, Franco S, Dermer SW, Cook P (1990) Regional brain metabolism during auditory hallucinations in chronic schizophrenia. Br J Psychiatry 157:562–570

David GB (1957) The pathological anatomy of the schizophrenias. In: Richter D (ed) Schizophrenia, somatic aspects. Pergammon, New York, pp 93–130

Dunlap CB (1924) Dementia praecox: some preliminary observations on brains from carefully selected cases and a consideration of certain sources of error. Am J Psychiatry 80:403–421

Ebmeier KP, Blackwood DHR, Murray C, Souza V, Walker M, Dingall N, Moffoot APR, O'Carroll RE, Goodwin GM (1993) Single photon emission tomography with 99_m Tc-exametazine in unmedicated schizophrenic patients. Biol Psychiatry (in press)

Feighner JP, Robins E, Guze SB, Woodruff A, Winokur G, Munoz R (1972) Diagnostic criteria for use in psychiatric research. Arch Gen Psychiatry 26:57–63

Ferrier IN, Johnstone EC, Crow TJ, Rincon-Rodriguez I (1983) Anterior pituitary hormone secretion in chronic schizophrenia. Arch Gen Psychiatry 40:755–761

Griesinger W (1857) New Sydenham Society, London

Häfner H, Riecher A, Maurer K (1989) How does gender influence age at first hospitalisation for schizophrenia? A translational case register study. Psychol Med 19:903–918

Hounsfield GN (1973) Computerised transverse axial scanning (tomography), part I: description of the system. Br J Radiol 46:1016–1022

Hunter R, MacAlpine I (1963) Three hundred years of psychiatry, 1535–1860. Oxford University Press, London

Jakob H, Beckmann H (1986) Prenatal development disturbances in the limbic allocortex in schizophrenics. J Neurol Transm 65:303–326

Johnstone EC (1992) Heterogeneity in Schizophrenia. In: Lindenmeyer JP, Kay SR (eds) New biological vistas on schizophrenia. Brunner Mazel, New York

Johnstone EC, Crow TJ, Frith CD, Husband J, Kreel L (1976) Cerebral ventricular size and cognitive impairment in chronic schizophrenia. Lancet ii:924–926

Johnstone EC, Owens DGC, Gold A, Crow TJ, Macmillan JF (1981) Institutionalisation and the defects of schizophrenia. Br J Psychiatry 139:195–203

Johnstone EC, Owens DGC, Frith CD, Crow TJ (1987) The relative stability of positive and negative features in chronic schizophrenia. Br J Psychiatry 150:60–64

Kovelman JA, Schiebel AB (1984) A neurohistological correlate of schizophrenia. Biol Psychiatry 19:1601–1621

Kraepelin E (1907) Lehrbuch der Psychiatrie. Macmillan, New York

Krawiećka M, Goldberg D, Vaughan M (1977) Standardised psychiatric assessment scale for rating chronic psychotic patients. Acta Psychiatr Scand 55:299–308

Lawrie SM, Ingle G, Roberts A, Johnstone EC, Santosh C, Best J (1993) Biological and clinial correlates of the heterogeneity in schizophrenia. Proceedings of annual meeting of Royal College of Psychiatry

Lewine RJ (1981) Sex differences in schizophrenia: timing or subtypes. Psychol Bull 90:432–444

Liddle PF, Friston KJ, Frith CD, Hirsch SR, Jones T, Frackowiak RSJ (1992) Patterns of cerebral blood flow in schizophrenia. Br J Psychiatry 160:179–186

May PRA, Dencker SJ, Hubbard LW, Midha KK, Liberman RP (1988) A systematic approach to treatment resistance in schizophrenic disorders. In: Dencker SJ, Kulhanek F (eds) Treatment resistance in schizophrenia. Braunschweig, Wiesbaden

Owens DGC (1992) Imaging aspects of the biology of schizophrenia. Currt Opin Psychiatry 5:6–14

Owens DGC, Johnstone EC (1980) The disabilities of chronic schizophrenia, their nature and the factors contributing to their development. Br J Psychiatry 136:384–395

Owens DGC, Johnstone EC, Frith CD (1982) Spontaneous involuntary disorders of movement: their prevalence, severity and distribution in chronic schizophrenics with the without treatment with neuroleptics. Arch Gen Psychiatry 39:452–461

Owens DGC, Johnstone EC, Crow TJ, Frith CD, Jagoe JR, Kreel LL (1985) Lateral ventricular size in schizophrenia. Relationships to the disease process and its clinical manifestations. Psychol Med 15:27–41

Reveley MA, Taylor CJA (1990) Recent advances in brain imaging. In: Dinan TG (ed) Principles and practice of biological psychiatry. Clinical Neuroscience Publishers, London

Shenton ME, Kikinis R, Jolenz FA, Pollak SD, Le May M, Wible CG, Hokama H, Martin J, Metcalf D, Coleman M, McCarley RW (1992) Abnormalities of the left temporal lobe and thought disorder in schizophrenia. N Engl J Med 327:604–612

Smith RC, Calderon M, Ravichandrian GK, Largen J, Vroulis G, Shrartsbund A, Gordon J, Schoolar LC (1984) Nuclear magnetic resonance in schizophrenia: a preliminary study. Psychiatr Res 12:137–147

Stevens JR (1982) Neuropathology of schizophrenia. Arch Gen Psychiatry 39:1131–1139

Withers E, Hinton J (1971) Three forms of the clinical tests of the sensorium and their reliability. Br J Psychiatry 119:1–8

Zec RF, Weinberger DR (1986) Relationship between CT scan findings and neuropsychological performance in chronic schizophrenia. Psychiatr Clin North Am 9:49–61

Zigun JR, Weinberger DR (1992) In vivo studies of brain morphology in schizophrenia. In: Lindenmayer JP, Kay SR (eds) New biological vistas on schizophrenia. Brunner Mazel, New York

Idiopathic and Organic Presentations of Schizophrenia

M.R. Trimble

Introduction

The term "idiopathic" has been used in medicine for many years as a shield to hide our ignorance regarding aetiology and pathogenesis of many diseases. It has been customary to draw distinctions between syndromes secondary to a known or assumed organic cause, and the idiopathic, synonyms for the latter being cryptogenic, essential, and endogenous. The presumption has been that causes in such cases are likely to have a substantial genetic basis.

Literally referring to disease that arises sui generis, idiopathic forms part of many classifications. For example, the latest classification of the epilepsies (ILAE 1985) dichotomises both localisation-related and generalised seizures into idiopathic and symptomatic groupings.

Applied to the syndromes of schizophrenia, idiopathic remains a euphemism for lack of knowledge, and, as with epilepsy, improved power of investigation will lead to an unravelling of the underlying organic substrates of the clinical expression in many cases.

The classification of cases into the two categories organic and idiopathic seems as artificial as that between the so-called organic and the functional psychoses. The latter actually has more substance, assuming that the term "functional" is used correctly, in a physiological as opposed to a psychological sense. The use of functional in the latter way clearly harks back to the philosophical discourses of Descartes, and has been reinforced in the early part of the twentieth century by the separation, particularly at a clinical level, between neurologists and psychiatrists.

Our present concept of schizophrenia was shaped by three exact contemporaries, Kraepelin (1856–1926), Freud (1856–1939) and Bleuler (1857–1939). Kraepelin put forward the concept of dementia praecox in 1896. He suggested it was due to severe and wide spread disease of the frontal, motor and temporal cortex (Kraepelin 1919). Bleuler, inspired by the ideas of Freud, renamed it schizophrenia and suggested psychological explanations of many of its features. He felt, however, that complete justice to all the known facts could only be done by assuming the presence of some kind of disturbance of the brain (Bleuler 1950). Freud too accepted the brain as the organ of the mind (Freud 1964).

Institute of Neurology, Queen Square, London, WC1N 3BG, UK

H. Häfner and W.F. Gattaz (Eds.)
Search for the Causes of Schizophrenia, Vol. III

In recent years a working hypothesis of a cerebral basis for schizophrenia has been adopted by most clinical and research psychiatrists, and some of the neurological features and underpinnings of schizophrenia have been established (Trimble 1988). In such a setting, the division of idiopathic and organic can no longer be maintained. The number of cases that may be regarded as idiopathic is shrinking, and in any case, perhaps the word "genetic" should be revived for some of these.

Neurological Disorders Presenting with Schizophrenia-like Symptoms

Very crudely, an initial level of discussion could involve a distinction between disorders of the white and grey matter of the brain, and the axis (laterality, anterior–posterior) influenced by any change of function or structure. Thus, as a general rule, white matter disorders appear less associated with psychotic symptomatology than grey matter disorders, and neurological disease of the anterior aspects of the brain, particularly the frontal and temporal lobes, are more involved than the posterior lobes.

The neurological conditions most often associated with a schizophrenia-like state are shown in Table 1. While it is the case that some white matter disorders, for example, multiple sclerosis, or metachromatic leucodystrophy, are occasionally reported to lead to schizophreniform states, by and large, these are isolated case histories. Often the presentation is acute, reminiscent of an organic brain syndrome, with subsequent resolution. Recent studies of multiple sclerosis suggest that when patients present with psychosis, white matter plaques in the temporal lobes are likely to be identified (Feinstein et al. 1992).

Of the grey matter disorders, it is basal ganglia, and temporal lobe lesions which stand out. Davison and Bagley (1969), after a comprehensive review of the literature, concluded that schizophrenia-like psychoses occur in Wilson's disease, and Huntington's chorea more frequently than by chance. However, although paranoid and schizophrenia-like presentations are commonly noted in the history of a patient with Huntington's chorea (Trimble 1981), recent systematic investigation of the psychopathology of Wilson's disease suggests

Table 1. Neurological disorders most offen cited as associated with schizophrenia-like states

White matter:	Multiple sclerosis Metachromatic leucodystrophy
Grey matter:	Epilepsy Huntington's chorea Cortical Lewy body disease Encephalitis lethargica Systemic lupus erythematosis

that psychoses are really quite rare, and are far outweighed by neurotic disturbance and personality disorder (Denning and Berios 1989).

Schizophrenia-like psychoses rarely occur in association with idiopathic Parkinson's disease, and likewise, are not reported in association with presumed basal ganglia disorders such as the dystonias. Thus, the association of neurological illness of the basal ganglia to classical schizophrenia-like presentations is perhaps less extensive than may originally been thought. These models nonetheless are revealing in the sense that Huntington's chorea, where the pathology is in the caudate nuclei, emerges as a disturbance of the balance between dopamine and GABA. The same medication, namely neuroleptics, are effective at treating both the psychoses and the movement disorder. In Parkinson's disease, psychoses are more likely driven by dopamine agonist therapies, and their phenomenology is quite distinctive. Often this resembles an organic brain syndrome, with variable clouding of consciousness, delusions, often paranoid, and comples visual hallucinations. Wilson's disease is now thought to be due largely to damage of the globus pallidus, while, the motor aspects of Parkinson's disease are more associated with the putamen.

Recent evidence from studies of the pathology of schizophrenia reveal that basal ganglia disturbance occurs (Bogerts et al. 1983). Clinico-pathological correlation of such studies suggest that the basal ganglia changes may be related more to negativism and catalepsy (Stevens 1986). Thus, while some structures of the basal ganglia link with psychosis, this may be quite restricted, and associated with changes of neurotransmitter, especially dopamine function.

The neurological disorder most commonly associated with the development of a schizophrenia-like psychosis is that of epilepsy. Some contributions of this literature to an understanding of schizophrenia will now be discussed.

Schizophrenia-like Psychoses of Epilepsy

The most extensive series of the schizophrenia-like psychoses of epilepsy came from Slater and Beard (1963). The most important points that Slater and Beard made were that these psychoses occurred in the setting of clear consciousness, that there was no evidence of a family history of schizophrenia in the probands, and that the psychoses tended to come on several years after the onset of a history of epilepsy.

Slater and Beard carefully studied the phenomenology of these patients. They made the point that the first-rank symptoms of Schneider were seen in all cases, and that motor symptoms and hebephrenic deterioration were rare.

There are a number of other groups that have now studied these psychoses (for full review see Trimble 1991). In general, the incidence of psychosis is higher in temporal lobe epilepsy (17%) than in other focal epilepsies (2%), and in studies of patients with epilepsy and psychosis, temporal lobe epilepsy is over-represented (217 out of 287 cases where the diagnosis could be established – Trimble 1991).

Studies carried out at the Institute of Neurology have clarified the phenomenology of these psychoses. Perez and Trimble (1980) described the mental state of patients with epilepsy and psychosis, prospectively referred, with psychoses lasting longer than a month in a setting of clear consciousness. They were evaluated using the Present State Examination of Wing, a semistandardised technique which allows categorisation of patients' symptoms by use of the Catego computer programme. In their sample, 50% of the patients with epilepsy and psychosis were categorised as having schizophrenic psychosis, 92% having a profile of nuclear schizophrenia based on the first-rank symptoms of Schneider.

The syndrome profile of patients with schizophrenia and epilepsy compared with those with schizophrenia and no epilepsy, as a comparison group, showed few differences, emphasising the similarities of the clinical presentation of these two disorders.

In their sample, 17 patients had complex partial seizures and a clinical diagnosis of temporal lobe epilepsy. In seven the seizure type was generalised. When the PSE sub-class diagnosis of the epileptic psychotic patients were examined by epilepsy diagnosis, a significant association was noted between the presentation of nuclear schizophrenia, and temporal lobe epilepsy. Paranoid psychoses were also over represented.

Similar data have now been reported by others (Toone et al. 1982). These data suggest that in a group of patients with psychosis and epilepsy a significant number have a schizophrenia-like presentation with many features similar to nuclear schizophrenia in the absence of epilepsy. Our group has emphasised the presentation of nuclear schizophrenia with first-rank symptoms in association with temporal lobe pathology. We confirm direct links therefore between certain types of epilepsy and certain patterns of clinical presentation, hinted at in the work of Slater and Beard (1963).

A number of authors have examined the relationship of age of onset of psychosis to the age of onset of seizures. A histogram showing these data is given in Fig. 1. There is an interesting homogeneity of the interval, most studies finding this in the region of eleven to fifteen years. Clinically, many patients with these problems have seizures in late childhood or early adolescence, and go on to develop a psychosis in their late teens and twenties. These data are of interest. First, they suggest that the chronic recurrent seizure disorder, with disturbances emanating from the temporal lobes, somehow influence the brain, priming it for later psychotic breakdown. Secondly, for the similarity which is now being appreciated in schizophrenia research, namely that lesions placed early on in specific brain areas appear to lead to psychotic break down many years later.

A final point for discussion is the effect of laterality. Although not all series have found clear cut results, there is a bias towards the left side of the brain, particularly left temporal lobe epilepsy being associated with a presentation of the schizophrenia-like psychosis. Again, our studies at the Institute of Neurology (Perez et al. 1985) have noted a much more exclusive association

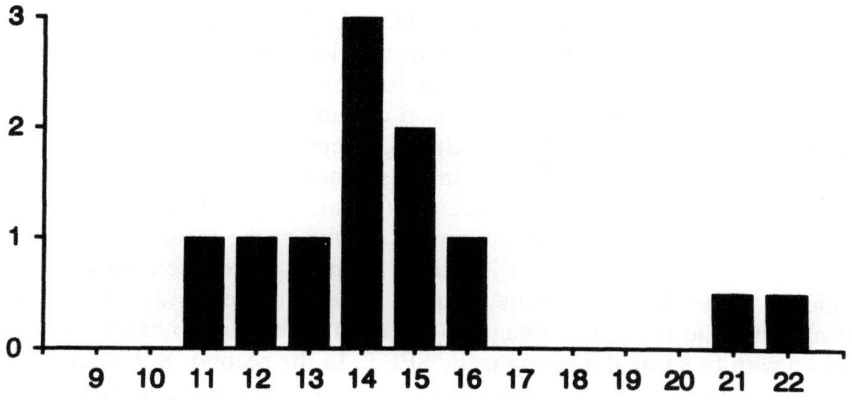

Fig. 1. Age of onset epilepsy and age of onset psychosis (11 studies)

between left hemisphere, in particular, left temporal lobe lesions and nuclear schizophrenia.

A summary of these data are that patients with temporal lobe epilepsy are at risk for the later development of psychosis, although this only occurs in a minority. The psychoses may take the flavour of a schizophrenia-like illness, with predominant first-rank symptoms of Schneider. The latter presentation comes on a number of years after the onset of the epilepsy, and is associated with dominant temporal lobe lesions.

As a final point, neuropathological studies have identified that the lesions underlying the epilepsy affect medial temporal lobe, in particular, limbic structures (Bruton 1988).

These data on epilepsy have in part been responsible for the acceleration of research on the temporal lobes in schizophrenia. This literature is reviewed by others, but it has become clear over the past few years that, although pathological changes may be found in various brain regions in schizophrenia, and identified by either neuropathological studies, or more recently, by structural imaging with MRI, and MRS, pathology is frequently reported in temporal structures. There is an increasing literature emphasising the importance of left hippocampal amygdala and parahippocampal gyrus lesions (for review see Trimble 1994). In addition, there is a considerable literature reflecting on frontal lobe changes in schizophrenia (Trimble and Rogers 1987). It should be noted, however, that underlying structural changes in the frontal lobes have been much more hard to identify than the lesions in the hippocampus and related structures, while techniques assessing frontal lobe function, particular neuropsychological and neuroimaging studies of function (for example, SPET or PET) have been more successful. Although there are well known neuroanatomical connections between medial temporal structures and frontal cortex, and orbital frontal cortex is often referred to as part of the limbic system, one important challenge in schizophrenia research at the present time is to link up changes of structure which have been identified, with changes of function. The latter not only include alteration of frontal lobe activity, but also changes of neurotransmitter activity, especially of dopamine. Thus, although the dopamine hypothesis of schizophrenia has caused considerable controversy, it has stood the test of time (Trimble 1988). Although reports are conflicting, at least one group has suggested that there is increased activity in basal ganglia dopamine receptors in drug naive schizophrenics (Wong et al. 1986) compatible with a hypothesis of over activity of dopamine activity in the disorder.

In the following two investigations, attempts have been made to use the epilepsy model to clarify some of these issues. In the first, structural imaging using magnetic resonance imaging (MRI) and spectroscopy (MRS) have been linked with functional imaging using SPET. In the second, SPET has been used with a dopamine ligand to identify post-synaptic dopamine activity in patients with psychosis.

Institute of Neurology Studies

MRI, MRS and SPET in Temporal Lobe Epilepsy

Thirteen patients with epilepsy, the majority being identified as having temporal lobe epilepsy, received MRI, SPET, using a GE/CGR triple-headed neurocam, and MRS (for details see Schmitz et al. 1994, in preparation). Correlations were sought between changes of structure, as revealed through MRI hippocampal relaxometry and ^{1H}MRS, and changes of function as revealed by ^{99mTc}HMPAO perfusion in various brain regions. In particular we wished to test a hypothesis that structural changes in medial temporal areas would lead to changes of frontal lobe function.

Our data revealed significant correlations between mesial temporal sclerosis as identified by MRI parameters, and decreased perfusion of the ipsilateral temporal cortex. The latter correlated well with spectroscopy data, but not relaxometry. Thus, the hypoperfusion related to neuronal loss, rather than gliosis. The areas of functional abnormality were far greater than the very restricted areas of identifiable structural damage. They extended to lateral temporal and orbital frontal regions. Further, correlations were shown between perfusion in the temporal lobes, and the frontal lobes, but not between the spectroscopy data and frontal lobe perfusion.

A summary of these data is that functional change in the frontal lobes is associated with functional change in the temporal lobes, but not directly with neuronal loss in the latter.

These data may have relevance for further understanding of the pathogenesis of schizophrenia-like states, emphasising that it is perhaps not the underlying neuronal loss in the limbic structures which is so relevant to the frontal lobe abnormalities, but that the latter are more associated with alteration of function in the temporal lobes. This raises important issues as to the potential for reversibility of these functional abnormalities, perhaps reflecting a more widespread area of disturbance of function in limbic structures. The recent reports that some neuroleptics, for example clozapine, may reverse some of the so-called negative symptoms of schizophrenia, which are thought by some to be inter-related with frontal lobe dysfunction (Trimble 1988), become more plausible.

Psychoses of Epilepsy Studied with SPET and ^{123}I-IBZM

Seven patients with temporal lobe epilepsy who became psychotic were compared with a matched group of seven patients who were not psychotic. Both groups received SPET using an SME 810 scanner. The radio ligand was ^{123}IBZM, selective for the D_2 receptor. This provides good visual images of the basal ganglia during the later stages of imaging following an injection when specific binding is recorded. All the psychotic patients were evaluated using the Present State Examination, and all had a schizophrenia-like psychosis. None was receiving neuroleptic medications, although both groups received anticonvulsant drugs. None was receiving vigabatrin.

The results of this study revealed that the psychotic group had decreased uptake of the exogenous ligand during the period of specific binding, suggesting that there was increased endogenous ligand available at the time of the scan (for details see Ring et al. 1994). These data suggest that the state of psychosis is associated with an increased release of dopamine, although other interpretations may also be valid. However, the results are not compatible with the hypothesis that the receptors are increased, or have higher affinity. They support the dopamine hypothesis of an association between the state psychosis and increased dopamine release, rather than a hypothesis that links increased receptor binding or affinity to psychosis.

Summary and Conclusions

Of all the neurological models of schizophrenia, epilepsy has proved the most interesting and most useful heuristically. Psychoses, particularly of a schizophreniform nature, are over represented in patients with temporal lobe epilepsy, there is a bias towards left sided lesions, and there is great similarity between the clinical presentation of the psychosis in epilepsy, and the phenomenology of schizophrenia in the absence of epilepsy. Other psychotic presentations, for example, manic and paranoid do occur, but research on the schizophreniform patients has been part of the growing evidence that medial termporal lobe limbic structures are somehow involved in the pathogenesis of certain psychotic phenomena. The data presented here emphasise that structural lesions, especially cell loss, at medial temporal sites, can lead on to functional alteration of frontal lobe function via associated temporal lobe decreased perfusion. This perhaps explains the findings of hypofrontality in schizophrenia. It can occur without any necessary underlying alteration of structure in the frontal lobes.

We have also shown that the psychotic state is not associated with increased activity of dopamine receptors at the D_2 post-synaptic receptor in basal ganglia structures. Increased release of dopamine is a more compatible explanation of our data.

We have tried in these studies to pull together three important concepts found in schizophrenia research. The first relates to the growing evidence of structural changes in the medial temporal structures, and the second to abnormalities of frontal lobe function. The third is the dopamine hypothesis. It seems that widespread functional change, including perhaps altered dopamine function, may be related to cell loss in medial temporal, namely limbic structures in epilepsy. The same may hold true for schizophrenia.

There has been as much confusion over the terms "structural" and "functional" as there has been over the terms "organic" and "idiopathic". As Rogers (1992) has pointed out: "Accepting psychoses associated with known cerebral disorder as equivalent to those in which cerebral disorder is presumed but not established, is arguably the single most important step foreward in research in this field". Organic has some resonance with structural, but cannot be seen as the antithesis of that which is functional, and always encroaches on the idiopathic.

References

Bleuler E (1950) Dementia praecox or the group of schizophrenias. International Universities Press, New York

Bogerts B, Hantsch J, Herzer M (1983) A morphometric study of the dopamine containing cell groups in the mesencephalon of normals, Parkinson patients and schizophrenics. Biol Psychiatry 18:951

Bruton CJ (1988) The neuropathology of temporal lobe epilepsy. Oxford University Press, Oxford (Maudsley monograph, no 31)

Davison K, Bagley CR (1969) Schizophrenia-like psychoses associated with organic disorders of the central nervous system: a review of the literature. In: Herrington RN (ed) Current problems in neuropsychiatry. Headley Brothers, Ashford, Kent (British Journal of Psychiatry, special publication no 4)

Denning TR, Berrios G (1989) Wilson's disease. Psychiatric symptoms in 195 cases. Arch Gen Psychiatry 46:1126–1134

Feinstein A, duBoulay G, Ron M (1992) Psychotic illness in multiple sclerosis. Br J Psychiatry 161:680–685

Freud S (1964) An outline of psychoanalysis. In: The complete psychological works of Sigmund Freud, vol 23. Hogarth, London, pp 141–207

International League against Epilepsy (1985) Proposal for classification of the epilepsies and epileptic syndromes. Epilepsia 26:268–278

Kraepelin E (1919) Dementia praecox and paraphrenia. Livingstone, Edinburgh

Perez MM, Trimble MR (1980) Epileptic psychosis – diagnostic comparison with process schizophrenia. Br J Psychiatry 137:245–249

Perez MM, Trimble MR, Murray MF, Reider I (1985) Epileptic psychosis: an evaluation of PSE profiles. Br J Psychiatry 146:155–163

Ring HA, Trimble MR, Costa DC, Moriarty J, Verhoeff NPLG, Ell PJ (1994) Striatal dopamine receptor binding in epileptic psychoses. Biol Psychiatr 35:375–380

Rogers D (1992) Motor disorder in psychiatry. Wiley, Chichester

Slater E, Beard AW (1963) The schizophrenia-like psychoses of epilepsy. Br J Psychiatry 109: 95–150

Stevens J (1986) Clinico-pathological correlations of schizophrenia. Arch Gen Psychiatry 43:115

Toone BK, Garralda ME, Ron MA (1982) The psychoses of epilepsy and the functional psychoses: a clinical and phenomenological comparison. Br J Psychiatry 141:256–261

Trimble MR (1981) Neuropsychiatry. Wiley, Chichester

Trimble MR (1988) Biological psychiatry. Wiley, Chichester

Trimble MR (1991) The psychoses of epilepsy. Raven, New York

Trimble MR (1994) Biological psychiatry, 2nd edn. Wiley, Chichester

Trimble MR, Rogers D (1987) Neurological disorder in schizophrenia. In: Henn FA, Delisi LE (eds) Schizophrenia, vol 2. Elsevier Science, Amsterdam, pp 439–465

Wong DF, Wagner HN, Tune LE et al (1986) Positron emission tomography reveals D_2 receptors in drug naive schizophrenics. Science 234:1558–1563

Associations Between Regional Brain Abnormalities and Specific Syndromes in Schizophrenia

P.F. Liddle

Introduction

The historical roots of schizophrenia are to be found in Kalhbaum's (1863) identification of three psychotic conditions of young and middle adult life: catatonia, hebephrenia and dementia paranoides. Catatonia is a disorder of the initiation of voluntary activity; hebephrenia is characterized by inappropriate speech, mood and behaviour, while the essence of dementia paranoides is a defect in evaluating reality. Kraepelin amalgamated these three disorders to form a single disease because of their overlapping clinical features and shared propensity to run a chronic course.

In accord with these historical roots, recent studies of the relationships between schizophrenic symptoms (Liddle 1984, 1987; Peralta et al. 1992) demonstrate that schizophrenic symptoms segregate into three syndromes: psychomotor poverty (poverty of speech, flat affect, decreased spontaneous movement); disorganization (formal thought disorder and inappropriate affect); and reality distortion (delusions and hallucinations). Most schizophrenic patients exhibit evidence of more than one of the syndromes. Thus it appears the diverse symptoms of schizophrenia reflect at least three distinguishable, but related pathological processes.

Neuropsychological studies indicate that each syndrome is associated with a specific impairment of the supervisory processes responsible for initiation, selection and evaluation of mental activity. Psychomotor poverty is associated with impaired initiation of activity, and in particular, with impaired word generation (Liddle and Morris 1991; Allen et al. 1993). Disorganization is associated with impaired suppression of inappropriate responses, including impaired performance in the Stroop test (Liddle and Morris 1991; McGrath 1992). In the Stroop task the subject is presented with a list of colour names printed in ink of a colour that is not congruent with the colour name. For example, the word "red" might be printed in green ink. The subject is asked to say the colour of the ink. Efficient performance of this task demands suppression of interference from the competing tendency to respond to the colour name.

Hammersmith Hospital, Du Cane Road, London, W12 0HS, UK

H. Häfner and W.F. Gattaz (Eds.)
Search for the Causes of Schizophrenia, Vol. III

In general, delusions or of hallucinations are not associated with wide-ranging neuropsychological impairment. It appears that the abnormality producing reality distortion symptoms involves a quite circumscribed aspect of mental processing. Some studies indicate that delusions and hallucinations are associated with impaired monitoring of internally generated activity. For example, in a study by Frith and Done (1989), patients played a computer game which involved aiming at a moving target. Patients with delusions of control were impaired in their ability to adjust their aim to allow for movement of the target under circumstances in which there was no visual feedback. Under these circumstances, accurate adjustment of aim depends on internal monitoring of the action of aiming. Although further studies are required, the available evidence supports the hypothesis that reality distortion reflects impaired monitoring of self-generated mental activity.

Patterns of Regional Cerebral Blood Flow

In a PET study of regional cerebral blood flow (rCBF) in patients with persistent, stable symptoms, Liddle et al. (1992a) found that each syndrome was associated with a particular pattern of rCBF in association cortex and related subcortical nuclei. Furthermore, for each syndrome, the cerebral sites affected included the sites maximally activated in normal individuals performing the type of supervisory mental process implicated in that syndrome.

Psychomotor poverty was associated with decreased rCBF in the left prefrontal cortex and parietal cortex and with increased rCBF in the caudate nuclei. As shown in Fig. 1, the area of decreased blood flow in left prefrontal cortex coincided with the site that is maximally activated in normal individuals during a word generation task (Frith et al. 1991). Thus patients with psychomotor poverty appear to have decreased activity at the cerebral site implicated in internal generation of words.

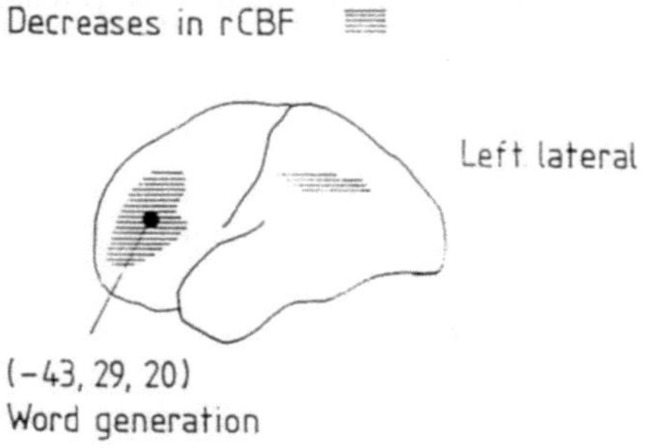

Fig. 1. Locus of maximal activation of the prefrontal cortex during the internal generation of words in normal subjects superimposed on the areas of decreased cortical blood flow associated with psychomotor poverty in schizophrenia. Coordinates refer to the reference frame defined in the atlas of Tailarach and Tournoux (1988), and specify distance in millimetres from the origin at the midpoint of the anterior commissure. (From Liddle et al. 1992b)

Disorganization was associated with decreased rCBF in the right ventral prefrontal cortex and insula, and with increased rCBF in the right anterior cingulate and in the thalamus. As shown in Fig. 2, the area of increased rCBF in the right anterior cingulate cortex associated with disorganization in schizophrenic patients includes the site maximally activated in normal individuals during the Stroop task (Pardo et al. 1990). In light of the evidence from animal studies that the ventral prefrontal cortex plays a role in the suppression of previously learned responses which are no longer appropriate to the circumstances, the finding that disorganization is associated with decreased rCBF in ventral prefrontal cortex and increased rCBF in anterior cingulate, suggests that patients with persistent disorganization syndrome have a malfunction of the ventral prefrontal cortex resulting in a tendency towards the intrusion of a inappropriate material into the current stream of mental processing, necessitating pathological overactivity of the anterior cingulate cortex.

Reality distortion was associated with increased rCBF in left parahippocampal gyrus, temporal pole and lateral prefrontal cortex, and with decreased rCBF in superior temporal gyrus bilaterally and in caudate nucleus and posterior cingulate cortex on the right side. As shown in Fig. 3, the area of positive

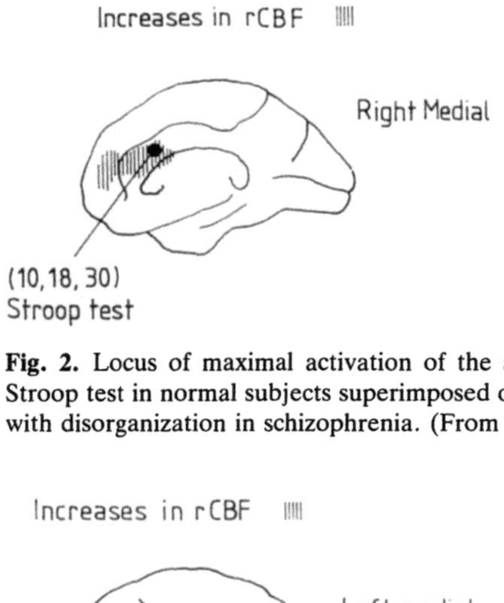

Fig. 2. Locus of maximal activation of the anterior cingulate cortex during performance of the Stroop test in normal subjects superimposed on the area of increased cortical blood flow associated with disorganization in schizophrenia. (From Liddle et al. 1992b)

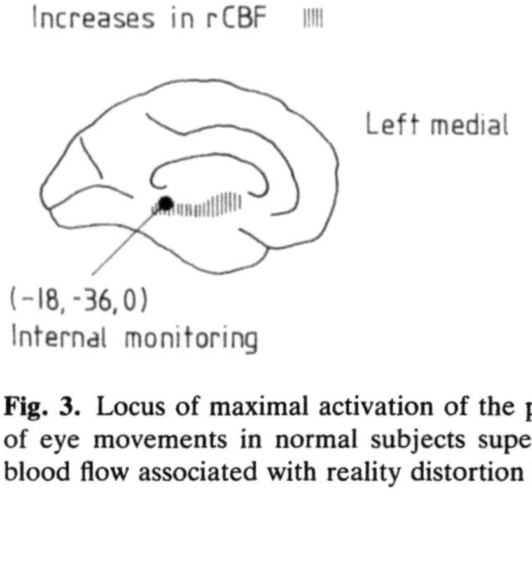

Fig. 3. Locus of maximal activation of the parahippocampal gyrus during the internal monitoring of eye movements in normal subjects superimposed on the areas of increased medial temporal blood flow associated with reality distortion in schizophrenia. (From Liddle et al. 1992b)

correlation between reality distortion and rCBF in the left medial temporal lobe includes the parahippocampal site activated in normal individuals during an eye movement task requiring internal monitoring of activity (Frith et al. 1992). Thus, the evidence supports the hypothesis that reality distortion entails a pathological increase in internal monitoring of self-generated activity.

Although the patients in this study were receiving antipsychotic medication, partial correlation analysis indicated that variation in antipsychotic dose could not account for the observed pattern of correlation between symptoms and rCBF. Subsequent studies of unmedicated patients using single photon emission tomography (Ebmeier et al. 1993) confirm this conclusion. In particular, Ebmeier et al. confirmed the negative correlation between psychomotor poverty and rCBF in prefrontal cortex, the positive correlation between disorganization and rCBF in the right anterior cingulate, and the negative correlation between reality distortion and rCBF in left lateral temporal cortex.

However, despite the negative correlation between psychomotor poverty and frontal rCBF, the group mean frontal rCBF in these unmedicated patients was greater than in normal controls. This suggests that some other clinical feature present in acute patients tends to produce frontal overactivity. Clinical evidence suggests that psychomotor poverty is at one pole of a bipolar dimension, with psychomotor excitation at the opposite pole. Although some acutely ill patients suffer from psychomotor poverty, if a larger proportion of the patients suffered psychomotor excitation, the group as a whole would be expected to exhibit hyperfrontality.

An examination of the aspects of the pattern of cerebral actvity common to psychomotor poverty, disorganization and reality distortion in the patients studied by Liddle et al. (1992a) revealed that all three syndromes are associated with increased rCBF in left parahippocampal gyrus and globus pallidus (Friston et al. 1992). Thus, it appears that the pathological processes generating each the three syndromes share features in common, in accord with Kraepelin's proposal that catatonia, hebephrenia and dementia paranoides should be considered as components of a single illness. It is also important to note that for each of the three syndromes, the pattern of cerebral actvity included areas of relative underactivity and areas of relative overactivity, implying a dynamic imbalance of function in different cerebral areas, rather than a fixed deficit of function.

Activation of Prefrontal Cortex

The seminal studies by Weinberger et al. (1986) of frontal activation during the performance of the Wisconsin Card Sorting Test demonstrated that schizophrenic patients produce lesser frontal activation than normal individuals. However, since the patients perform less well than normal individuals in this task, Weinberger's studies do not answer the question of whether or not frontal activation in schizophrenic patients is less than normal when the patients perform a "frontal" task at the same rate as normal individuals. To address this question, my colleagues and I (Liddle et al. 1994) measured the changes in rCBF during the generation of words at a fixed rate of one word every three seconds, compared with a baseline condition in which subjects repeated words provided by the experimenter at the same rate. We studied four groups of subjects: schizophrenic patients with poverty of speech, schizophrenic patients exhibiting odd speech, schizophrenic patients with unimpaired verbal fluency, and normal individuals.

Patients with poverty of speech exhibited a maximal level of activation in the left prefrontal cortex that was similar in magnitude to that occurring in patients with unimpaired verbal fluency and those with odd speech, and also similar to that occurring in normal individuals. Thus, the magnitude of frontal activitation in schizophrenic patients during word generation is similar to that in normal individuals generating words at the same rate.

However, the patients produced an abnormal pattern of activation in other cerebral areas. Whereas in normal individuals, frontal activation during word generation was associated with suppression of activity in lateral temporal cortex, and an increase in thalamic activity, in the schizophrenic patients the frontal activation was accompanied by increased activity in lateral left temporal lobe and parietal lobe, decreased activity in the posterior cingulate cortex and no significant change in the thalamus. The pattern was similar in all three groups of schizophrenic patients, though the extent of abnormal left temporo-parietal activation was greatest in the patients with poverty of speech (Fig. 4).

While in general, the patients exhibited more extensive cortical activation than the normal controls, it should be noted that the patients produced ab-

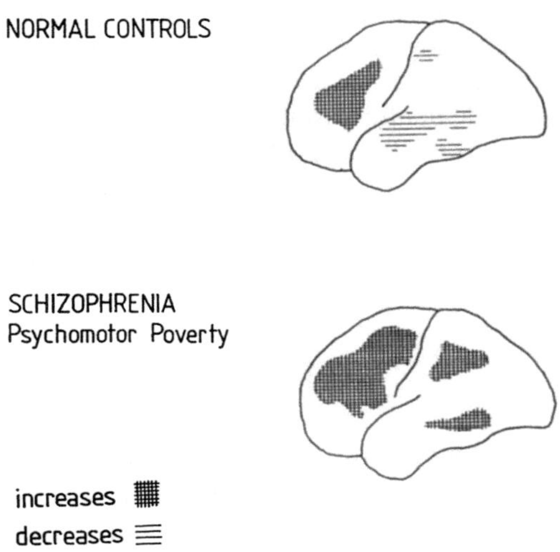

Fig. 4. Area of significant change in blood flow in left lateral cerebral cortex during the generation of words in normal subjects (*upper figure*) compared with that in schizophrenic patients with poverty of speech (*lower figure*). In both groups of subjects, the rate of word generation was paced at the rate of one word every 3 s

normal suppression of activity in posterior cingulate gyrus, suggesting that the observed pattern of activity cannot be attributed merely to more extensive activation due to greater effort. Rather, the pattern of activation suggests an abnormal pattern of functional connectivity between cerebral areas in schizophrenia.

Conclusion

The symptoms of schizophrenia segregate into three syndromes in a manner suggesting that they arise from at least three distinguishable but related pathological processes.

Neuropsychological studies indicate that each of the syndromes reflects abnormality of a specific aspect of the supervisory mental processes responsible for the initiation, selection and monitoring of mental activity. Measurements of rCBF confirm that the syndromes are associated with three different patterns activity in a multimodal assocation cortex and related subcortical nuclei.

In accord with the view that these syndromes are part of a single illness, there is overlap between these rCBF patterns, especially in left parahippocampal gyrus and globus pallidus. Furthermore, the evidence suggest that areas of relative cerebral underactivity, such as the prefrontal activity associated with psychomotor poverty do not reflect irreversible loss of function, but rather, a dynamic imbalance of activity in different cerebral regions reflecting an abnormal pattern of functional connectivity between regions.

References

Allen HA, Frith CD, Liddle PF (1993) Negative features, retrieval processes and verbal fluency in schizophrenia. Br J Psychiatry (in press)

Ebmeier KP, Blackwood DHR, Murray C, Souza V, Walker M, Dougall N, Moffoot APR, O'Carroll RE, Goodwin GM (1993) Single emission tomography with 99mTc-exametazime in unmedicated schizophrenic patients. Biol Psychiatry (in press)

Frith CD, Done DJ (1989) Experiences of alien control in schizophrenia reflect a disorder in the central monitoring of action. Psychol Med 19:359–363

Frith CD, Friston KJ, Liddle PF, Frackowiak RSJ (1991) Willed action and the prefrontal cortex in man: a study with PET. Proc R Soc Lond [Biol] 244:241–246

Frith CD, Friston KJ, Liddle PF, Frackowiak RSJ (1992) PET imaging and cognition in schizophrenia. J R Soc Med 85:222–224

Friston KJ, Liddle PF, Frith CD, Hirsch SR, Frackowiak RSJ (1992) The left medial temporal region and schizophrenia: a PET study. Brain 115:367–382

Kahlbaum K (1863) Die Gruppierung der psychischen Krankheiten. Kafeman, Danzig

Liddle PF (1984) Chronic schizophrenic symptoms, cognitive function and neurological impairment. Thesis, Royal College of Psychiatrists, London

Liddle PF (1987) The symptoms of chronic schizophrenia: a re-examination of the positive-negative dichotomy. Br J Psychiatry 151:145–151

Liddle PF, Morris DL (1991) Schizophrenic syndromes and frontal lobe performance. Br J Psychiatry 158:340–345

Liddle PF, Friston KJ, Frith CD, Jones T, Hirsch SR, Frackowiak RSJ (1992a) Patterns of cerebral blood flow in schizophrenia. Br J Psychiatry 160:179–186

Liddle PF, Friston KJ, Frith CD, Frackowiak RSJ (1992b) Cerebral blood flow and mental processes in schizophrenia. J R Soc Med 85:224–227
Liddle PF, Herold S, Fletcher P, Friston KJ, Silbersweig D, Frith CD (1994) A PET study of word generation in schizophrenia. Schizophr Res 11:168
McGrath JJ (1992) The neuropsychology of thought disorder. Schizophr.Res 6:157
Pardo JV, Pardo PJ, Janer KW, Raichle ME (1990) The anterior cingulate mediates processing selection in the Stroop attentional conflict paradigm. Proc Natl Acad Sci USA 87:256–259
Peralta V, deLeon J, Cuesta MJ (1992) Are there more than two syndromes in schizophrenia? A critique of the positive-negative dichotomy. Br J Psychiatry 161:335–343
Talairach J, Tournoux P (1988) A co-planar stereotactic atlas of the human brain. Thieme, Stuttgart
Weinberger DR, Berman KF, Zec RF (1986) Physiologic dysfunction of dorsolateral prefrontal cortex in schizophrenia. 1. Regional cerebral blood flow evidence. Arch Gen Psychiatry 43: 114–124

Genetic Heterogeneity and Phenotype Variation of Schizophrenia

W. Maier

It is well proven that the manifestation of schizophrenia is under genetic control (Kendler and Diehl 1993). It is however less clear:

1. How relevant environmental factors are, and how they operate and interact with genetic factors.
2. If the genetic susceptibility is identical across the total population of schizophrenics (or at least across all subjects with familial schizophrenia), or if the specific genetic components vary across families.
3. What is transmitted in families of schizophrenics and how co-familial traits are related to genetic and genetic risk factors; are only disorders and symptoms that belong to the schizophrenic sepctrum transmitted or are neurophysiological, neuropsychological, or biochemical correlates also transmitted (independent of the occurrence of the disease)?

These controversial issues might be summarized under the heading of genetic and phenotypic heterogeneity of schizophrenia. This paper discusses a selection of relevant aspects.

The Relative Contribution of Genetic and Environmental Factors to Schizophrenia

Twin and adoption studies clearly demonstrate that most of the variance in familial patterns of schizophrenia is due to genetic determinants and that familial environment is of reduced impact (McGue et al. 1986). The concordance rates between monozygotic twins are far from being perfect (about 50%). Therefore it was concluded that only a subgroup of schizophrenia is genetic (creating the excess concordance rates in twins) and another subgroup of schizophrenia is environmental (McGuffin et al. 1987). This distinction is also referred to by the familial/sporadic dichotomy (Murray et al. 1985). However, the validity of the genetic/environmental discrimination is not convincingly evidenced: current empirical evidence does not exclude that the occurrence of schizophrenia may always require the presence of genetic factors,

Department of Psychiatry, University of Mainz, Untere Zahlbacher Str. 8, 55101 Mainz, Germany

H. Häfner and W.F. Gattaz (Eds.)
Search for the Causes of Schizophrenia, Vol. III

but expression of the genetic predisposition is dependent on additional environmental risk factors or the absence of protective factors; variation of individual environmental factors may induce imperfect monozygotic concordance rates under this condition. This view is supported by investigations in offspring of monozygotic twins discordant for schizophrenia (Gottesman and Bertelsen 1989). The risk for offspring of the unaffected twin (with a schizophrenic monozygotic co-twin) to develop schizophrenia was observed to be equal to the offspring of the affected index case. Simultaneously, the risk to offspring of the unaffected monozygotic twin of a schizophrenic is substantially higher than the risk in offspring of the unaffected twin with a dizygotic schizophrenic co-twin. A similar study by Kringlen (1991) reported divergent results as far as schizophrenia is concerned; however, when the disease definition is extended by affective and anxiety disorders, conclusions are very similar.

The Copenhagen high-risk study supplements this view by demonstrating that familial susceptibility to schizophrenia might be either expressed as schizophrenia or schizotypal personality disorder. Whereas both conditions are familial, schizophrenia occurs preferentially in the presence of an additional risk factor such as birth or prenatal complication (Cannon et al. 1990). The Copenhagen Study (Cannon et al. 1990) and the Helsinki Adoption Study (Tienari 1991) clearly demonstrate that the presence of both genetic and environmental etiological factors is necessary for the manifestation of most cases of schizophrenia. Environmental factors may be of a somatic (infection during pregnancy, prenatal complications or both) or of a psychological (rearing styles, home environment) nature. Thus, there is agreement with studies in offspring of discordant monozygotic twins that the "pure" environmental subgroup is only of limited size.

On the other hand, some studies comparing subjects with vs. without a somatic risk factor demonstrated that the development of schizophrenia does not require a familial loading (Table 1); an additional recent prospective study reported that in the presence of a somatic risk factor familial loading was negatively correlated with the manifestation of schizophrenia (McNeil et al. 1993); at variance with Done et al. (1991). More generally, deviant prenatal development has been postulated to present an environmentally determined etiological subtype of schizophrenia operating independently of genetic forces (Castle and Murray 1991). However, an inverse relationship between obstetric and birth complications and familial loading has not been found consistently (Table 1). In addition, at least some of the prenatal morphological deviations

Table 1. Obstetric complications associated with a lack of familial load?

Positive	Negative
Lewis et al. (1987)	Nimigaonkar et al. (1988)
Schwarzkopf et al. (1989)	Heun and Maier (1993)
O'Callaghan et al. (1990)	

which are assumed to be linked to the development of schizophrenia might also have genetic sources (Nowakowski 1991).

Empirical investigations focusing on a valid distinction between genetically and environmentally determined cases of schizophrenia are severely limited by sample size. Given this limitation, there is evidence that genetic susceptibility operates more broadly than proposed by the 50% concordance rate for monozygotic twins and that "pure" environmental cases represent a minority.

Heterogeneity of the Genetic Liability to Schizophrenia

The interfamilial heterogeneity of the genetic basis of diseases is a well-known phenomenon in human genetics. Even in monogenic diseases, different mutant alleles may give rise to the same disease phenotype. This phenomenon is called genetic heterogeneity (Risch 1990a). The different alleles may occur at the same locus (intra-locus heterogeneity such as Duchenne's and Becker's muscular dystrophy) or at different loci (inter-locus heterogeneity such as neurofibromatosis). Evidence for genetic heterogeneity can be obtained by:

1. Starting with identified genotypes: genetic heterogeneity is proposed by strong, replicable evidence for *linkage* to a particular region in some families and strong, replicable evidence against linkage to the same region in other families.
2. Starting with the familial phenotypes: genetic heterogeneity is proposed by:
   a. Evidence for distinct *clinical subtypes* of the same disease which run true within families.
   b. Evidence for distinct *patterns of inheritance* (e.g., dominant in some families and recessive in others).

Heterogeneity Evidenced by Linkage Studies?

Linkage studies explore if a genetic polymorphism (marker) is transmitted together with the phenotype in families. A marker and a disease are linked if their genetic loci are closely located on the same chromosome (therefore only interlocus but not intralocus heterogeneity can be detected by linkage analysis). Did linkage studies in schizophrenia provide evidence for genetific heterogeneity?

It was claimed that schizophrenia is linked to a marker on chromosome 5. One study in a heterogeneous sample of families demonstrated linkage. This claim could not be substantiated in any of the numerous replication trials although it was initially established in a heterogeneous population (McGuffin et al. 1990). The meaning of these negative findings might either be non-replication, since the initial finding was a false-positive finding, or indication of heterogeneity. It is generally recognized that the first possibility is more likely than the second in this particular instance (McGuffin et al. 1990).

Mutations of the amyloid–precursor–protein gene (APP gene) are causative for early onset of Alzheimer's disease in a minority of cases. The APP gene has also been screened in a sample of chronic schizophrenics, and unexpectedly a point mutation in the expressed part of the gene was found in a single patient with a familial history of schizophrenia (Jones et al. 1992). Subsequent association studies were not able to detect any further cases with the same APP gene mutation (Nöthen et al. 1993). Again, it remains undecided if non-replications of a rare etiological relationship of a genetic polymorphism with schizophrenia (genetic heterogeneity) or a random co-occurrence of two rare events in one instance explains this diversity of findings.

Tentative evidence for genetic heterogeneity was found by tracing a balanced translocation on chromosome 11 through a pedigree loaden with schizophrenia and other mental diseases (St. Clair et al. 1990). Linkage was found between the broadly defined phenotype (schizophrenia, schizoaffective and recurrent major depression being considered as affected) and the breaking point on chromosome 11 which is located in close proximity to the D_2 receptor gene (Fletcher et al. 1993). Owing to the limited size of a single extensive family, the strength of linkage was only modest. However, two other smaller families have been described with two independent translocations on the long arm of chromosome 11 (Smith et al. 1989; Holland and Gosden 1992). Unfortunately, the breaking point was not so accurately defined that the relative position to the D_2 receptor gene could be identified.

As the observed breaking point in St. Clair's family is located close to the D_2 receptor gene, linkage to the D_2 receptor gene should also include linkage to the breaking point on chromosome 11. The functional relevance of D_2 receptors for the activity of anti-psychotic drugs additionally increased the a priori odds for linkage to the D_2 receptor gene or to a locus in the adjacent region which can be expected to include regulatory genes for the expression of the D_2 receptor gene. D_2 receptor gene polymorphisms were tested in a series of studies for linkage to schizophrenia (using a variety of disease models), but all these studies were negative (Table 2). Thus, there is no strong indication for linkage of schizophrenia to markers in the region of the breaking point of chromosome 11 in any other family or family sample. Again, given the only modest evidence for linkage in the families with translocations, the positive linkage might reflect a false-positive finding; or, alternatively, the translocation events might indicate a distinct genetic subtype for a small minority of families. However, replications of linkage observed by St. Clair et al. (1990) in another family or families would be required in order to exclude a false-positive finding.

Other variants of the dopamine receptor are also of functional relevance for the activity of neuroleptic drugs and, consequently, they represent candidate genes. A series of linkage studies have explored linkage to D_1, D_2, and D_3 to D_5 receptor genes without providing definite or tentative evidence for linkage (e.g., Coon et al. 1993).

Tentative but controversial evidence for genetic heterogeneity emerged from linkage studies of schizophrenia and a marker DXYS14 in the pseudo-

Table 2. Linkage between the region of the D_2 receptor gene and schizophrenia?

Translocations of chromosome 11q[a]	Linkage studies to 11q (preferably region of the D_2 receptor gene)
	Negative
St. Clair et al. (1990)	Moises et al. (1990)
Smith et al. (1989)	Wang et al. (1993)
Holland and Gosden (1990)	Su et al. (1993)
	Gill et al. (1993)
	Coon et al. (1993)
	Hallmayer et al. (1993)
	Positive
	None

[a] Breakpoint sites near the site of the D_2 receptor gene

autosomal region of the X chromosome. Crow (1988) proposed this region as a candidate area for linkage studies as some family and twin studies showed excess same-sex concordance of affected siblings (in particular when the disease is transmitted from the paternal side). This finding would be the consequence if a major locus were located in this area. The current state of linkage studies to candidates in this area is controversial as Table 3 shows. The early finding of linkage between schizophrenia and the genetic variation at the DXYS14 locus by Collinge et al. (1991), who applied the sibling pair method, has been independently replicated by another study using the same technique. However, two published linkage studies in multiplex families and a third not yet published study from our group excluded linkage to this particular locus and also to most of the residual pseudoautosomal region. Three putative explanations have to be discussed for an explanation of these discrepant findings:

1. The linkage reports using the sibling pair method are false positives. While this possibility cannot be excluded, independent replications argue against it.

Table 3. Genetic association studies to pseudoautosomal region (DXYS 14) by study type

	Sibling pair	Linkage	
Positive			
DXYS 14	Collinge et al. (1991)	None	
DXYS 14	d'Amato et al. (1992)		
Negative			
DXYS 14	Asherson et al. (1992)	DXYS 14	Asherson et al. (1992)
		DXYS 14	Wang et al. (1993)

2. The sibling pair method is more sensitive than the classical lod score technique. While simulation studies argue against this possibility, it should be remembered that linkage of IDDM to HLA markers was detected by applying the sibling pair technique, whereas the lod score method in multiplex families provided misleading results.
3. There may be heterogeneity across families with some of them but not others, contributing to linkage. Intrafamilial heterogeneity may contribute to failure to detect linkage in multiplex families. Recently, evidence was obtained from simulation studies that, assuming genetic heterogeneity, multiplex families are at a substantially higher risk for intrafamilial heterogeneity than patients from less highly loaded families (as sibling pair samples) (Greenberg 1993). Family-specific evaluation of our linkage study lends some support to this possibility of interfamilial heterogeneity.

Given the published literature, there is currently no fully convincing evidence for genetic heterogeneity of schizophrenia from linkage studies. Among the loci tested for linkage, all results were either negative or putatively false positive; only linkage to the pseudoautosomal region for a minority of familial cases remains a possibility. Thus, linkage studies are currently not able to provide an equivocal answer to the question of genetic heterogeneity of schizophrenia, and we also still have to rely on the analysis of familial phenotypes.

Schizophrenic Symptoms Associated with Genetic Diseases

Genetic heterogeneity of schizophreniform syndromes is proposed by the excess risk for schizophreniform symptoms in subjects with a variety of genetic diseases transmitted in a monogenic (Mendelian) mode in families. Propping (1989) has compiled a long list of monogenic diseases associated with excess morbidity for schizophrenic symptoms. These monogenic diseases are distributed widely across the human genome. Although the quantification of the relative risks is difficult as the estimation of the frequency of co-occurrence of two rare disorders is prone to extensive error. It could be convincingly demonstrated that psychotic syndromes are more common if genetic variations causing monogenic diseases are present.

For example, Wilson's disease and Marfan's disease – two monogenic diseases with established linkage findings – are associated with an increased risk for schizophrenic symptoms but also with an elevated risk for affective disorders (Bowers and Swinger 1987; Propping 1989). The most interesting example is fragile x syndrome; schizophrenia and schizophrenic traits as well as affective disorders (Reiss and Freund 1991; Reiss et al. 1988; Mendlewicz and Hirsch 1991) have been found to occur more often among subjects with expression of the fragile x syndrome on a cytogenetic level but without the complete phenotype (i.e., without mental retardation). These psychopathological features were considered as "soft signs" of the psychopathological phenotype, which presents as mental retardation in its fully expressed form. The mutation of the gene on the long arm of the x chromosome causing the fragile x syndrome has recently been identified as have premutations anticipating the mutation in parental generations (Oberlé et al. 1991). Recent research demonstrated that schizophrenic and affective symptoms only occur when the gene is extensively mutated and methylated, but not when only premutations are present (Reiss et al. 1993).

Schizophrenic syndromes can apparently arise from a variety of chromosomal aberrations. The relative frequency of those cases of schizophrenia among the total population of schizophrenics is, however, very low (DeLisi et al. 1988a); thus, these genetic deviations cannot explain a major proportion of the total genetic variance of schizophrenia.

Usually the occurrence of schizophrenic traits among carriers of monogenic disease genes requires the expression of the phenotype of the underlying genetic disease; classification schedules might therefore label the psychopathological syndromes as organic psychotic disorders. Despite this convention, these examples propose that schizophrenia-like symptoms may be caused by a variety of monogenic aberrations. It is noteworthy at this point that nearly all monogenic diseases associated with schizophrenic syndromes are also at a higher risk for affective syndromes. From these considerations we can propose:

1. Multiple single gene effects mediate schizophrenic or related psychotic syndromes; if this finding extends to schizophrenia as defined in DSM-III-R, genetic heterogeneity would be established.
2. The multiple single genes mediating psychotic disorders simultaneously express other phenotypic traits.
3. The relative risk for schizophrenia among these monogenic diseases is elevated, but still relatively low. Therefore the genes involved operate as susceptibility genes (Greenberg 1993) modulating the risk but not causing schizophrenia.

Evidence for Distinct Clinical Subtypes Running Time Within Families?

The classical tool to subdivide schizophrenia is by psychopathological features. Clinical subtypes might indicate a distinct genetic subtype if the subtype is breeding true (i.e., if the subtype aggregates in families of probands with this particular subtype but not in families of probands with other subtypes or disorders). The familial aggregation of a variety of different subtypes has been explored in order to find specific patterns of aggregation: the distinction by paranoid/non-paranoid type, by hebephrenic/catatonic/paranoid type, by negative/positive symptoms, by age at onset, and by clinical outcome (overview: McGuffin et al. 1987; Dworkin and Lenzenweger 1984; Kendler and Tsuang 1988; Kendler et al. 1987). Some intrafamilial similarity of symptom patterns was found, but breeding time of distinct subtypes was missing. In particular, Murray (1991) has proposed that early onset schizophrenia occurring more frequently in male compared to female subjects is etiologically distinct from late onset schizophrenia in females. He postulated that late onset schizophrenia in females represents a familial genetic variant of affective disorders. This hypothesis did not hold true in independent replications (Maier et al. 1993a; Pulver et al. 1990).

Most promising to identify etiological subtypes are indicators with high causal priority. It is well established that schizophrenics are more likely to be born during winter and early spring than other subjects. Differential impact of genetic factors for various periods of birth was proposed (Boyd et al. 1986). Empirically the hypothesis that the season of birth defines distinct familial subtypes was not substantiated (Pulver et al. 1992). Subtypes defined by neurobiological indicators which are breeding true have not yet been identified.

In conclusion, the apparent variation of clinical symptomatology and subtypes of schizophrenia does not propose distinct entities with regard to genetic or familial transmission. The clinical variation (e.g., by age at onset or clinical subtypes) might partly reflect quantitative differences in genetic–familial loading, but not qualitative differences without proposing convincingly genetic heterogeneity. Whereas some clinical correlates of varying familial risk are identified, the neurobiological determinants still remain obscure.

Patterns of Inheritance/Segregation Analysis

A variety of modes of genetic transmissions of schizophrenia have to be taken into consideration (Table 4). Segregation analyses were previously used to identify the most compatible pattern of familial transmission. There are several limitations to these studies:

1. In the absence of associated genetic markers the power of segregation analysis is too low to identify a single most appropriate model of transmission. These methods are only able to reject the most inappropriate modes of transmission leaving multiple modes of transmission as equally compatible with the patterns of familial aggregation.
2. The published segregation analyses mainly refer to pooled samples of different family and twin studies that have used divergent and, from a modern point of view, insufficient tools for recruitment, assessment, and case identification.
3. The variable expression of the familial phenotype is not sufficiently taken into account.

Classical Segregation Analysis

Most classical segregation analyses on samples of systematically recruited families of schizophrenic probands excluded the fact that the disorder is transmitted in a uniform monogenic mode. The most crucial argument against monogenic transmission is the relatively high recurrence risk in monozygotic co-twins of schizophrenics compared to nonidentical siblings (the ratio of risks to be

Table 4. Modes of transmission and segregation analysis

Modes of transmission	Segregation analysis
Single gene (with reduced penetrance)	–
Multiple single genes (with reduced penetrance)	–
e.g., neurofibromatosis	
Oligogenic transmission	
(with major/modest effects in each family)	
Without epistasis	–
With epistasis	+
e.g., IDDM	
Polygenic transmission	+
(without genes with major/modest effect)	
e.g., intelligence	
Mixed models	+
(genes with major/modest effect and polygenic background, homogeneous across families)	
e.g., hypertonus	
Heterogenous models	+
[proportion of families with single gene transmission(s) and another proportion with polygenic/mixed transmission]	
e.g., mental retardation	

–, no support; +, not excluded.

expected by assumption of monozygotic transmission is 2.0 but the observed ratio is about 50%/10% = 5). Segregation studies simultaneously argue against a monogenic model with different single genes in different families (O'Rourke et al. 1982; McGue and Gottesman 1991; Tsuang et al. 1991). These results are not an artefact due to aggregation across a variety of families with schizophrenic index cases: even in extended, multiplex families with a very high density of cases and very similar environments for all family members (as the North Swedish pedigree originally identified by Böök 1953), multilocus transmission is more appropriate than monogenic transmission.

It was claimed that the phenotype of schizophrenia or the schizophrenic spectrum is too broadly defined. Extension of the phenotype by defining a latent trait comprising schizophrenia and impairment of smooth pursuit eye movement and by assuming conditional independence between both manifest traits was proposed to fit into a monogenic dominant mode of transmission (Holzman et al. 1988). However, extension of the phenotype can only account for reduced penetrance but not for the excessive drop-off of recurrence risks between different degrees of familial relationship (McGue and Gottesman 1989; Risch 1990a,b).

The classical segregation analyses in samples of families nearly equivocally propose the involvement of multiple genes (e.g., as polygenic or multifactorial transmission), each contributing only a small proportion of the genetic variance. Some genes may substitute each other or some may partly be substituted by environmental factors (susceptibility genes in contrast to necessary genes; Greenberg 1993). However, there is still space for major genes. Most of these classical segregation analyses are compatible with genes contributing a major or modest component to the genetic variance (mixed models) in so far as a polygenic background is the main genetic source. Subgroups of multiplex families with single genes causing schizophrenia (heterogeneous models) are also compatible with these analyses in so far as the majority of genetic causes are transmitted in a polygenic manner (McGue and Gottesman 1991).

In this respect, Meehl's (1990) concept of schizotype is of particular interest. Meehl was the first to suggest that both a major gene and a polygenic background play a role in the etiology of schizophrenia. According to Meehl's theory, inheritance of a single gene gives rise to a neural integrative deficit termed schizotaxia that is expressed at the personality level as schizotypy; the polygenic background operates as a potentiator in favor of schizophrenia. Expression of clinical schizophrenia among individuals who inherit the single gene defect is postulated to be a function of status on a host of polygenical and environmentally influenced potentiators including anxiety, anhedonia, and social introversion.

Meehl (1990) collected evidence to show that, in the absence of the postulated schizotype gene, schizophrenia is rare; polygenetic potentiators alone account only for 10% of all cases, whereas the schizotaxia gene is present in 90% of the cases with schizophrenia. While these proposed estimates might overestimate the relevance of the major gene (given the results of classical segregation analysis), the postulate of two components of a major gene and a polygenic background is in full accordance with the biometrics of familial aggregation of schizophrenia.

Meehl (1990) also acknowledges the very low penetrance of the single gene for schizotype; the phenotype "schizophrenia" is expressed only in the presence of a substantial adversive load (polygenic and/or environmental). Alternative manifestations of the gene for schizotype are other signs of a neurointegrative deficit such as eye movement disorders or attentional deficits.

Non-classical Analysis

Classical segregation analyses do not take into account putative interactions between genes. A very powerful type of analysis addressing the interaction between segregating genes was proposed by Risch (1990c). This technique (Table 5) is able to identify genetic diseases transmitted in a non-monogenic manner with various contributing genes operating non-additively (i.e., interaction between genes, also called epistasis). Reanalysis of aggregated classical family study data of schizophrenia, including first, second, and third degree relatives, excluded (as the classical segregation analyses) monogenic transmission and demonstrated the involvement of multiple, interacting genes (Risch 1990). Simultaneously, additive models with multiple, non-interacting genes

Table 5. Familial risk of schizophrenia and related illnesses

Disorders aggregating in families of probands with schizophrenia diagnosis	Excess familial risk for schizophrenia as function of type of probands
Schizophrenia	Schizophrenia
Schizoaffective disorders	Schizoaffective disorders
Other non-affective psychoses	Other non-affective psychoses
Psychotic affective disorders	Psychotic affective disorders
Non-psychotic unipolar depression	–
Anxiety disorders	–

No familial relationship established to non–psychotic bipolar disorder and alcoholism/drug dependence.

were found to be unlikely to be appropriate for the mode of transmission of schizophrenia. As a limitation of this analysis, putative bias inherent in the methodology of these classical family studies and the aggregation of family data from multiple sources should be taken into account; most of these studies are non-blind assessments of familial aggregation without interviewing distant relatives directly. The application of this method to an ongoing family study in Mainz results in similar conclusions.

What is Transmitted in Families of Schizophrenics (Variation of Familial Phenotype)?

Co-segregation of Schizophrenia and Other Disorders

It is widely acknowledged that less severe conditions similar to schizophrenia in psychological features, such as schizotypal personality disorders and chronic schizoaffective disorder, are genetically related to schizophrenia; other psychotic disorders, paranoid and schizoid personality disorders are also discussed in this respect (Kendler and Diehl 1993). Together with schizophrenia these syndromes are summarized under the heading of "schizophrenic spectrum". The concept of schizophrenic spectrum implies two hypotheses:

1. Schizophrenia aggregates in family members of probands recruited by spectrum disorders.
2. Spectrum disorders aggregate in families of schizophrenics.

The first condition is more restrictive than the second one. In a most carefully conducted case register-based family study Kendler et al. (1993) demonstrated that schizophrenia aggregates not only in families of probands with schizophrenia and schizoaffective disorders, but also in families of probands with schizotypal personality disorder and with affective disorders with psychotic features. A series of well-designed family studies demonstrated the reverse relationship, i.e., an elevation of the risk for these disorders in families of schizophrenics (overview: Kendler and Diehl 1993). This familial relationship is at least partly supported by twin studies (Kendler and Diehl 1993). These results reveal lack of specifity of the genetic basis of schizophrenia. By contrast, they suggest that the familial liability to schizophrenia is, at least in part, a liability to develop psychosis, as the risk for schizophrenia was substantially increased in relatives of all proband groups with non-schizophrenic psychotic disorders.

A familial relationship to other psychiatric disorders and behavioral traits has also been observed. The relationship to schizophrenia, however, is less strong and less consistent. Among psychiatric disorders, unipolar depression and also anxiety disorders were more common than in control samples by some, but not all of the recently conducted controlled family studies fulfilling methodological criteria (e.g., blindness, personal interviews, best estimate diagnoses) (see Table 6 and overview by Taylor 1992).

The familial relationship of unipolar depression to schizophrenia is less strong than the relationship spectrum disorders to schizophrenia as an elevation of schizophrenia in families of probands with unipolar depression has not been reported (Table 5).

Schizophrenia is the most severe among the non-organic psychiatric diseases, and it might therefore also be proposed that schizophrenia represents just the extreme of a continuum of liability to all kinds of psychiatric diseases graded by severity with a strong intrafamilial correlation of liability scores. According to this assumption, elevations of the risk to all psychiatric disorders is to be expected. Recently conducted family studies do not support this general relationship. By contrast, there is overwhelming evidence that (non-psychotic) bipolar affective disorder and schizophrenia are neither genetically nor familially linked (Kendler and Diehl 1993; Taylor 1992). Substance abuse disorders were also not linked to schizophrenia (Gershon et al. 1988; unpublished data from the Mainz family study).

The familial relationship between two disorders does not prove that both conditions emerge from the same etiological basis. In particular, a "true" familial relationship requires:

Table 6. Aggregation of non-psychotic unipolar affective disorder in families of schizophrenics

Positive		Negative
	Tsuang et al. (1980)	Guze et al. (1983)
Gershon et al. (1988)		Frangos et al. (1985)
Maier et al. (1990)[a]		Baron et al. (1982)
Taylor et al. (1993)		Coryell and Zimmermann (1988)
Farmer et al. 1987[b]		
Kringelen and Cramer 1989[b]		

[a] 1. Not due to co-morbidity in probands.
2. Not secondary to schizophrenia in relatives.

[b] Controlled studies with direct interviews.

1. That the aggregation of non-psychotic disorders in families of schizophrenics is not due to the co-morbidity between schizophrenia and this disorder in the index proband with two disorders segregating in the family.
2. That the aggregation of the non-psychotic disorder in families of schizophrenics is not secondary (e.g., a reaction due to living with a schizophrenic) to the manifestation of schizophrenia in the index case.

The Mainz family study (Maier et al. 1993b) is not in favor of any of these alternative explanations with regard to the relationship of schizophrenia and unipolar depression. In particular, chronic schizophrenics without co-morbidity with any affective disorder were also associated with excess of familial risk for unipolar depression. An aggregation of unipolar depression in relatives was also observed in first onset cases of schizophrenia without an additional secondary case for schizophrenia in the family. Finally, unipolar depression is unlikely to represent a precursor of schizophrenia in relatives as 85% of the relatives diagnosed as having unipolar depression were older than 35 years, i.e., they have passed most of the lifetime risk for schizophrenia.

A preliminary evaluation of an ongoing controlled family study in schizophrenia conducted in Mainz in first, second, and third degree relatives of schizophrenics confirmed previous findings with regard to an elevation of the risk for unipolar depression in first degree family members of schizophrenics by a factor of about 2.0. Although more attenuated, the risk for, unipolar depression was also elevated in second and third degree relatives (Figs. 1, 2). Third degree relatives are usually unlikely to share environmental risk factors or to react with psychological impairment to the manifestation of a disorder in the third degree relatives. Therefore, a sharing of genetic factors influcencing both schizophrenia and unipolar depression is suggested.

This new family study also reaffirms previous observations that bipolar disorder and schizophrenia do not at all or only modestly overlap in families (Fig. 3). Given these results, the substantial overlap between these two conditions (schizophrenia, bipolar disorder) in samples of multiplex families (e.g., Blackwood et al. 1991) is more likely to result from the particular recruitment strategy applied in linkage studies (the likelihood to include a family increases with the number of severe disorders in a pedigree).

The ongoing Mainz family study also allows the exploration of another alternative putative explanation for the co-segregation of schizophrenia and unipolar depression: schizophrenia might either be misclassified in index cases (e.g., owing to limited reliability and validity of assessment and diagnostic tools), or there might be a lack of specificity of the clinical concept of schizophrenia as defined by DSM-III-R. If these possibilities were to explain the excess of unipolar depression in relatives of schizophrenics, the risk of schizophrenia should be reduced in families that include family members with a lifetime diagnosis of non-psychotic affective disorders. According to this assumption, schizophrenic probands with a family history of affective disorder should bear a reduced familial risk for schizophrenia compared with the residual group of schizophrenic probands. Given the very limited family sizes in Northern Europe, this type of analysis would not work very well in families

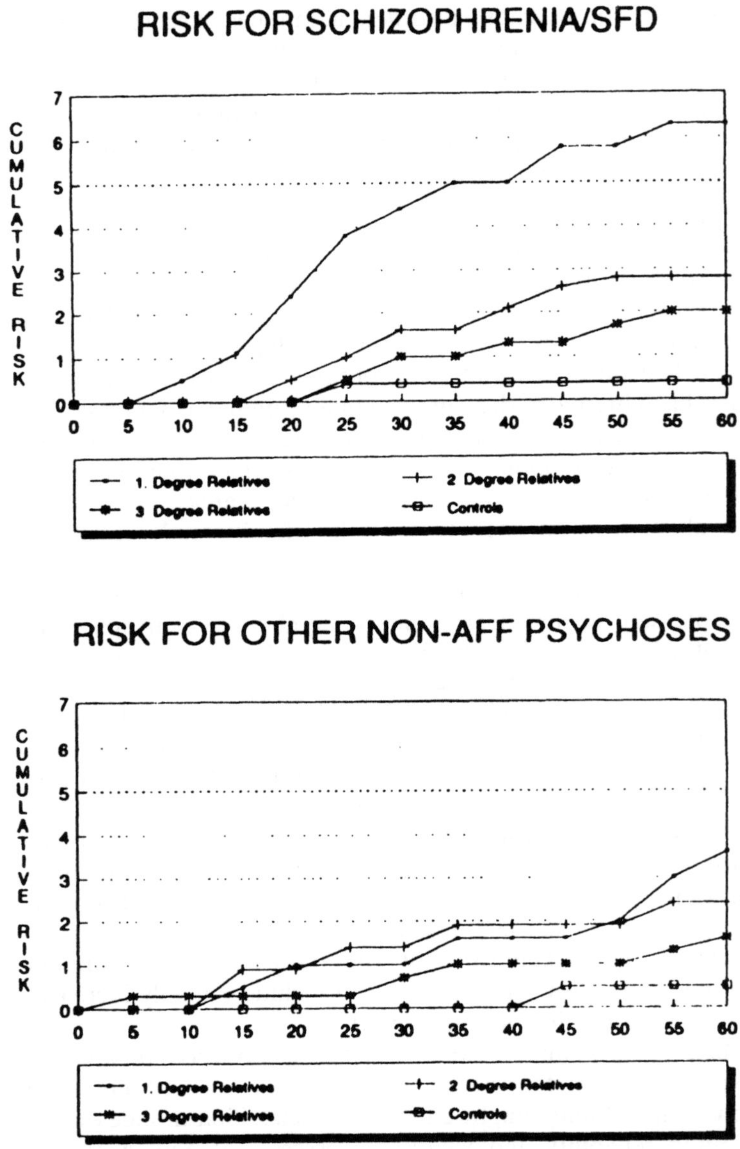

Fig. 1. Cumulative lifetime risks of psychotic disorders (DSM-III-R) – relatives of schizophrenics by degree of relationship and controls

with first degree relatives only. Extension to second and third degree relatives would, however, provide sufficient power (mean number of living first degree relatives: 4–5; and mean number of living second and third degree relatives: 14–15). Table 6 demonstrates the results for our ongoing family study in Mainz.

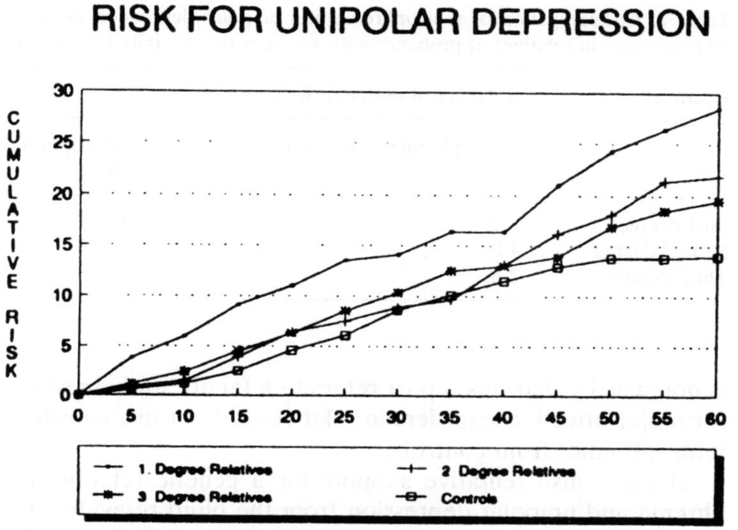

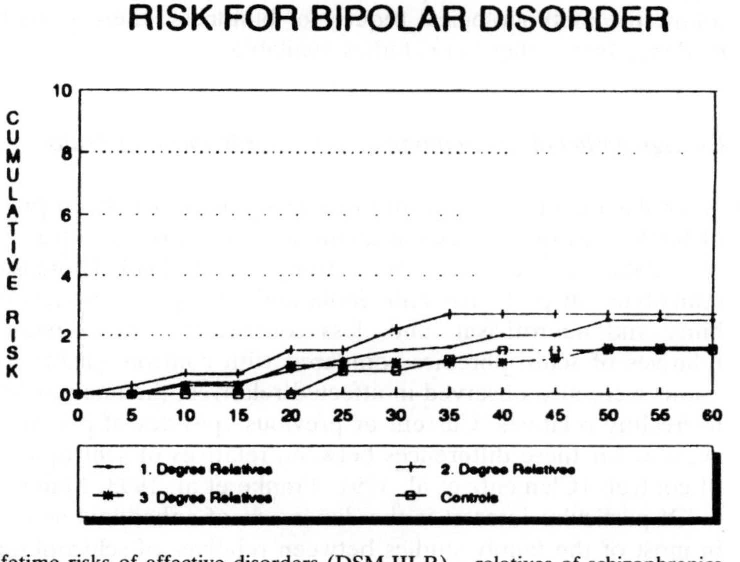

Fig. 2. Cumulative lifetime risks of affective disorders (DSM-III-R) – relatives of schizophrenics by degree of relationship and controls

The results shown in Table 7 suggest that schizophrenia aggregates particularly in families that are additionally loaded with unipolar depression. It may be concluded that the familial liability to schizophrenia is associated with non-psychotic unipolar depression. This conclusion is in agreement with a previous report by Baron and Gruen (1991) which was limited to first degree relatives. It

Table 7. Co-segregation of schizophrenia and unipolar depression in families: Lifetime risks of schizophrenia in relatives of probands with versus without family load of affective disorders

Relatives	Relatives of probands with	
	Schizophrenia + affective disorder (%)	Schizophrenia − affective disorder (%)
First degree	8.2	4.1
Second degree	4.0	2.3
Third degree	2.6	2.6

is noteworthy that this report refers to a family study in which the morbid risks for major affective disorders in relatives did not distinguish the whole group of schizophrenics from controls.

There is also tentative support for a genetic relationship between schizophrenia and unipolar depression from the blind re-evaluation of the Maudsley series of monozygotic twins with schizophrenic index cases by Farmer et al. (1987). Major depression was observed in a substantial proportion of monozygotic co-twins of schizophrenics. As nearly all the other twin studies did not follow the methodological request of blindness, there is no further published evidence from other twin studies available.

Co-segregation of Schizophrenia and Other Behavioral Traits

A similar familial relationship to schizophrenia has been proposed by a series of family and twin studies in quantitative psychological traits, mainly from the personality sphere, that have a strong genetic load. Dimensional measures of schizotypy, in particular anhedonia and perceptual aberration, but also irritability and neuroticism, and, less consistently, introversion score higher in relatives of schizophrenics compared with controls (Table 8). As these deviations were also observed in affected relatives, the analysis has to be restricted to healthy relatives. Current or previous episodes of psychiatric illness cannot account for these differences between relatives of schizophrenics and relatives of controls (Clementz et al. 1991; Franke et al. 1993; Maier et al. 1993c).

Of particular interest is the dimension of anhedonia as it discriminates best in most of the family studies between relatives of schizophrenics and relatives of controls in comparison with other measures of schizotypy (Clementz et al. 1991; Franke et al. 1993). Furthermore, a genetic relationship with schizophrenia was demonstrated by twin studies (Berenbaum et al. 1990). However, this link is not specific to schizophrenia as anhedonia seems to predispose to a broad variety of psychological impairments (Berenbaum et al. 1990, 1992). In agreement with the most recent modification of Meehl's concept of schizotypy, twin and high-risk studies suggest that anhedonia does not predispose to psychosis in the absence of other risk factors, with the most important one being

Table 8. Selected behavioral traits under genetic control cosegregating with schizophrenia

Personality measures
Neuroticism
Anhedonia[a]
Perceptual aberration
Schizotypal traits[a]
Attention and perception
Attentional deficits
Span of apprehension
Visual backward masking
Continuous performance
Psychophysiology
Orientation reaction/habituation
Neurophysiological traits
SPEM abnormalities[a]
P 300 amplitude, latency
P 50 inhibition

[a] Relationship to schizophrenia established by twin studies.

the relationship to a relative with schizophrenia. In line with this suggestion, there is no evidence for an elevated risk of schizophrenia in subjects identified by elevated anhedonia scores in non-patient samples (Berenbaum et al. 1993).

The perceptual aberration dimension has received less attention in families of schizophrenics. A single study (Dworkin et al. 1990) reports a modestly increased risk for treated psychosis in families of students with elevated psychosis proneness scores. Given the limited power inherent in this study, this result may reflect a chance finding. Furthermore, a strong familial relationship to schizophrenia as proposed by this family study can hardly be reconciled with studies that have demonstrated a lack of familial and genetic association between cognitive perceptual features and schizophrenia, or at least a less strong association than for the emotional–interpersonal features of schizotypy (Togersen 1985).

A series of neuropsychological, neurophysiological, and psychophysiological measures that characterize schizophrenics and convey a strong genetic component have also been found to differentiate healthy relatives of schizophrenics from controls. The most discriminative measures are listed in Table 7; they are more extensively described in recently published overviews by Braff (1993) and Holzman (1992).

Attentional deficits are also of particular interest as they have been found to be associated with anhedonia (Franke et al. 1993; Erlenmeyer-Kimling et al. 1993). By contrast, personality measures and attentional functioning do not covary in control samples. Thus, covariation in relatives of schizophrenics is likely to be mediated by a common factor influencing both measures. The common factor is putatively the gradation of vulnerability to schizophrenia with a broad variation in the relatives of schizophrenics, but less so in the general population. The relationship also holds true for attentional impairment in childhood and adolescence (Erlenmeyer-Kimling et al. 1993).

The smooth pursuit eye movement (SPEM) disturbances have received most attention. In particular, Holzman (1989) found a true genetic relationship between SPEM disturbances and schizophrenia. However, the mode of co-segregation between the two traits remains, obscure.

Most of the dimensional measures in Table 7, as well as non-psychotic disorders aggregating in schizophrenia, are also associated with patients with a broad variety of other diagnoses, and they also aggregate in families of probands with other disorders (e.g., with alcoholism or bipolar disorders) that do not by themselves show a familial or genetic relationship to schizophrenia. The correlation of these risk factors to other liability dimensions apparently goes together with a lack of elevation of risk to schizophrenia in relatives of probands defined by deviations in these measures. They mainly represent indicators of increased risk for psychopathology in general. High-risk studies have claimed that at least some of these personality deviations and attentional deficits predispose to the subsequent development of psychotic disorders in high-risk samples (Cannon et al. 1990). However, this specific predictive role seems to depend on an additional risk factor, such as being a family member of a schizophrenic patient.

Models for Phenotype Variation

Continuum Models for the Schizophrenia Phenotype

The elevated risk for schizophrenia – such as conditions in families of schizophrenics – motivated the idea that the behavioral traits aggregating in families are due to the same familial risk factors. Twin and adoption studies supported this view (e.g., Kety et al. 1975). Consequently, it was proposed that the transmission of behavioral traits in families of schizophrenics proceeds along a continuum of liability to schizophrenia reflecting interindividual quantitative differences in genetic familial loading (e.g., Baron and Gruen 1991). The liability dimension is thought to be highly positively correlated to the severity of the expressed schizophrenia – like symptomatology with chronic schizophrenia (Kraepelin) – as the most severe form. In addition, the model assumes multiple thresholds on the continuum, each reflecting diagnostic boundaries with schizophrenia referring to the maximal liability scores on the continuum; next is schizotypal personality disorder and schizoaffective disorder, followed by less severe schizophrenia-like conditions with transition to healthy states (i.e., no lifetime diagnosis). The model assumes a normal distribution of the latent (i.e., assumed and non-observable) dimension of liability to schizophrenia. This model can well be combined with a polygenic mode of inheritance of schizophrenia and related traits which is compatible with results of segregation studies; additional familial environmental risk factors may also be integrated.

This specification postulates a strong intrafamilial correlation of the individual position of a subject on the continuum of liability (multithreshold model according to Falconer).

Although this model is able to integrate many results from epidemiological family, twin, adoption, and segregation studies, doubts have been raised regarding its explanatory power (Häfner 1990):

1. Plausibility of the model decreases with increasing distance from the position of strictly defined schizophrenia as the specificity and variety of disturbances increase with lowering degrees of vulnerability (e.g., in the area of subclinical schizophrenia form traits).
2. The psychopathological dimension corresponding to the dimension of familial vulnerability may not be as homogenous by content as assumed in the model.

These doubts were even more strongly supported after it became evident in recent years that the bulk of heterogeneous disorders, symptom signs, and deviations are co-familial with schizophrenia, and (at least most of them) are also genetically related to schizophrenia (as described in the previous section). This model also excludes the putative impact of genes with a major, or at least a modest, effect on the phenotype variation and excludes gene–environment interaction (environment may enter only as an additive component). The continuum model therefore needs to be modified in order to integrate the heterogeneity of the familial phenotype and and unspecificity of co-familial behavioral traits, and to leave space for genes with unequal contributing effects and gene–environment interaction.

Figure 3 shows the modification of the continuum concept of schizophrenia. For heuristic reasons, the model proposes a dichotomic distinction among the co-familial traits to schizophrenia:

1. Conditions which not only aggregate in families of schizophrenics and which carry an elevated familial risk to "core" schizophrenia.
2. Conditions which only aggregate in families of schizophrenics and which do not carry an elevated familial risk to "core" schizophrenia.

This discrimination might not be as distinct in reality as it is proposed here. A further complication factor is that it is not known for each co-familial trait whether relatives of carriers of this trait are at an elevated risk for schizophrenia. However, this distinction is heuristically justified by another observation discussed in previous sections: most of the co-familial traits which are listed as not carrying (by themselves) an elevated familial risk to schizophrenia are also related to other disorders with strong genetic components. For example, most of these traits also occur in association with bipolar affective disorders and/or alcoholism/substance dependence (e.g., Squires-Wheeler et al. 1989). Most of them can also be observed to aggregate among relatives of probands with bipolar disorder and/or alcoholism/substance dependence. Therefore, the interval on the liability continuum beyond the interval representing schizophrenia, other spectrum disorders, and other psychotic or related disorders is

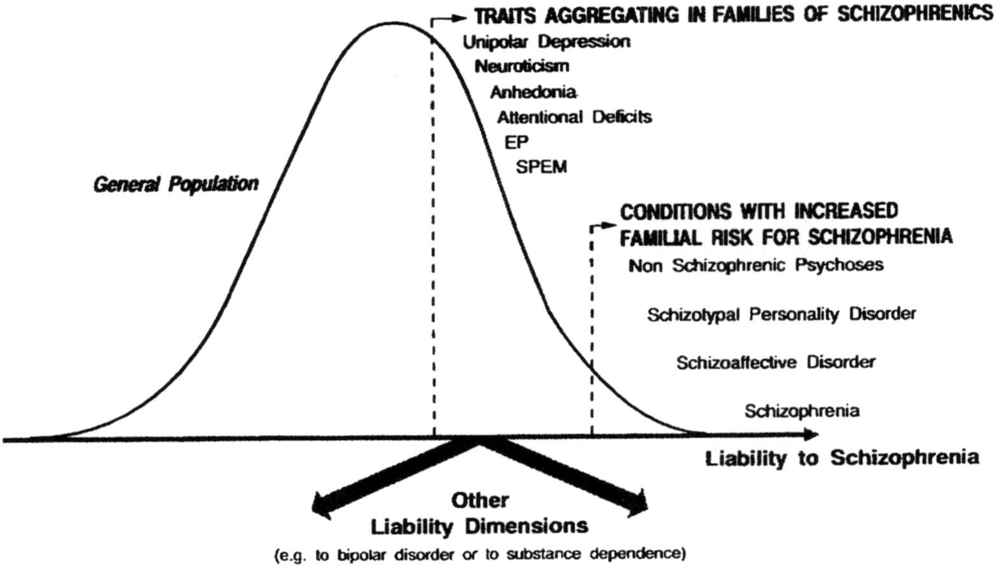

Fig. 3. Continuum model for family liability to schizophrenia

crossed by other genetic–familial liability dimensions (in particular for bipolar disorder and for substance dependence) which are independent or at best modestly correlated with the liability dimension accounting for schizophreniform disorders (Maier and Merikangas 1992).

This model allows for the interaction of familial risk factors. Although all behavior traits included in this model are under genetic control, the genes involved may not overlap between two selected traits. However, the simultaneous occurrence of two risk factors (e.g., anhedonia and having a schizophrenic relative) may increase the risk of developing psychoses in a super-additive manner. In particular, the proposed model is in agreement with the schizotype concept of Meehl (1990). This model also allows for population-specific correlations (e.g., between anhedonia and attentional deficits in relatives of controls).

How Might Hypothesized Genes and Environmental Risk Factors Contribute to Liability in This Model?

A quantitative variation of the phenotype across various heterogeneous behavioral traits is assumed. It is well known from genetic research in animals (inbred–recombinant experiments) and from genetic epidemiology of human behavior (e.g., intelligence) that evidence for the impact of major genes on quantitative traits (which are determined by genetic factors) is lacking or at best rate (Plomin 1990). Therefore, polygenetic background for phenotype variation along the schizophreniform liability dimension is even more likely than it has already been proposed by segregation analysis based on more restrictively defined phenotypes. As some behavioral traits on the psychopathological continuum corresponding to the liability dimension for schizophrenia are not necessarily associated with an elevated risk for schizophrenia in their relatives, additional either genetic or environmentally determined risk factors are usually necessary to show that schizophrenia is occurring. Currently, it cannot be decided if, beyond a polygenetic background, major environmental risk factos or major genes are more relevant. Therefore, two alternatives are proposed:

1. Either multiple genes with equal or unequal effects or major genes which are heterogeneous or homogeneous across multiplex families may define additional genetic risk factors which substantially increase the expression of psychotic illness. This model includes Meehl's concept of the origin of a schizotype additional major gene. The results of the segregation analysis of schizophrenia by Risch (1990a–c) propose that multiple interacting genes are additionally required for psychosis to become manifest.
2. Additional environmental risk factors are required to shift the polygenically mediated liability to manifest psychopathological syndromes in general to an increased liability to manifest psychoses or even schizophrenia. This mode of gene–environmental interaction was earlier proposed by Crow (1986). Given current empirical evidence, the environment risk factors are obstetric or gestation complications or an adverse (individual) psychosocial environment. This additional assumption is in agreement with recent research on the impact of environmental factors on schizophrenia. As reviewed in the first section, recent adoption and high-risk studies have demonstrated that the recurrence risk of schizophrenia spectrum disorders is increased in biological relatives of schizophrenics if additional, specified environmental risk factors occur to the subject at genetic risk; this conclusion is in agreement with this model. Most recently, an association study by Wright et al. (1993) also points very convincingly in this direction. It can also be predicted from this model that the simultaneous occurrence of environmental risk factors and non-psychotic symptomatology in relatives of schizophrenics is unlikely to occur; subjects carrying a strong polygenetic component for general psychopathology (e.g., indicated by the diagnosis of major depression in relatives of schizophrenics) would be at a higher risk of developing psychosis if adverse environmental effects are also operative. Empirical evidence to support this prediction convincingly is currently not available. Only Matthysse et al. (1992) collected casuistic evidence in support of this prediction.

How Does This Concept of Phenotype Variation Relate to Crow's Concept of Unitary Concept?

The proposed continuum considers a subgroup of affective disorders (but not all affective disorders) on a continuum of genetic/familial liability with schizophrenia. Previously, Crow (1991) argued in this direction and proposed a unitary concept of affective disorders and schizophrenia. However, beyond this overall similarity, there are some differences:

1. Crow (1991) assumes a single gene (or at best multiple single genes) to express the whole phenotype variation. We consider a polygenic model to constitute the link between the components of the broadly, quantitatively, and qualitatively varying phenotype. Segregation analyses are in support of this suggestion. However, major genes might be operating in addition, and particularly so in the case of major genes for the more severe types of disorder.
2. Only particular subtypes of affective disorders are located on a continuum with schizophrenia according to the proposed model. In particular, subtypes of unipolar depression and less so of bipolar disorder are linked to schizophrenia by sharing of common familial factors. Bipolar disorder may define its own liability dimension crossing with the schizophrenia dimension. The model proposed in this paper assumes that only a subtype of unipolar depression is co-familial with schizophrenia, whereas other subtypes might be co-familial with other traits or may be environmental in origin.

Sex-Specific Effects

Two well-established findings (Table 9) on the impact of gender on familial risks are a challenge to all concepts of familial transmission of schizophrenia. An excess of same sex concordance among sibling pairs with schizophrenia was found in a series of independent family studies. Some studies were able to demonstrate that an excess of same sex concordance occurs preferentially if the disease is transmitted by the father. Two putative explanations for this finding are:

1. Siblings of the same sex share more environmental factors; if any of these shared factors increases the risk to schizophrenia, excess of same sex concordance would be the consequence.
2. A gene in the pseudoautosomal region might contribute to the risk of schizophrenia; the likelihood of this explanation depends critically on the contribution of the paternal origins of the disease to the excess of same sex concordance.

Current research is not able to exclude one of these two plausible explanations. Both possibilities are compatible with the proposed models of transference.

Previous as well as recently conducted family studies showed that the gender of the proband has an impact on recurrence risks in relatives. Female probands carried an elevated familial risk (Table 9). Ignoring the fact that a broad variety of artefacts may serve as putative explanations for this gender effect (Maier et al. 1993a), it should be noted that either differential fertility (female schizophrenics being more fertile than male schizophrenics with the con-

Table 9. Impact of gender on recurrence risk

Excess of recurrence risk in female probands with schizophrenia		Same sex concordance among affected siblings	
Positive	Bellodi et al. (1986)	Positive	Crow (review, 1988)[a]
	Goldstein et al. (1990)		Crow (1990b)[a]
	Pulver et al. (1992)		Gorwood et al. (1992)[a]
	Maier et al. (1993d)		Asherson et al. (1992)
			Maier et al. (1993)
		Negative	Sturt and Shar (1985)
Increased fertility in female schizophrenics?		Shared sex-specific environment?	
Differential impact of genes?		Pseudoautosomal locus?	

[a] Only for paternal transmission.

sequence of a higher relative weight of parents – with a lower risk for schizophrenia than children – in families of male schizophrenics) or a stronger environmental influence of a schizophrenic mother to a child may increase mean familial risk in female schizophrenics.

Does this observation indicate that differential etiological factors operate in male compared to female schizophrenics? For example, Murray et al. (1992a) proposed linking the finding (of excess familial risk in female schizophrenics) to the postulated "schizophrenic" effect of influenza; the environmental risk factor is supposed to transfer an increased risk to schizophrenia not of itself, but preferably if the mother is carrying additional risk factors in order to induce schizophrenia in the infant. This additional risk factor is preferably an elevated risk for schizophrenia or the disorder itself. This explanation is speculative; however, this suggested mechanism would fit nicely into the proposed model of phenotype variation.

Implications on the Search for Genes Contributing to Schizophrenia

Linkage studies in multiplex families have been proposed as the king's road to unravelling the mode of transmission of complex diseases in families and to identifying genes making a major contribution to the genetic variance (Ott 1990; Gershon et al. 1988). A major advantage of this strategy is that it is not compromised by the putative operation of more than one contributing major gene or by a polygenetic background (Martinez and Goldin 1990). As false cases are deleterious, complications in following this strategy arise from the broad variation of the familial phenotype. The model presented for phenotype variation proposes that the definition of the phenotype in linkage studies should not be overinclusive. In order to be convincing, exclusion and (hopefully sometimes) identification of putatively contributing genes should be demonstrated by using a phenotype definition restricted to psychotic illness. Two reasons argue for this proposal:

1. As extensively argued in the previous section, the genetic relationship between schizophrenia and "fringe" cases (i.e., phenotypes not associated with an increased familial risk to schizophrenia) might be preferentially mediated by the polygenetic background, but not by a putative major gene.
2. Overinclusive phenotypes increase the risk of false-positive cases with increasing general population prevalence; thus, non-genetic cases might become more frequent.

Implication on Recruitment Strategies

Recent experience with successful linkage studies (e.g., in epilepsy, Greenberg 1992) indicates that the index proband of a multiplex family should fit stringent diagnostic criteria as well as other family members in order to increase the likelihood of a genetic etiology. This argument is in agreement with the proposed model of phenotype variation. In addition, the relevance of major genes probably increases with increasing severity of the expressed phenotype. Accordingly, Matthysse et al. (1979) provided biometrical support to the thesis that whether a gene has a major or minor effect depends on the population being studied; the magnitude of a gene of effect increases the more extreme the phenotypes of index cases are (if a polygenetic or oligogenic transmission model with unequal effects of the various genes is assumed).

Linkage Versus Association Studies

Linkage studies can only identify major genes; their power to detect linkage exponentially decreases with the decreasing magnitude of the contribution of a particular gene. The decrement in power is less pronounced for association studies; whereas those studies are likely to be less successful in identifying major genes, they have greater power to detect the contribution of minor genes (McGuffin et al. 1992). Given current evidence of genetic epidemiological studies and experience with linkage studies, the possible operation of a polygenetic background should be taken very seriously. Thus, association studies are necessary tools which cannot be substitued by linkage studies. Unfortunately, an association study to a genetic locus cannot address the presence or absence of association to a broad region on the chromosome surrounding the candidate locus as economically as linkage studies are able to do. Presumably 30000 genes are expressed in the brain and all may serve as putative candidate genes. Thus, if no major genes account for schizophrenia, the identification of contributing genes may need a long period of time.

References

Asherson P, Parfitt E, Sargeant M, Tidmarsh S, Buckland P, Taylor C, Clements A, Gill M, McGuffin P, Owen M (1992) No evidence for a pseudoautosomal locus for schizophrenia. Linkage analysis of multiply affected families. Br J Psychiatry 161:63–68

Baron M, Gruen RS (1991) Schizophrenia and affective disorder: are they genetically linked? Br J Psychiatry 159:267–270

Baron M, Gruen R, Asnis L, Kane J (1982) Schizoaffective illness, schizophrenia and affective disorders: morbidity risk and genetic transmission. Acta Psychiatr Scand 65:253–262

Bellodi L, Bussoleni C, Scorza-Smeraldi R, Grassi G, Zacchetti L, Smeraldi E (1986) Family study of schizophrenia. Schizophr Bull 12:120–128

Berenbaum H (1992) Posed facial expressions of emotion in schizophrenia and depression. Psychol Med 22:929–937

Berenbaum H, Oltmanns TF, Gottesman II (1990) Hedonic capacity in schizophrenics and their twins. Psychol Med 20:367–374

Blackwood DHR, St Clair DM, Muir WJ, Duffy JC (1991) Auditory P300 and eye tracking dysfunction in schizophrenic pedigrees. Arch Gen Psychiatry 48:899–909

Böök JA (1953) A genetic and neuropsychiatric investigation of a North-Swedish population. Acta Genet Stat Med 4:1–100

Bowers MB, Swigar ME (1987) Acute psychosis and plasma catecholamine metabolites. Arch Gen Psychiatry 44:190

Boyd JH, Pulver AE, Sewart W (1986) Season of birth: schizophrenia and bipolar disorder. Schizophr Bull 12:173–185

Braff DL (1993) Information processing and attention dysfunctions in schizophrenia. Schizophr Bull 19:223–259

Cannon TD, Mednick SA, Parnas J (1990) Antecedents of predominantly negative- and predomiantly positive-symptom schizophrenia in a high-risk population. Arch Gen Psychiatry 47:622–632

Castel DJ, Murray RM (1991) The neurodevelopmental basis of sex differences in schizophrenia. Psychol Med 21:565–575

Clementz BA, Grove WM, Katsanis J, Iacono WG (1991) Psychometric detection of schizotypy: perceptual aberration and physical anhedonia in relatives of schizophrenics. J Abnorm Psychol 100:607–612

Collinge J, DeLisi LE, Boccio A, Johnston EC, Lane A, Larkin C, Crow TJ (1991) Evidence for a pseudoautosomal locus for schizophrenia using the method of affected sibling pairs. Br J Psychiatry 158:624–629

Coryell W, Zimmerman M (1988) The heritability of schizophrenia and schizoaffective disorder. Arch Gen Psychiatry 45:323–327

Coon H, Byerley W, Holik J, Hoff M, Myles-Worsley M, Lannfelt L, Sokoloff P, Schwartz J-C, Waldo M, Freedman R, Plaetke R (1993) Linkage analysis of schizophrenia with five dopamine receptor genes in nine pedigrees. Am J Hum Genet 52:327–334

Crow TJ (1986) Secular changes in affective disorder and variations in the psychosis gene. Arch Gen Psychiatry 43:1013–1014

Crow TJ (1988) Sex chromosomes and psychosis. The case for a pseudoautosomal locus. Br J Psychiatry 153:675–683

Crow TJ (1990a) The continuum of psychosis and its genetic origins. The Sixty-fifth Maudsley Lecture. Br J Psychiatry 156:788–797

Crow TJ (1990b) Nature of the genetic contribution to psychotic illness – a continuum view point. Acta Psychiatr/Scand 81:401–408

Crow TJ (1991) The demise of the Kraepelinian binary concept and the aetiological unity of the psychoses. In: Marneros A, Andreasen NC, Tsuang MT (eds) Negative versus positive schizophrenia. Springer, Berlin Heidelberg New York, pp 425–440

D'Amato T, Campion D, Gorwood P, Jay M, Sabate O, Petit C, Abbar M, Malafosse A, Leboyer M, Hillaire D, Clerget-Darpoux F, Feingold J, Waksman G, Mallet J (1992) Evidence for a pseudoautosomal locus for schizophrenia. II: Replication of a non-random segregation of alleles at the DXYS14 locus. Br J Psychiatry 161:59–62

DeLisi LE, Reiss A, White B, Gershon E (1988a) Cytogenetic studies of males with schizophrenia screening for fra x chromosome and othe chromosomal abnormalities. Schizophr Res 1:277–281

DeLisi LE, Dauphinais ID, Gershon ES (1988b) Perinatal complications and reduced size of brain limbic structures in familial schizophrenia. Schizophr Bull 14:185–191

Done DJ, Johnstone EC, Frith CD, Golding J, Shepherd PM, Crow TJ (1991) Complications of pregnancy and delivery in relation to psychosis in adult life: data from the British perinatal mortality survey sample. Br Med J 302:1576–1580

Dworkin RH, Lenzenweger MF (1984) Symptoms and the genetics of schizophrenia: implications for diagnosis. Am J Psychiatry 141:1541–1546

Dworkin PH, Lenzenweger MF, Moidin SO, Skillings GF, Levick SE (1988) A multidimensional approach to the genetics of schizophrenia. Am J Psychiatry 145:1077–1083

Erlenmeyer-Kimling L, Cornblatt BA, Rock D, Robert S, Bell M, West A (1993) The New York high-risk project: anhedonia, attentional deviance, and psychopathology. Schizophr Bull 19:141–153

Farmer AE, McGuffin P, Gottesmann II (1987) Twin concordance for DSM-III schizophrenia: scrutinizing the validity of the definition. Arch Gen Psychiatry 44:634–641

Fletcher JM, Evans K, Baillie E, Byrd P, Hanratty D, Leach S, Julier C, Gosden JR, Muir W, Porteous DJ, St Clair D, van Heyningen V (1993) Schizophrenia-associated chromosome IIq21 translocation: Identification of flanking markers and development of chromosome IIq fragment hybrids as cloning and mapping resources. Am J Hum Genet 52:478–490

Frangos E, Athanassenas G, Tsitourides S, Katsanou N, Alexandrou P (1985) Prevalence of DSM III schizophrenia among the first-degree relatives of schizophrenic probands. Acta Psychiatr Scand 72:382–386

Franke P, Maier W, Hardt J, Hain C (1993) Cognitive functioning and anhedonia in subjects at risk for schizophrenia. Schizophr Res 10:77–84

Gershon ES, DeLisi LE, Hamovit J, Nurnberger JI Jr, Maxwell ME, Schreiber J, Dauphinais D, Dingman CW II, Guroff JJ (1988) A controlled family study of chronic psychoses: schizophrenia and schizoaffective disorder. Arch Gen Psychiatry 45:328–336

Gill M, McGuffin E, Parfitt E, Mant R, Asherton P, Collier D, Vallada H, Powell J, Shaikh S, Taylor C, Sargeant M, Clements A, Nanko S, Takazawa T, Llewellyn D, Williams J, Whatley S, Murray R, Owen M (1993) A linkage study of schizophrenia with DNA markers from the long arm of chromosome 11. Psychol Med 23:27–44

Goldstein JM, Faraone SV, Chen WJ, Tolomiczencko GS, Tsuang MT (1990) Sex differences in the familial transmission of schizophrenia. Br J Psychiatry 156:819–826

Gorwood P, Leboyer M, d'Amato T, Jay M, Campion D, Hillaire D, Mallet J, Feingold J (1992) Evidence for a pseudoautosomal locus for schizophrenia. I: A replication study using phenotype analysis. Br J Psychiatry 161:55–58

Gottesman II, Bertelsen A (1989) Confirming unexpressed genotypes for schizophrenia. Arch Gen Psychiatry 46:867–872

Greenberg DA (1992) There is more than one way to collect data for linkage analysis. What a study of epilepsy can tell us about linkage strategy for psychiatric disease. Arch Gen Psychiatry 49:745–750

Greenberg S (1993) Linkage analysis of "necessary" disease loci versus "susceptibility" loci. Am J Hum Genet 52:135–143

Guze SB, Cloninger CR, Martin RL, Clayton PJ (1983) A follow-up and family study of schizophrenia. Arch Gen Psychiatry 40:1273–1276

Häfner H (1990) New perspectives in the epidemiology of schizophrenia. In: Häfner H, Gattaz WF (eds) Search for the causes of schizophrenia, vol II. Springer, Berlin Heidelberg New York, pp 408–431

Hallmaier J, Maier W, Minges J, Schwab S, Ertl M, Ackenheil M, Wildenauer DB (1993) No linkage of the dopamine D_2 receptor gene to schizophrenia in 15 German families. Psychiatry Res (in press)

Heun R, Maier W (1993) The role of obstetric complications in schizophrenia. J Nerv Ment Dis 181:220–226

Holland T, Gosden C (1990) A balanced chromosomal translocation partially cosegregating with psychotic illness in a family. Psychiatry Res 32:1–8

Holzman PS (1992) Behavioral markers of schizophrenia useful for genetic studies. J Psychiatr Res 26:427–445

Holzman PS, Kringlen E, Matthysse S, Flanagan SD, Lipton RB, Cramer G, Levin S, Lange K, Levy DL (1988) A single dominant gene can account for eye tracking dysfunctions and schizophrenia in offspring of discordant twins. Arch Gen Psychiatry 45:641–647

Jones CT, Morris S, Yates CM, Moffoot A, Sharpe C, Brock DJH, St Clair D (1992) Mutation in codon 713 of the β amyloid precursor protein gene presenting with schizophrenia. Nature Genet 1:306–309

Kendler KS, Diehl SR (1993) The genetics of schizophrenia: a current, geneticepidemiologic perspective. Schizophr Bull 19:261–285

Kendler KS, Tsuang MT (1988) Outcome and familial psychopathology in schizophrenia. Arch Gen Psychiatry 45:338–346

Kendler KS, Tsuang MT, Hays P (1987) Age at onset in schizophrenia. A familial perspective. Arch Gen Psychiatry 44:881–890

Kendler KS, McGuire M, Gruenberg AM, O'Hare A, Spellman M, Walsh D (1993) The Roscommon family study. I. Methods, diagnosis of probands and risk of schizophrenia in relatives. Arch Gen Psychiatry 50:527–540

Kety SS, Rosenthal D, Wender PH et al (1975) Mental illness in the biological and adoptive families of adopted individuals who have become schizophrenic: a preliminary report based on psychiatric interviews. In: Fieve R, Rosenthal D, Brill H (eds) Genetic research in psychiatry. Johns Hopkins University Press, Baltimore MD, pp 147–165

Kringlen E (1991) Adoption studies in functional psychosis. Eur Arch Psychiatry Clin Neurosci 240:307–313

Kringelen E, Cramer G (1989) Offspring of monozygotic twins discordant for schizophrenia. Arch Gen Psychiatry 46:873–877

Lenzenweger MF, Loranger AW (1989) Detection of familial schizophrenia using a psychometric measure of schizotypy. Arch Gen Psychiatry 46:902–907

Lewis SW, Own MJ, Murray RM (1987) Obstetric complications and schizophrenia: methodology and mechanisms. In: Schulz SC, Tamminga CA (eds) Schizophrenia: scientific progress. Oxford University Press, New York, pp 56–68

Maier W, Merikangas KR (1992) Cotransmission and comorbidity of affective disorders, anxiety disorders and alcoholism in families. In: Mendlewicz J, Hippius H (eds) Genetic research. Springer, Berlin Heidelberg New York, pp 207–224

Maier W, Hallmayer J, Minges J, Lichtermann D (1990) Affective and schizoaffective disorders: Similarities and differences. In: Marneros A, Tsuang MT (eds) Morbid risks in relatives of affective, schizoaffective, and schizophrenic patients. Results of a family study. Springer, Berlin Heidelberg New York, pp 201–207

Maier W, Lichtermann D, Minges J, Heun R, Hallmayer J (1993a) The impact of gender and age at onset on the familial aggregation of schizophrenia. Eur Arch Psychiatry Clin Neurosci 242:279–285

Maier W, Lichtermann D, Minges J, Hallmayer J, Heum R, Benkert O, Levinson DF (1993b) Continuity and discontinuity of affective disorders and schizophrenia: results of a controlled family study. Arch Gen Psychiatry (in press)

Maier W, Minges J, Lichtermann D, Heun R, Franke P (1993c) Personality variations in healthy relatives of schizophrenics. Schizophr Res (in press)

Maier W, Lichtermann D, Minges J, Franke P, Heun R, Hallmayer J (1993d) Concordance for gender in sib pairs affected with schizophrenia and related disorders. Schizophr Res 9:71–76

Martinez MM, Goldin LR (1990) Power of the Linkage Test for a heterogeneous disorder due to two independent inherited causes: a simulation study. Genet Epidemiol 7:219–230

Matthysse S, Lange K, Wagener DK (1979) Continuous variation caused by genes with unequal effects. Proc Natl Acad Sci USA 76:2862–2865

Matthysse S, Levy DL, Kinney D, Deutsch C, Lajonchere C, Yurgelun-Todd D, Woods B, Holzman PS (1992) Gene expression in mental illness: a navigation chart to future progress. J Psychiatr Res 26:461–473

McGue M, Gottesman II (1989) A single dominant gene still cannot account for the transmission of schizophrenia. Arch Gen Psychiatry 46:478–479

McGue M, Gottesman II (1991) The genetic epidemiology of schizophrenia and the design of linkage studies. Eur Arch Psychiatry Clin Neurosci 240:174–181

McGue M, Gottesman II, Rao DC (1986) The analysis of schizophrenia family data. Behav Gene 16:75–87

McGuffin P, Farmer A, Gottesman II (1987) Is there really a split in schizophrenia? The genetic evidence. Br J Psychiatry 150:581–592

McGuffin P, Sargeant M, Hetti G, Tidmarsh S, Whatley S, Marchbanks RM (1990) Exclusion of schizophrenia susceptibility gene from the chromosome 5qII-q13 region: New data and a reanalysis of previous reports. Am J Hum Genet 47:524–535

McGuffin P, Owen M, Gill M (1992) Molecular genetics of schizophrenia. In: Mendlewicz J, Hippius H, Bondy B, Ackenheil M, Sandler M (eds) Genetic research in psychiatry. "2. Münchner Genetikgespräche", 12–15 Sept 1991, C.I.N.P. President's Workshop. Springer, Berlin Heidelberg New York, pp 25–43

McNeil TF, Cantor-Graae E, Cardenal S (1993) Prenatal cerebral development in individuals at genetic risk for psychosis: head size at birth in offspring of women with schizophrenia. Schizophr Res 10:1–5

Meehl PE (1990) Toward an integrated theory of schizotaxia, schizotypy, and schizophrenia. J Person Disord 4:1–99

Mendlewicz J, Hirsch D (1991) Bipolar manic depressive illness and the fragile X syndrome. Biol Psychiatry 29:298–299

Moises HW, Gelernter J, Giuffra LA, Zarcone V, Wetterberg L, Civelli O, Kidd KK, Cavalli-Sforza LL (1990) No linkae between D2 dopamine receptor gene region and schizophrenia. Arch Gen Psychiatry 47:643–647

Murray RM (1991) The neurodevelopmental basis of sex differences in schizophrenia. Psychol Med 21:565–575

Murray RM, Lewis SW, Reveley AM (1985) Towards an aetiological classification of schizophrenia. Lancet 4:1023–1026

Murray RM, Jones P, O'Callaghan E, Takei N, Sham P (1992a) Genes, viruses, and neurodevelopmental schizophrenia. J Psychiatr Res 26:225–235

Murray RM, O'Callaghan E, Castle D, Lewis SW (1992b) A neurodevelopmental approach to the classification of schizophrenia. Schizophr Bull 18:319–332

Nimgaonkar VL, Wessely S, Murray RM (1988) Prevalence of familiality, obstetric complications, and structural brain damage in schizophrenic patients. Br J Psychiatry 153:191–197

Nöthen MM, Erdmann J, Lanczik M, Rietschel M, Körner J, Maier W, Albus M, Ertl MA, Wildenauer DB, Propping P (submitted) Mutation in the β amyloid precusor protein gene and schizophrenia. Biol Psychiatry (in press)

Nowakowski RS (1991) Neuronal migration and differentiation during normal and genetically perturbed development of the hippocampal formation. In: Mednick SA et al (eds) Developmental neuropathology of schizophrenia. Plenum, New York, pp 29–47

O'Callaghan E, Larkin C, Waddington JL (1990) Familial, obstetric, and other clinical correlates of minor physical anomalies in schizophrenia. Am J Psychiatry 148:479–483

O'Rourke DH, Gottesman II, Suarez BK et al (1982) Refutation of the single locus model in the aetiology of schizophrenia. Am J Hum Genet 33:630–649

Oberlé I, Rousseau F, Heitz D, Kretz C, Devys D, Hanauer A, Boue J, Bertheas MF, Mandel JL (1991) Instability of a 550-base pair DNA segment and abnormal methylation in fragile X syndrome. Science 252:1097–1102

Ott J (1990) Invited editorial: cutting a gordian knot in the linkage analysis of complex human traits. Am J Hum Genet 46:219–221

Plomin R (1990) The role of inheritance in behavior. Science 248:183–188

Propping P (1983) Genetic disorders presenting as "schizophrenia": Karl Bonhoeffer's early view of the psychoses in the light of medical genetics. Hum Genet 65:1–10

Propping P (1989) Psychiatrische Genetik: Befunde, Konzepte. Springer, Berlin Heidelberg New York

Pulver AE, Brown CH, Wolyniec P, McGrath J, Tam D, Adler L, Carpenter WT, Clarke B (1990) Schizophrenia: age at onset, gender and familial risk. Acta Psychiatr Scand 82:344–351

Pulver AE, Liang KY, Brown CH, Wolyniec P, McGrath J, Adler L, Tam D, Carpenter WT, Childs B (1992) Risk factors in schizophrenia: season of birth, gender, and familial risk. Br J Psychiatry 160:65–71

Reiss AL, Freund L (1990) Fragile X syndrome. Biol Psychiatry 27:223–240

Reiss AL, Hagerman RJ, Vinogradov S, Abrahams M, King R (1988) Psychiatric disability in female carriers of the fragile X chromosome. Arch Gen Psychiatry 45:25–30

Reiss AL, Freund L, Abrahams MT, Boehm C, Kazazian H (1993) Neurobehavioral effects of the fragile X premutation in adult women: a controlled study. Am J Hum Genet 52:884–894

Risch N (1990a) Genetic linkage and complex diseases. With special reference to psychiatric disorders. Genet Epidemiol 7:3–16

Risch N (1990b) Genetic linkage and complex diseases. A response. Genet Epidemiol 7:41–45

Schwarzkopf SB, Nasrallah HA, Olson SC, Coffman JA, McLaughlin JA (1989) Perinatal complications and genetic loading in schizophrenia: preliminary findings. Psychiatry Res 27:233–239

Smith M, Wasmuth J, McPherson JD, Wagner C, Grandy D, Civelli O, Potkin S et al (1989) Cosegregation of an 11q22. 3–9 p22 translocation with affective disorder: proximity of the dopamine D_2 receptor gene relative to the translocation breakpoint. Am J Hum Genet 45:A220

Squires-Wheeler E, Skodol AE, Bassett A, Erlenmeyer-Kimling L (1989) DSM-III-R schizotypal personality traits in offsprings of schizophrenic disorder, affective disorder, and normal control parents. J Psychiatr Res 23:229–239

St Clair D, Blackwood D, Muir W, Carothers A, Walker M, Spowart G, Gosden C, Evans HJ (1990) Association within a family of a balanced autosomal translocation with major mental illness. Lancet 336:13–16

Sturt E, Shur E (1985) Sex concordance for schizophrenia in proband-relative pairs. Br J Psychiatry 147:44–47

Su Y, Burke J, O'Neill A, Murphy B, Nie L, Kipps B, Bray J, Shinkwin R, Nuallain MN, MacLean CJ, Walsh D, Diehl SR, Kendler KS (1993) Exclusion of linkage between schizophrenia and the D2 dopamine receptor gene region of chromosome 11q and 112 Irish multiplex families. Arch Gen Psychiatry 50:205–211

Taylor MA (1992) Are schizophrenia and affective disorder related? A selective literature review. Am J Psychiatry 149:22–32

Taylor MA, Berenbaum SA, Jampala VC, Cloninger CR (1993) Are schizophrenia and affective disorder related? Preliminary data from a family study. Am J Psychiatry 150:278–285

Tienari P (1991) Genes, family environment or interaction. Findings from an adoption study. In: Kringlen E, Lavik NJ, Torgersen S (eds) Etiology of mental disorder. Department of Psychiatry, University of Oslo, Oslo, pp 33–49

Torgersen S (1985) Relationship of schizotypal personality disorder to schizophrenia: genetics. Schizophr Bull 11:554–563

Tsuang MT, Winokur G, Crowe RR (1980) Morbidity risks of schizophrenia, mania, depression and surgical conditions. Br J Psychiatry 137:497–504

Tsuang MT, Gilbertson MW, Faraone SV (1991) The genetics of schizophrenia. Current knowledge and future directions. Schizophr Res 4:157–171

Wang ZW, Black D, Andreasen N, Crowe RR (1993) Pseudoautosomal locus for schizophrenia excluded in 12 pedigrees. Arch Gen Psychiatry 50:199–204

Wright P, Murray RM, Donaldson PT, Underhill JA (1993) Do maternal HLA antigens predispose to schizophrenia? Lancet 342:117–118

Schizophrenia: Disease or Syndrome? Discussion

R.M. Murray and P. Jones

Back to the Future in Schizophrenia Research

Many ideas proposed by clinicians working around the time of the first descriptions of schizophrenia are now being re-researched and honed by the tools of modern psychiatric research: neuroimaging, genetics and epidemiology. These four chapters are excellent examples.

Johnstone reports results from the first neuropathological study of schizophrenia to have extensive antemortem information. Inclusion of cases with focal neuropathology, previously excluded (Bruton et al. 1990), raises the possibility of two distinct pathological pathways to schizophrenia. Abnormalities of development are implicated in the first, in which decreased brain size is associated with poor premorbid function and limited academic achievement, prominent negative symptoms and poor cognitive function; a clinical pattern we have termed "neurodevelopmental schizophrenia" (Murray et al. 1992a,b).

The second pathological phenotype proposed by Johnstone is more common in women, and is not associated with small brains or impaired premorbid function but rather with gliosis and focal brain damage. Johnstone concludes that when schizophrenia occurs in women it may sometimes be on the basis of pathology separate from, or additional to, developmental abnormality. This echoes the view that we have outlined elsewhere (Castle and Murray 1991; Murray et al. 1992b) that schizophrenia comprises two main entities: the first a severe, early onset condition associated with premorbid dysfunction, and commoner in males; the second a milder, adult onset disorder occurring particularly in women.

Johnstone's findings on neurodevelopmental impairment were presaged over a century ago by one of her predecessors as Professor of Psychiatry in Edinburgh, Clouston (1892). Clouston described developmental or adolescent insanity, a psychotic condition which, he believed, had its onset between 15 and 24 years, was more common in men, frequently had a family history, and proceeded to a secondary dementia (or defect state) in 30% of cases. Furthermore, Clouston anticipated the modern interest in the frequency of minor physical abnormalities, including palatal abnormalities (Green et al. 1987), by

Department of Psychological Medicine, Institute of Psychiatry and King's College Hospital, De Crespigny Park, London, SE5 8AF, UK

H. Häfner and W.F. Gattaz (Eds.)
Search for the Causes of Schizophrenia, Vol. III

noting that 55% of patients with adolescent insanity had a deformed palate compared with 19% of the general population. He considered that this was a marker for a neurodevelopmental failure, and that adolescent insanity was "the last cortical developmental disease".

Physiological Basis of Schizophrenic Symptoms

Trimble points out that Kraepelin, whose concept of dementia praecox was both broader and more influential than Clouston's ideas of developmental insanity, also regarded the disorder as secondary to cortical disease. Unfortunately, over the subsequent century, psychiatrists gradually lost sight of what Clouston and Kraepelin knew. As Ron and Harvey (1990) have concluded, "To have forgotten that schizophrenia is a brain disease will go down as one of the great aberrations of twentieth century medicine."

Trimble also reminds us of the important role played by Slater when he redirected interest to brain dysfunction in schizophrenia by suggesting that the psychosis associated with temporal lobe epilepsy might provide a useful model of the disorder. Following Slater's presentation to the Royal Society of Medicine (Slater and Beard 1963), the English neurologist, Sir Charles Symmond commented: "The schizophrenia-like aural symptoms of temporal lobe epilepsy suggest that the temporal lobe includes within its boundaries circuits concerned with the physiological basis of the psychological disorder which we call schizophrenia . . . it is not loss of neurones in the temporal lobe that is responsible for the psychosis, but the disorderly activity of those that remain."

Trimble and his colleagues (Perez et al. 1985) subsequently demonstrated that Schneider's First Rank Symptoms were characteristic of the psychosis associated with temporal lobe epilepsy. A further demonstration of this link between epilepsy and first rank symptoms has been provided by Jablensky et al. (1992) who noted that among psychotic cases studied by the World Health Organization, a history of childhood febrile convulsions was particularly associated with nuclear schizophrenia, a PSE/Catego category based particularly on the presence of Schneider's First Rank Symptoms.

Liddle has shown previously that schizophrenic symptoms reliably segregate into three syndromes which correlate with neuropsychological impairments and, importantly, have meaning to the clinician (Liddle 1984, 1987; Peralta et al. 1992). Liddle's syndrome of reality distortion has much in common with Schneider's First Rank Symptoms. It is, therefore, satisfying that this syndrome is associated with increased blood flow (rCBF) in the left medial temporal structures, including those involved in the internal monitoring activities thought to go awry in reality distortion (Frith et al. 1992).

Liddle's evidence that psychomotor poverty correlates with both impaired word generation and with decreased rCBF in the left prefrontal cortex is reminiscent of Kleist's (1934a) observation over half a century ago that traumatic damage to the left forebrain" . . . is characterised by a lack of productivity of thought and by an incapacity to carry through a chain of thought". Kleist points out that such patients show poverty of thought and are incapable of producing coherent cognitions from single ideas or perceptions, and that these deficits match closely those that occur in schizophrenia.

We (van Os et al. 1993) have attempted to relate Liddle's three syndromes to the growing evidence of premorbid deficits in schizophrenia. In a study of 166 psychotic patients, we found that a negative symptom factor akin to Liddle's psychomotor poverty syndrome, was associated with male sex, single status, low IQ and childhood psychiatric attendance. The disorganisation syndrome was strongly associated with poor premorbid social adjustment. Thus, both these syndromes may develop from childhood deficits, and the findings confirm our previous suggestion that negative symptoms are the adult manifestation of childhood developmental impairments (Murray et al. 1988).

We found no such relationships for the reality distortion syndrome of delusions and hallucinations. This is not surprising; these symptoms reflect state related changes in brain function occurring not only in schizophrenia, but also in affective psychosis and drug-related psychosis.

In view of this, we (McGuire et al. 1993) examined the temporal relationship between verbal hallucinations and regional cerebral blood flow (rCBF). SPET was used to define rCBF in 12 male acute schizophrenic patients at the precise moment they reported verbal auditory hallucinations; the scan was later repeated when the patients were no longer hallucinating. In the hallucinated state, rCBF was greater in Broca's area and, to a lesser, non-significant extent in the left anterior cingulate cortex and parts of the left temporal lobe.

These findings suggest that the production of verbal auditory hallucinations is associated with increased activity in cortical areas specialised in language. The activation of Broca's area is particularly interesting because it is implicated in the production of inner speech. Frith has suggested that schizophrenics may misinterpret their own inner speech as alien, and perceive it instead as verbal hallucinations.

Such studies show how functional imaging can be used to elucidate the pathophysiology underlying schizophrenic symptoms. Using such an approach, Weinberger et al. (1986, 1988) and Berman et al. (1986) have suggested there is an irreversible loss of prefrontal function in schizophrenia. However, Liddle now demonstrates quite clealry that patients with the psychomotor poverty syndrome are able to activate the prefrontal cortex with an appropriate cognitive challenge. Indeed, they produce more widespread cortical activation than controls, and Liddle suggests that they "require greater cortical tone to achieve a specified level of performance".

This finding is reminiscent of Asarnow's evidence that children with schizophrenia show a delay in cognitive processing as if they were less able to mobilise their processing resources rapidly (see chapter by R.F. Asarnow et al., this volume). Similarly, we (Birkett et al. 1994) have found that the relatives of schizophrenics also show impairment on a word fluency task, but only for the first 15 s. These findings again point to the importance of language dysfunction, not only in frankly psychotic patients, but also in the premorbid phase (Done et al. 1993; C. Hollis and E. Taylor, personal communication), and in relatives who may share a similar genetic risk for the disorder.

Genes and Environment

Johnstone has only scant data on the occurrence of birth trauma and family history in her patients and, wisely, she avoids speculation on aetiology. Liddle is similarly reticent, and Trimble confines himself to reminding us that psychosis associated with temporal lobe epilepsy is rarely accompanied by a family history of schizophrenia, thus establishing the existence of phenocopies.

Maier considers the contentious area of the relative contribution of genes and environment to liability to schizophrenia. The calculation of heritability is dependent not only on unproven assumptions about the mode of transmission but also on the information available concerning the sample and the diagnostic criteria employed. These problems are well illustrated by a very recent analysis of the Maudsley Twin series, originally investigated by Gottesman and Shields (1972). Farmer et al. (1987) had previously applied the DSM-III criteria for schizophrenia to the case summaries prepared when Gottesman and Shields examined the sample in the mid-1960s and reported a heritability of 0.85. However, we (Marshall et al. 1993) followed up the twins for a further 25 years, and were able to apply the DSM-III criteria to much more extensive information. The heritability estimate plunged to 0.29! When broader diagnostic criteria for schizophrenia were applied, the estimate rose, to 0.55 for DSM-IIIR and 0.70 for ICD-10.

There are two main explanations. Firstly, such calculations may be a futile exercise. Secondly, and more optimistically, it may be that very restrictive criteria for schizophrenia tend to identify a less genetic disorder, and that as one broadens the criteria one includes more genetic cases. Indeed, as Marshall et al. (1993) broadened the criteria to include not only DSM-III schizophrenics but successively atypical psychotics and all patients with affective disorder, the ratio of monozygotic to dizygotic twin concordance rates rose from 1.60 to 3.04 and 3.99, respectively; this implies that genetic influence is greater on "broad" than "narrow" schizophrenia.

Such findings are congruent with the evidence that remitting cases of schizophrenia, variously termed good outcome or schizo-affective, are highly familial. Such cases are more commonly female, and there is growing evidence that schizophrenia is more familial in women than in men (e.g., Sham et al. 1993). We have also recently shown (van Os et al. 1993) that insidious onset schizophrenia is associated with a low morbid risk of disease in first degree relatives (2%); an intermediate morbid risk was found in relatives of subjects with acute onset schizophrenia, while the highest morbid risk (11%) was found in the relatives of acute schizophrenics who broke down following adverse life events. Thus, male schizophrenics with an insidious onset should show least evidence of a genetic effect.

Might the strategy of dividing schizophrenics into familial and sporadic cases (Murray et al. 1985) illuminate this issue? Maier considers that it is too crude a strategy to be productive, and the long debate regarding familial risk and cerebral ventricle size (Lewis 1990) might, at first sight, be taken as supporting him. However, recent analyses present a clearer picture.

Vita et al. (1992) carried out a meta-analysis of existing studies and demonstrated that ventricle-brain ratio (VBR) was greater, on average, in sporadic schizophrenics than in familial schizophrenics. More importantly, in an analysis of their own series, Vita et al. found that among men with schizophrenia, sporadic cases had significantly larger lateral ventricles than familial cases; no such differences were found in females. We (Jones et al. 1993) have replicated these findings in a series of 121 consecutive admissions for schizophrenia. Once again, the association of larger ventricles and negative family history was confined to men; male sporadic cases had significantly larger lateral ventricles then male familial cases who were not different from women, with or without a positive family history.

It remains uncertain as to which environmental factors cause the structural brain changes found in schizophrenia. Much work (see McGrath and Murray 1993) has implicated pre- and peri-natal complications (PBCs), particularly in severe, early onset schizophrenia in men. For instance the Danish high-risk sample demonstrated a convincing link between PBCs and chronic, though remitting, schizophrenia and an association between PBCs and large cerebral ventricles. However, our own recent findings do not (Jones et al. 1993).

It is possible that ischaemic damage, perhaps mediated by release of excitotoxic amino acids, may contribute to severe early onset neurodevelopmental schizophrenia, which we have noted earlier to be commoner in males. The reason for the greater frequency of milder, good outcome, schizophrenia in females remain uncertain. The main possibilities appear to be:

1. Oestrogens modify the severity of the disorder and delay its onset, through their anti-dopaminergic properties. This view, proposed by Häfner et al. (1991a,b), is readily tested; if true, male and female schizophrenics with a similar aetiology should have different ages at onset. We (Walsh et al. 1993) have examined this among schizophrenics from multiply affected families, where all cases are presumed to be genetic in origin. There was no difference in age of onset in men and women, providing no evidence for any delaying effect of oestrogens.
2. Schizophrenia in women may have closer links with affective disorders, a view expressed by Castle et al. (1993).
3. The putative increased risk of schizophrenia mediated by perinatal exposure to influenza appears to be greater in women than men (Takei et al. 1993) and may be associated with a milder, more paranoid disease (Mednick et al. 1993).
4. Finally, as Johnstone tentatively implies, a higher proportion of female schizophrenia may result from acquired focal brain damage, though why this should be so is quite unclear.

References

Asarnow RF, Asarnow JR, Strandburg R (1989) Schizophrenia: a developmental prospective. In: Tackle D, Cicchetti L (eds) The emergence of a discipline: Rochester symposium on development psychopathology, vol 1. Erlbaum, Hillsdale, New Jersey

Berman KF, ZEC RF, Weinberger DR (1986) Physiologic dysfunction of dorsolateral prefrontal cortex in schizophrenia. 2. Role of neuroleptic treatment, attention, and mental effort. Arch Gen Psychiatry 42:126–135

Birkett PBL, Griffiths TD, Frangou S, Sharma T, Sigmundson T, Murray RM, Reveley A (1994) Neuropsychology in familial schizophrenia. Schizophr Bull (in press)

Bruton CJ, Crow TJ, Frith CD, Jonstone EC, Owens DGC, Roberts GW (1990) Schizophrenia and the brain: a prospective clinico-neuropathological study. Psychol Med 20:285–304

Castle DJ, Murray RM (1991) The neurodevelopmental basis of sex differences in schizophrenia. Psychol Med 21:565–575

Castle DJ, Abel K, Takei N, Murray RM. The relationship between schizophrenia in women and affective disorder. Schizophr Bull (submitted)

Clouston TS (1892) Clinical lectures on mental diseases, 3rd edn. Churchill, London

Done DJ, Crow TJ, Johnstone EC, Sacker A (1993) Childhood antecedents of schizophrenia and affective illness: social adjustment at ages 7 & 11. Br Med J (in press)

Farmer A, McGuffin P, Gottesman II (1987) Twin concordance for DSM-III schizophrenia. Arch Gen Psychiatry 44:634–641

Frith CD, Friston KJ, Liddle PF, Frackowiak RSJ (1992) The left medial temporal region and schizophrenia; a PET study. Brain 115:367–382

Gottesman II, Sheilds J (1972) Schizophrenia and genetics: a twin study perspective. Academic, London

Green MF, Satz P, Soper HV et al (1987) Relationship between physical anomolies and age of onset of schizophrenia. Am J Psychiatry 144:666–669

Häfner H, Behrens S, deVry J, Gattaz WF (1991a) Oestradiol enhances the vulnerability threshold for schizophrenia in women by an early effect in dopaminergic neurotransmission. Eur Arch Psychiatry Clin Neurosci 241:65–68

Häfner H, Behrens S, de Vry J, Gattaz WF (1991b) An animal model for the effects of estradiol on dopamine-mediated behaviour – implications for sex differences in schizophrenia. Psychiatry Res 38:125–134

Jablensky A, Sartorius N, Ernberg G, Anker M, Korten A, Cooper JE, Day R, Bertelsen A (1992) Schizophrenia: manifestations, incidence and course in different cultures. A World Health Organization ten-country study. Psychol Med Monogr Suppl 20:40

Jones PB, Harvey I, Lewis SW, Toone BK, Van Os J, Williams M, Murray RM. Cerebral ventrical dimensions as risk factors of schizophrenia and effective psychosis. An epidemiological approach to analysis. Psychol Med (submitted)

Kleist K (1934a) Gehirnpathologie. Barth, Leipzig

Kleist K (1934b) Kriegverletzungen des Gehirn in ihre Bedeutung für Hirnlokalisation und Hirnpathologie. Barth, Leipzig

Lewis SW (1990) X-Ray CT in schizophrenia: fifteen years on. Br J Psychiatry 157(9):16–24

Liddle PF (1984) Chronic schizophrenic symptoms, cognitive function and neurological impairment. Thesis, Royal College of Psychiatrists, London

Liddle PF (1987) The symptoms of chronic schizophrenia: a re-examination of the positive-negativce dichotomy. Thesis, Royal College of Psychiatrists. Br J Psychiatry 15:145–151

Marshall EJ, Revely AM, Coid B, Birkett PBL, Murray RM (1993) The Maudsley schizophrenic twin series. A 25 year follow-up: concordance and heritability for operationally defined schizophrenia. Arch Gen Psychiatry (submitted)

McGrath J, Murray RM (1994) Early risk factors for schizophrenia: from conception to birth. In: Weinberger D, Hirsch S (in press)

McGuire PK, Shah GMS, Murray RM (1993) Increased blood flow in Broca's area during auditory hallucinations in schizophrenia. Lancet 342:703–706

Mednick S (1993) Presentation to NATO Advanced Study Insitute on fetal brain development and schizophrenia. 22 Sept 1 Oct. Il Ciocco Hotel, Italy
Murray RM, Lewis SW, Reveley AM (1985) Towards an aetiological classification of schizophrenia. Lancet i:1023–1026
Murray RM, Lewis SW, Owen M, Foerster A (1988) The neurodevelopmental origins of dementia praecox. In: Bebbington P, McGuffin P (eds) Schizophrenia: the major issues. Heinmann, London, pp 90–106
Murray RM, Jones PB, O'Callaghan E, Takei N, Sham P (1992a) Genes, viruses and neurodevelopmental schizophrenia. J Psychiatric Res 26:225–235
Murray RM, O'Callaghan E, Castle DJ, Lewis SW (1992b) A neurodevelopmental approach to the classification of schizophrenia. Schizopr Bull 18:319–332
Peralta V, deLeon J, Cuesta MJ (1992) Are there more than two syndromes in schizophrenia? A critique of the positive-negative dichotomy. Br J Psychiatry 161:335–343
Perez MM, Trimble MR, Murray MF, Reider I (1985) Epileptic Psychosis: An evaluation of PSE profiles. Br J Psychiatry 146:155–163
Ron M, Harvey I (1990) The brain in schizophrenia. J Neurol Neurosurg Psychiatry 53:725–726
Sham PC, Jones PB, Russel A, Foerster A, Gilvary, Bebbington P, Lewis SW, Toone BK, Murray RM (1993) Age at onset, sex and familial psychiatric morbidity in schizophrenia. Report from the Camberwell Collaborative Psychosis Study. Br J Psychiatry (submitted)
Slater E, Beard AW (1963) The schizophrenia-like psychoses of epilepsy. Br J Psychiatry 109:95–150
Takei N, O'Callaghan E, Sham P, Glover G, Murray RM (1994) Does pre-natal influenza divert susceptible females from later depressive psychosis to schizophrenia? Acta Psychiatr Scand (in press)
van Os J, Fahy T, Jones P, Harvey I, Lewis S, Toone B, Maurray RM et al (1994) Psychopathological syndromes in the functional psychoses and their prognostic significance. (submitted)
van Os J, Fahy T, Bebbington P, Jones P, Wilkins SW, Sham P, Russell A, Gilvary K, Lewis S, Toone B, Maurray RM (1994) The influence of life events on the subsequent course of psychotic illness. Psychol Med (in press)
Vita et al (1992) Familial and sporadic schizophrenia: a meta-analysis. Presentation to the Association of European Psychiatrists, Barcelona, Nov 1992
Walsh C, Asherson T, Sham P, Castle D, Williams J, Taylor C, Clements A, Watt D, Sargeant M, Owen M, Gill M, McGuffin P, Murray RM (1994) Age of onset of schizophrenia in multiply affected families is early and shows no sex difference. (submitted)
Weinberger DR, Berman KF, Zec RF (1986) Physiologic dysfunction of dorsolateral prefrontal cortex in schizophrenia. 1. Regional cerebral blood flow evidence. Arch Gen Psychiatry 43: 114–124
Weinberger DR Berman KF Illowsky BP (1988) Physiological dysfunction of dorsolateral prefrontal cortex in schizophrenia. 3. A new cohort and evidence for a monoaminergic mechanism. Arch Gen Psychiatry 45:609–615

Part III
Brain, Receptors and Schizophrenia

Multiple Dopamine Receptors: Implications for Schizophrenia

P. Sokoloff[1], M.-P. Martres[1], D. Levesque, J. Diaz[2], C. Pilon[1], N. Griffon[1], and J.-C. Schwartz[1]

Introduction

The classical dopamine (DA) hypothesis of schizophrenia owed its birth to the serendipitous discovery in the early 1950s of the potent antipsychotic properties of chlorpromazine (Delay et al. 1952). The ability of this compound, as well as other related "neuroleptics", to increase animal brain levels of DA metabolites in rough proportion to their clinical efficacy had been attributed to blockade of, at this time still undiscovered, brain DA receptors (Carlsson and Lindqvist 1963). Conversely, it was hypothesized that overstimulation of these receptors could be involved in the etiology of the disorder. In support were the observations that DA-stimulant drugs such as amphetamine were able to induce psychosis-like manifestations in healthy individuals or to exacerbate psychotic symptoms in schizophrenics. The subsequent discovery that therapeutic potencies of antipsychotics were correlated with their affinity for certain brain DA receptors (Creese et al. 1976) has added arguable value to the hypothesis.

However, the DA hypothesis of schizophrenia has progressively been challenged in view of clinical and physiological observations that did not fit with an excessive DA transmission (for a review see Jaskiw and Weinberger 1992). Cognitive deficits and impairments of affective relationships, which are prominent in some patients and often mark the evolution of the disease to a chronic status, generally do not improve with neuroleptic treatment. The evidence for a primary increase of DA activity is controversial. Levels of DA metabolites in body fluids, which constitutes an indirect index of DA neurotransmission, has never unequivocally supported an enhanced DA activity in schizophrenics (Crow 1982). The consistent increase in striatal DA receptors in post-mortem brain may have been due in most cases, but not all (Seeman et al. 1984; Wong et al. 1986), to previous neuroleptic medication, since this treatment is known to cause DA receptor upregulation in animals (Creese and Sibley 1980). In addition, the idea that brain DA receptors are increased in schizophrenics has been recently weakened by position emission tomography studies (Farde et al. 1987; Martinot et al. 1990). On the other hand, non-invasive

[1] Unité de Neurobiologie et de Pharmacologie de l'INSERM, Centre Paul Broca, 2ter rue d'Alésia, 75014 Paris, France
[2] Laboratoire de Physiologie, Université René Descartes, 4 Avenue de l'Observatoire, 75006 Paris, France

H. Häfner and W.F. Gattaz (Eds.)
Search for the Causes of Schizophrenia, Vol. III

neuroimaging methods have provided eivdence for anatomical abnormalities such as enlargement of ventricles and hypofunction of medial temporal lobes (Wagner et al. 1988), accompanied by structural changes (Roberts 1990), which might result in the clinical symptoms that form the core of schizophrenia.

Even though DA hyperactivity might not be the *primary* etiological event in schizophrenia but subsequential to developmental abnormalities, the DA theory, in its most parsimonious acceptation, is still operative to explain the relative efficacy of DA receptor antagonists, i.e., neuroleptics, to control some symptoms of schizophrenia such as productive symptoms (hallucinations, delusions). However, the favorable therapeutic effect of neuroleptics is often impaired by severe motor and/or endocrine side effects, leading to patients' non-compliance, which limits their use. Since the dual classification of DA receptors into D_1 and D_2, based on their opposite effects on adenylate cyclase (Spano et al. 1978; Kebabian and Calne 1979), it has been shown that the effects of neuroleptics are mostly due to a blockade of the D_2 receptor, because their affinity for the D_1 receptor is variable and, generally, much lower. The use of molecular biology in the study of receptors has now introduced a revolution in the field, showing that both categories of DA receptor were more heterogeneous than previously suspected from classical pharmacological studies. The possibility that the blockade of these various receptors results in different clinical effects, for instance, favourable versus adverse effects, affords opportunities to select more efficient and safer drugs. From the present review of the DA receptors so far identified, it emerges that the pharmacological and anatomical properties of the D_3 receptor, as well as aspects of its regulation and of its genetics, suggest that this particular receptor plays an important role in the antipsychotic effects.

Molecular Biology of DA Receptors

The first cloning of a D_2 receptor (Bunzow et al. 1988) paved the way for the cloning of a series of other DA receptor genes, based on the significant sequence homology between these receptors and the D_2 receptor. As anticipated, the D_1 receptor, which is nearly as abundant as the D_2 receptor in brain, was the first to follow (Zhou et al. 1990; Dearry et al. 1990; Sunahara et al. 1990; Monsma et al. 1990). Then came the genes of a series of less abundant and less expected receptors that markedly extend the DA receptor family: the rat and human D_3 (Sokoloff et al. 1990; Giros et al. 1990), the human D_4 (van Tol et al. 1991) and human D_5 receptors (Sunahara et al. 1991) as well as the rat counterpart of the D_5, termed D_{1B} (Tiberi et al. 1991). The amino acid sequences of all these receptors, as deduced from their established nucleotide sequences, reveal that they belong to a larger superfamily, that of receptors coupled to their intracellular transduction system by a GTP-binding (G) protein. Their structure, deduced from structural studies of rhodopsin "the receptor for light" (Henderson and Unwin 1975; Saibil 1990), is characterized by the occurrence of seven streches of 20–25 hydrophobic aminoacids, postulated to form transmembrane α-helices, connected by alternating extracellular and cytoplasmic loops composed of hydrophilic residues (see Fig. 1 in the case of the D_3 receptor). The N terminal part constitutes a glycosylated extracellular domain. The transmembrane helices form the ligand-binding domain, particularly three amino acid residues thought to interact with catecholamines: an aspartate (Asp^{110}) in the third transmembrane domain that forms an ion pair with the protonated amine group of DA and two serines (Ser^{193} and Ser^{196}) in the fifth transmembrane domain that presumably form a hydrogen-bonding interaction with the two phenol groups of DA. This latter interaction, specific for DA and its agonists, could cause a conformational change in the helix that would be transmitted to the third cytoplasmic loop, responsible to the coupling to G proteins.

The various genes of the DA receptor family can be classified into two groups according to their organization (Table 1): (a) intronless genes, i.e., those of the D_1 and D_5 receptor, in which the coding nucleotide sequence is

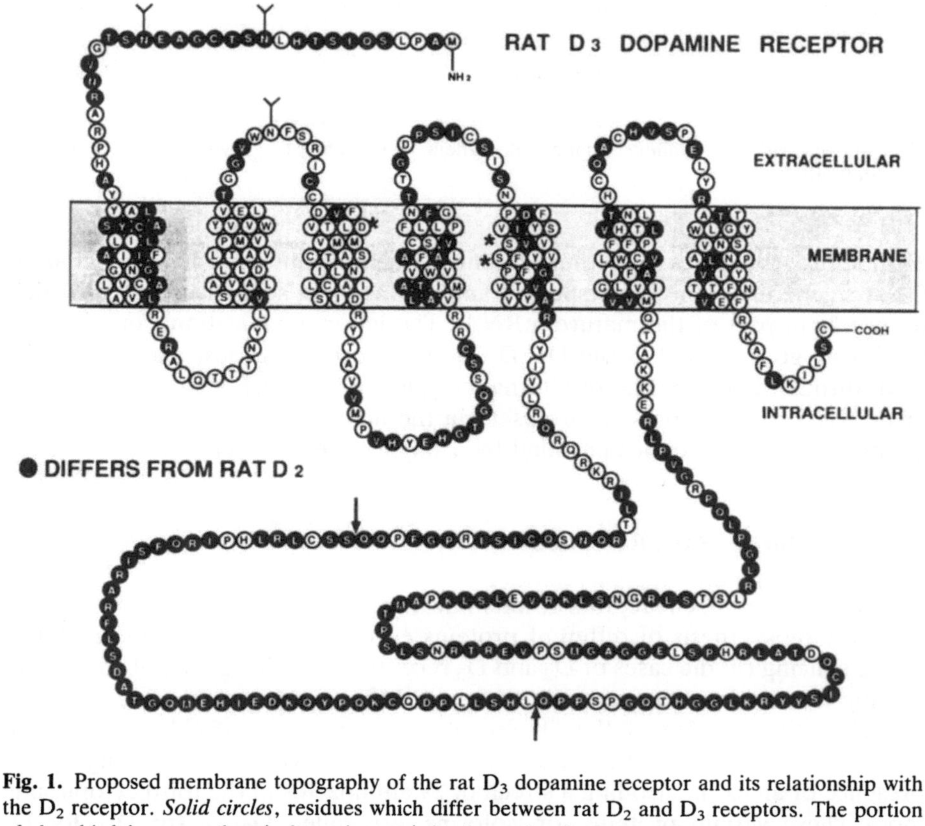

Fig. 1. Proposed membrane topography of the rat D_3 dopamine receptor and its relationship with the D_2 receptor. *Solid circles*, residues which differ between rat D_2 and D_3 receptors. The portion of the third intracytoplasmic loop (*arrows*) is absent in the human D_3 receptor, otherwise its structure is highly homologous to that of the rat D_3 receptor. The overall homology with the D_2 receptor is 50% and 75% if only transmembrane domains are considered

Table 1. Synopsis of the DA receptor family

	D_1-like receptors		D_2-like receptors		
	D_1	D_5	D_2	D_3	D_4
Coding sequence	446 a.a.	477 a.a.	D_{2A}: 443 a.a. D_{2B}: 414 a.a.	400 a.a.	387 a.a.
Chromosome localization	5q31–q34	4p16.3	11q22–q23	3q13.3	11p
Highest brain densities	Neostriatum	Hypothalamus, Hippocampus	Neostriatum	Isl. Calleja Nucleus accumbens	Medulla Frontal cortex
Pituitary	No	No	Yes	No	Yes
DA affinity	μM	μM	Sub-μM	nM	Sub-μM
Characteristic agonists	SKF 38393	SKF 38393	Bromocriptine	7-OH-DPAT	?
Characteristic antagonists	SCH 23390	SCH 23390	Haloperidol	UH 232	Clozapine
Adenylate cyclase	Stimulates	Stimulates	Inhibits	?	Inhibits

a.a., amino acid; Isl. Calleja, islands of Calleja; 7-OH-DPAT, 7-hydroxy-N,N'-di-*n*-propyl aminotetralin.

continuous; (b) genes with their coding sequence contained in discontinuous DNA segments (exons) interspersed among uncoding sequences (introns) that do not form part of the mature mRNA. The latter organization, found in the rhodopsin gene as well as the D_2, D_3, and D_4 receptor genes, may potentially lead, through a mechanism of alternative splicing (in which a given exon in the pro-mRNA is either present or absent in the final mRNA) to the biosynthesis of several distinct proteins encoded by a single gene (see below).

Various Isoforms of DA Receptors

The diversity of DA receptors results not only from gene multiplicity but also from the biosynthesis of different proteins encoded by a single gene by alternative splicing (in the cases of D_2 and D_3 receptors) or from gene polymorphism leading to sequence repetition (in the case of the D_4 receptor).

Alternative splicing of the D_2 receptor pro-messenger RNA produces, in several animal species, two mature messenger RNA (mRNA) differing by the presence of an exon encoding a strech of 29 aminoacids in the third cytoplasmic loop. These variant proteins, or isoforms, are called $D_{2(444)}$ (or D_{2L} for D_2 long, or D_{2A}) and $D_{2(415)}$ (or D_{2S} for D_2 short, or D_{2B}). They have different expression patterns among cerebral areas, although $D_{2(444)}$ predominates

(Giros et al. 1989, Monsma et al. 1989; Dal Toso et al. 1989). In addition, treatment with the antipsychotic haloperidol predominantly enhances the abundance of the smaller isoform in brain (Martres et al. 1992) and pituitary (Autelitano et al. 1989), suggesting a different mode of regulation. The two isoforms might functionally differ at the second messenger level, the coupling of which presumably involves the third cytoplasmic loop. While their stimulation activates the same second messenger pathways (Dal Toso et al. 1989; Piomelli et al. 1991), they may differ in their interaction with G proteins (Montmayer et al. 1993), thus endowing neurons with the ability to rapidly adapt intracellular responses. Initial pharmacological studies have indicated that the two D_2 receptor isoforms do not differ markedly in their recognition by several antipsychotics (Giros et al. 1989), but interest has been recently renewed when convergent observations were made that some "atypical" neuroleptics, namely among substituted benzamides, showed a preference for the shortest isoform (Castro and Strange 1993; Malmberg et al. 1993).

The D_3 receptor gene produces truncated isoforms also generated by alternative splicing, which correspond to short proteins because of early termination codons introduced by the deletions (Giros et al. 1991; Snyder et al. 1991; Fishburn et al. 1993). Nevertheless, a particular isoform of these still has a structure compatible with seven transmembrane helices, but no DA binding activity could be detected when the corresponding cDNA was transfected into cultured cells (Giros et al. 1991). Hence the truncated D_3 receptor isoforms appear inactive, which does not exclude their participation in the regulation of active isoforms, since they are concomitantly produced.

A polymorphism of the D_4 receptor gene is characterized by the presence of a stretch of 16 aminoacids in coding sequence, repeated up to eight times, in the third intracytoplasmic loop of the receptor. The variants might differ in their sensitivity to the sodium effect on recognition by some neuroleptics, which may underlie individual differences in responsiveness to antipsychotics (van Tol et al. 1992).

Dopamine Receptor Subtypes Use Different Intracellular Signalling Systems

Mammalian cell lines that do not normally express DA receptors, such as modified fibroblasts, can be transiently or permanently transfected with the various DA receptor subtypes providing a means to directly study the pharmacology of these artificially expressed receptors and to identify their signalling systems. The obvious advantage of transfected cells is that they usually express a single receptor subtype at high density, but one drawback is that they may differ from normal DA-responsive neurons in their natural components of the intracellular signalling system, particularly the G proteins.

Such tools allowed the confirmation of the positive coupling of the D_1 receptor to adenylate cyclase through the G protein known as Gs (Zhou et al. 1990; Dearry et al. 1990; Sunahara et al. 1990) and the negative coupling of the D_2 receptor to this enzyme through one of the G proteins termed Gi (Neve et al. 1989; Dal Toso et al. 1989; Sokoloff et al. 1990). This agrees with the known properties of these receptors in DA-responsive neurons and indicates that the two DA receptor subfamilies, which were initially distinguished from their structural properties, correspond to functional differences. These features suggest that the members of the "D_1-like" subfamily of intronless receptors, i.e., D_1 and D_5, would be similarly coupled to Gs and those of receptors with introns, i.e., D_2, D_3 and D_4, to Gi proteins. Actually, the picture is somewhat more complex, because a single receptor can couple to different second messenger pathways and because crosstalk between receptors affects the final response. The activation of D_1-like receptors not only triggers activation of adenylate cyclase, but also that of the inositol-phosphates' pathway through phospholipase C activation in some type of cells (Mahan et al. 1990). The complexity of coupling mechanisms of the D_2 receptor reflects the versatility of Gi/Go proteins in mediating various intracellular signals (Taylor 1990; Simon et al. 1991). The D_2 receptor inhibits adenylyl cyclase but also activates phospholipase A2 (see below) and increases intracellular Ca^{2+} in some transfected cells (Valler et al. 1990); however, intracellular Ca^{2+} is decreased through activation of K^+ channels and phospholipase C inhibited in other cells (Vallar et al. 1990).

Activations of D_1 and D_2 receptors have generally opposite effects but, nevertheless, act synergistically to produce many electrophysiological and behavioral responses (Clark and White 1987). A molecular basis for such D_1/D_2 receptor synergism may lay in the ability of D_1 receptor to potentiate the D_2 receptor-mediated increase in arachidonate release from transfected cells (Piomelli et al. 1991). The D_2 receptor-mediated arachidonate release, which is independent of inhibition of cyclic AMP formation, is of potential functional importance because it occurs with low concentrations of D_2 receptor agonists and arachidonic acid is a precursor for a variety of biologically active eicosanoids presumably involved in neurotransmission (Piomelli and Greengard 1990). As arachidonate and its cytochrome P_{450} metabolites are potent inhibitors of ($Na^+ + K^+$) ATPase in neurons of the striatum, their formation may well account for the D_1/D_2 synergistic responses in vivo, which may involve this transporter (Bertorello et al. 1990). Crosstalk at molecular level between D_1 and D_2 receptors was also revealed by examination of ligand-binding interactions in human post-mortem brain (Seeman et al. 1989). Interestingly, the D_1-mediated non-competitive inhibition by dopamine of D_2 receptor binding, present in brains from control, Alzheimer and Parkinson patients, was abolished in half of the brains tested from schizophrenics and Huntington patients, suggesting that impairment of such a link might occur in some of the hyperdopamine signs common in these latter diseases (Seeman et al. 1989). All these theories about D_1/D_2 interactions imply that both subtypes are coexpressed in the same neurons, a controversial issue, at least when striatum is considered (Gerfen et al. 1990; Le Moine et al. 1991).

The sequence homologous to that of the D_2 receptor and probably similar structure of the D_3 and D_4 receptors suggested that, like the D_2 receptor, they may be coupled to a Gi/Go protein. In agreement, D_4-mediated inhibition of adenylate cyclase was observed in mouse retina (Cohen et al. 1992). In contrast, none of the classical responses mediated by Gi proteins occurs upon stimulation of D_3 receptor in fibroblastic cells (Sokoloff et al. 1990, 1992; Piomelli et al. 1991; Seabrook et al. 1992). Nevertheless, when transfected in a neuroblastoma X glioma hybrid cell line, the D_3 receptor, while still without action on adenylate cyclase or phospholipase C, activates these proto-oncogene *c-fos* and induces mitogenesis through a pertussis toxin-sensitive Gi/Go protein (Pilon et al. 1994). These effects may be relevant for antipsychotic actions since Fos expression, thought to be a marker for neuronal activity (Morgan and Curran 1989) is activated in limbic areas of rat brain expressing the D_3 receptor, particularly the "atypical" neuroleptics (Robertson and Fibiger 1992). In addition, in immortalized mesencephalic neurons (MN9D cells) transfected with the D_3 receptor, prolonged application of dopamine or quinpirole induced a proliferation of neurites, whereas this effect is marginal and absent in cells expessing the D_2 receptor and the D_4 receptor, respectively (Todd et al. 1992). These studies using cultured cells suggest that the D_3 receptor may be involved in neuronal cell multiplication and plasticity through primary biochemical events which remain to be identified. Such long-term regulation of cell activities is also consistent with a prolonged occupancy of the D_3 receptor, at which DA displays a very high affinity, implying a slow dissociation of the endogenous ligand.

Dopamine Receptor Subtypes Differ Pharmacologically

The D_1-like and D_2-like subfamilies of DA receptors can be distinguished not only on genetic and molecular grounds, but also pharmacologically. Thus the "D_1-selective" agonist SKF 38393 and antagonist SCH-23390 are almost equipotent at D_1 and D_5 receptors, but 10^2–10^4 times less potent at D_2-like receptors; conversely, the antagonist sulpiride is approximately 100-fold more potent at D_2, D_3, or D_4 receptors than at D_1 or D_5 receptors (Table 2).

Within each subfamily the various subtypes differ in their apparent affinity for DA, which binds with micromolar Ki values at D_1 and D_2 receptors but with values five to ten times lower at D_4 and D_5 receptors and nearly 100-fold lower at D_3 receptors (Table 2). The unusually high affinity of DA for the D_3 receptor found, including on native receptors present on brain membranes (Levesque et al. 1992) is rather intriguing but it may reflect the ability of the neurotransmitter to act at greater distance or to remain bound for longer periods than at other receptors. Among DA agonists, only apomorphine displays similar affinities at all receptor subtypes; other agonists seem relatively specific for only a subset of receptors, e.g., quinpirole at D_3 and D_4 or bromocriptine at D_2 and D_3 receptors. This selectivity profile is likely to have therapeutic implications regarding efficacy and side effects for these agents when used to treat Parkinson's disease.

Until recently, it was generally assumed that neuroleptics derived their antipsychotic activity from D_2 receptor blockade (Seeman 1980). Nevertheless,

Table 2. Pharmacology of DA receptor subtypes

	D_1-like receptors		D_2-like receptors		
	D_1	D_5	D_2	D_3	D_4
Agonists					
Dopamine[a]	2000	250	2000	30	540
7-OH-DPAT	5300	–	60	0.7	650
Apomorphine	700	400	70	70	4[b]
Bromocriptine	700	500	5	7	300
Pergolide	1400	900	20	2	–
Quinpirole	14000	–	1400	40	50
SKF 38393	150	100	10000	5000	10000
Antagonists					
Haloperidol	30	40	0.6	3	5
Pimozide	–	–	10	11	40
Thioproperazine	–	–	0.5	1	50
(–)Sulpiride	40000	80000	10	20	50
UH 232	–	–	40	10	–
Clozapine	140	250	70	500	9
SCH 23390	0.3	0.3	1000	1000	3500

[a] Values in presence of guanyl nucleotide.
[b] High-affinity component. Values shown are inhibition constants expressed in nM, and were obtained with transfected cell lines expressing a single human DA receptor subtype (van Tol et al. 1991; Sunahara et al. 1991; Sokoloff et al. 1992), except for values of 7-OH-DPAT which are from the rat (Levesque et al. 1992).

the discovery of several D_2-like receptors, well recognized by these drugs, raises the question: what is(are) the cerebral target(s) responsible for this antipsychotic activity? Positron emission tomography allows the direct measurement of D_2 receptor occupancy in the brain of human subjects receiving various antipsychotics at therapeutic doses (Farde et al. 1989; Baron et al. 1989). From these data we have calculated what the occupancy by neuroleptics of D_3 and D_4 receptors would be, taking into account their respective in vitro affinities measured on cloned receptors (Table 3). Antipsychotic effects are obtained while 40%–80% of D_2 receptors are occupied, which also corresponds to a significant occupancy of D_3 receptors (Table 3). This holds true even in the case of haloperidol, which displays a rather high preference for the D_2 receptor, and it therefore cannot be excluded that the antipsychotic activity is related to occupancy of D_3 receptors, a hypothesis consistent with their distribution (see below). This may not hold true in all cases for the D_4 receptor for which several very potent antipsychotics, among which thioproperazine, have about a 100-fold lower affinity. Hence the D_2, and/or the D_3 receptor, remains the most likely common target for most antipsychotics.

Nevertheless, differential blockade of D_2, D_3, and D_4 receptors may account for the therapeutic profile of some antipsychotics. As clozapine displays significantly higher affinity at D_4 than at D_2 or, even more, D_3 receptors (Table 2),

Table 3. DA receptor subtype occupancy in human brain in vivo

	Receptor occupancy		
	D_2 receptor (%)	D_3 receptor %	D_4 receptor %
Haloperidol (3 mg)	87	58	44
Pimozide (4 mg)	77	75	43
Chlorpromazine (100 mg)	80	61	20
Sulpiride (400 mg)	74	59	27
Raclopride (4 mg)	72	57	2
Clozapine (300 mg)	65	21	93

The occupancy of D_2 receptors is a direct evaluation derived from positron emission tomography scan using [^{11}C]raclopride (Farde et al. 1989). Taking into account these data and the Ki values of neuroleptics measured in vitro on cloned receptors, the theoretical occupancies of D_3 and D_4 receptors were then calculated.

its "atypical" properties may derive from a rather selective blockade of D_4 receptors as proposed (van Tol et al. 1991). Note, however, that under optimal conditions, clozapine displays a similar affinity at D_4 and the short isoform of the human D_2 receptor (Malmberg et al. 1993). On the other hand, pimozide, which is almost as potent at D_3 as at D_2 receptors and much less potent at D_4 receptors (Table 2), belongs to a group of antipsychotics with a "disinhibitory" therapeutic profile. Such a profile might be more pronounced with the new compounds AJ 76 and UH 232, the only antagonists with higher affinity at D_3 than D_2 receptors, which have behavioral stimulant properties in animals (Svensson et al. 1986).

Dopamine Receptor Subtypes Are Variously Localized to Dopamine-Responsive Brain Areas

Northern blot and in situ hybridization studies have established that the genes for the D_1 and D_2 receptors are the most abundantly expressed in brain, the corresponding mRNAs being detected easily in all main DA-responsive areas. These localizations are essentially consistent with the results of membrane-binding studies and autoradiography (Dawson et al. 1988; Bouthenet et al. 1987), which support the view that a major fraction of these receptors is expressed in areas of cell bodies or on short neurons.

The ubiquitous expression of D_1 and D_2 receptors contrasts with the lesser abundance of the other DA receptors. Localization studies of the D_3 receptor gene transcripts, using Northern or PCR analysis (Sokoloff et al. 1990), as well as in situ hybridization histochemistry (Bouthenet et al. 1991), or of the D_3 receptor protein, using autoradiography with [^{3}H]7OH-DPAT (Levesque et al. 1992) indicate that the distribution of this receptor is markedly more restricted than that of the D_1 or D_2 receptors. For instance, only a weak D_3 receptor hybridization signal or [^{3}H]7OH-DPAT binding (Fig. 2) was detected in some restricted parts of the striatum, while the whole striatum contains the highest densities of dopamine axons and D_1 and D_2 receptor mRNAs. By contrast, the D_3 receptor mRNA and protein are highly expressed in the olfactory tubercle islands of Calleja complex, the anterior and shell parts of nucleus accumbens, and the bed nucleus of stria terminalis (Fig. 2). These areas constitute, with the ventral and ventromedial parts of the caudate putamen, the "ventral striatum" (Heimer and Wilson 1975), a territory receiving afferents from the prefrontal or allocortex and amygdala and its major dopamine inputs from the A10 cell

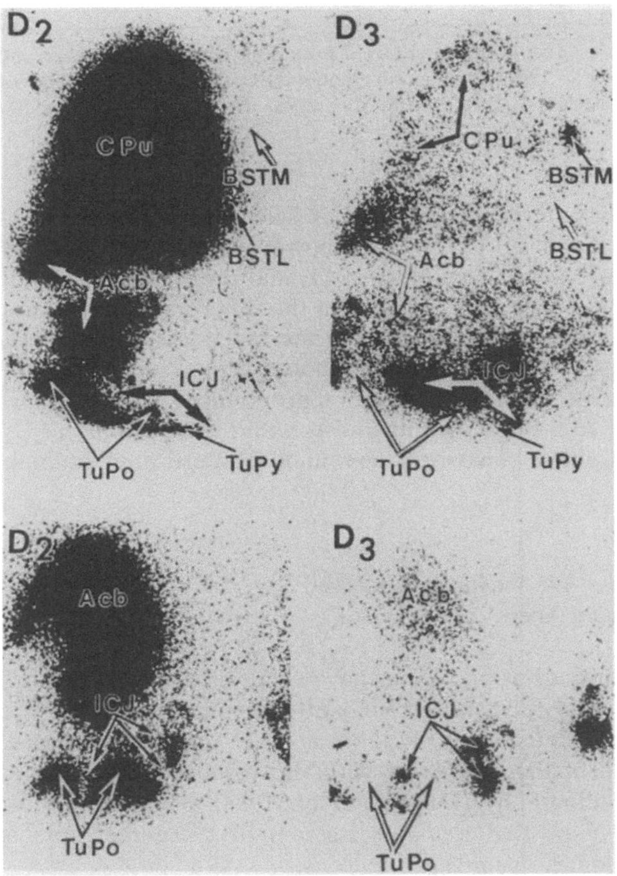

Fig. 2. Compared distributions of D_2 and D_3 receptor mRNAs established by in situ hybridization in sagittal (*top*) and frontal (*bottom*) sections of rat telencephalon. Note the non-overlapping complementary distributions of the two transcripts in the ventral striatum, particularly at the level of olfactory tubercle islands of Calleja and bed nucleus of the stria terminalis. *Acb*, accumbens nucleus; *BSTL*, *BSTM*, bed nucleus of the stria terminalis, lateral or medial part; *CPu*, caudate putamen; *ICj*, islands of Calleja; *TuPo*, *TuPy*, polymorph and pyramidal layers of the olfactory tubercle

group in the ventral tegmental area. It projects to ventral pallidum and the latter to the mediodorsal thalamic nucleus which selectively innervates the prefrontal cortex (Björklung and Lindvall 1984). This connectivity has led to the designation of this territory as the "limbic" part of the striatal complex, in which D_3 receptors may, therefore, mediate a large part of dopamine signals. The remaining of the striatal complex, which is mainly innervated by dopamine projections from the substantia nigra, receives its cortical inputs from the somatic neocortex and is highly enriched in D_1 or D_2 receptors.

D_3 receptor signals were also detected in other limbic areas such as the hippocampus, septum, or mammillary nuclei in the hypothalamus. D_1 and D_2 receptor mRNAs (Bouthenet et al. 1991; Fremeau et al. 1991; Meador-Woodruff et al. 1989) and, to a lesser extent, D_4 and D_5 mRNAs (Tiberi et al. 1991; van Tol et al. 1991; Sunahara et al. 1991) are also expressed in these areas but there is no strict overlap with D_3 receptor mRNA distribution. For instance, the highest levels of D_3 receptor mRNA in brain are detected in the islands of Calleja, in which the D_2 receptor signal is absent, whereas a reverse situation is found in the olfactory tubercles (Fig. 2). A more complete comparison of D_2 and D_3 receptor mRNA distributions reveals a seemingly mutually exclusive expression among brain regions (Bouthenet et al. 1991), suggesting that the two receptors are not expressed by the same cells.

Interestingly, no specific D_3 receptor signal could be detected by Northern and PCR analyses in the pituitary (Sokoloff et al. 1990; Bouthenet et al. 1991), which contains both D_2 and D_4 receptors. This allows the prediction that selective D_3 receptor ligands, when available in therapeutics, will not affect, like the currently used neuroleptics, the activity of mammotrophs.

In conclusion, localization studies of DA receptors support a major role for the D_3 receptor in dopaminergic neurotransmission in the phylogenetically old parts of the brain, namely the limbic areas, which are known to be involved in the control of cognitive, emotional, and reward processes. A variety of approaches indicate that these areas are more likely to be involved in the pathogenesis or treatment of schizophrenia than other dopaminoceptive areas, namely those implicated in motor control, which contain high levels of D_1 and D_2 receptors. Hence, the D_3 receptor distribution in the brain also supports a role of this receptor as a major target for antipsychotic action.

Dopamine Autoreceptors Belong to Diverse Subtypes

Dopamine autoreceptors are a functional class of receptors found in dendrites, cell bodies or axons of DA neurons and mediating inhibitions of these neuron activity or of neurotransmitter synthesis and release (Wolf and Roth 1987). These autoreceptors had been identified pharmacologically as D_2 receptors. Accordingly, DA neurons in the substantia nigra pars compacta and ventral tegmental area express D_2 mRNA (Mansour et al. 1990; Bouthenet et al. 1991), as confirmed by 6-hydroxydopamine lesions (Sokoloff et al. 1990), a localization which is likely to reflect autoreceptor biosynthesis. Then in substantia nigra pars compacta, particularly the lateral part, and in the ventral tegmental area, DA neurons were shown to express D_3 receptor mRNA, indicating that the D_3 receptor is also an autoreceptor (Sokoloff et al. 1990; Bouthenet et al. 1991).

D_2 and D_3 autoreceptors might exert their control on different sets of dopaminergic neurons. This latter hypothesis is supported by the non-overlapping localizations of the two autoreceptors in substantia nigra (Bouthenet et al. 1991). D_3 receptor mRNA is mainly, although faintly, expressed in the lateral part containing dopaminergic neurons projecting to colliculus and amygdala (Vankova et al. 1992), whereas the D_2 receptor signal is strongly expressed all over the whole compacta, where dopaminergic cells innervate various areas of the dorsal mesostriatal system (Björklund and Lindvall 1984). Finally, the question as to whether a single cell expresses both D_2 and D_3 receptors remains to be answered, namely by in situ hybridization studies at the cellular level.

According to the theory that schizophrenia might be due to DA hyperactivity, autoreceptor-selective DA agonists have been proposed as an alternative to neuroleptic medication (Tamminga et al. 1978; Carlsson 1988). However, confronted by the diversity of DA autoreceptors and by the fact that these receptors have their pharmacologically identical counterpart on postsynaptic neurons, the existence of autoreceptor-selective DA agonists may be questioned (Drukarch and Stoof 1990). Nevertheless, there is converging evidence suggesting that some D_3 receptor-selective agonists exert their main effects by reducing DA activity, although it is not clear whether pre- or postsynaptic receptors are involved. DA, as well as agonists such as TL99, quinpirole, or quinerolane, display very high affinities at D_3 receptors (Table 1). DA and DA agonists systemically administered in very low dosages reduce the activity of dopaminergic neurons (Skirboll et al. 1979), and several DA agonists, among those displaying selectivity for the D_3 receptor, act preferentially at presynaptic autoreceptors as judged in animal models, such as reversion of the butyrolactone-induced increase of dopamine synthesis (Martin et al. 1982). In agreement, AJ76 and UH232, the only antagonists exhibiting (limited) D_3 receptor selectivity, have behavioral stimulating properties in animals, presumably attributable to autoreceptor blockade (Svensson et al. 1986). Accordingly, D_3 receptor-selective agonists may be used in schizophrenia with the aim of decreasing dopaminergic output. However, preliminary open studies with postulated autoreceptor-selective DA agonists, among which are authentic D_3 receptor-selective compounds, yield inconsistent results, with only transient psychopathological improvement in some cases (Benkert et al. 1992). Nonetheless, the interest of D_3 receptor-selective agonists in schizophrenia might be revived when more selective compounds become available.

Regulation of DA Receptors

The changes in responsiveness or number of receptors, which occur mainly as a result of their blockade or neuronal input ablation, is a general phenomenon

(Cannon and Rosenblueth 1949) indicating that receptors are not fixed targets but rather macromolecules in a dynamic state. Prolonged interruption of DA neurotransmission by chronic treatment with neuroleptics or by lesions of dopaminergic neurons results in animals in behavioral supersensitivity to DA agonists and increased number of receptors (for reviews see Schwartz et al. 1980; Creese and Sibley 1980). Quantitation of DA receptors mRNAs provided direct evidence that the increases in responsiveness via increase in DA receptor number results from an increased rate of synthesis through activation of gene transcription. In accordance, chronic neuroleptic treatment enhances D_2 receptor mRNAs in various brain structures as assessed by Northern blot analysis (Autelitano et al. 1989; Martres et al. 1992) or in situ hybridization (Bernard et al. 1991; Kopp et al. 1992).

It is generally recognized, however, that no tolerance to the antipsychotic activity of neuroleptics develops in schizophrenic subjects, suggesting that D_2 receptors, in the brain structures where they are upregulated, do not constitute the target of neuroleptics regarding this antipsychotic activity. In contrast, prolonged treatment with these compounds elicits tardive dyskinesia with a prevalence ranging up to 40%. This condition has been attributed to development of hypersensitivity attributable to upregulation in the number of postsynaptic D_2 receptors in the nigrostriatal pathway (Klawans 1973).

We have recently compared the effects of a chronic haloperidol treatment on the level of D_2 and D_3 receptor binding and mRNA in various rat brain regions (Table 4). The drug was given at a high dosage, consistent with its blockade not only of the D_2 receptors, but also D_3 receptors, for which the drug displays 20-fold lower potency in the rat (Sokoloff et al. 1990). After a 2-week treatment haloperidol enhanced both D_2 receptor mRNA and D_2 receptor binding, the effect being specially marked in nucleus accumbens (+66%). In contrast, neither D_3 receptor binding nor mRNA changed in any brain region, suggesting that no upregulation develops at this receptor. This observation might be regarded as further support to the idea that the D_3 receptor is an important target for antipsychotics.

Dopamine Receptor Genes as Candidate Genes in Psychiatric Diseases

Family, twin, and adoption studies suggest that genetic factors contribute to the etiology of psychiatric disorders, although the mode of inheritance of these diseases in unknown (Gottesman 1991; Gershon 1990). Since dysfunctions of DA neurotransmission have been proposed as participating in these disorders, DA receptor genes are obvious candidate genes for genetic studies in psychiatry. The recent cloning of these genes and identification of polymorphisms in their sequence provide tools for linkage studies (in which the hypothesis of co-transmission of a given allele follows the transmission of the disease is tested within families) and for association studies (in which the frequency of a given allele is compared in an unaffected population versus a diseased population).

Table 4. Effects of a chronic haloperidol treatment on D_2 and D_3 receptor binding and mRNA

Brain area	D_2 receptor		D_3 receptor	
	mRNA (%)	Binding (%)	mRNA (%)	Binding (%)
Striatum	–	+24 ($p < 0.001$)	–	+8 (N.S.)
Nucleus accumbens	+44 ($p < 0.05$)	+66 ($p < 0.001$)	−3 (NS)	+7 (NS)
Olfactory tubercle	–	+11 ($p < 0.01$)	+2 (NS)	+3 (NS)

Rats were treated for 2 weeks by haloperidol (20 mg/kg s.c., twice a day) and killed 4 h or 60 h after the last injection for receptor mRNA or binding measurement, respectively. The mRNA levels were measured by quantitative polymerase chain reaction, using competition with an internal standard (Siebert and Larrick 1992). Respective D_2 and D_3 mRNA levels were normalized to β-actin mRNA level. Bindings to D_2 and D_3 receptors were measured using [^{125}I]iodosulpiride (Martres et al. 1985) and [^{3}H]7-hydroxy-N,N′-di-*n*-propylaminotetralin (Levesque et al. 1992), respectively. [^{125}I]iodosulpiride binding actually corresponds to binding to both D_2 and D_3 receptors, but the latter was found negligible in the studied areas (Levesque et al. 1992). Data shown were obtained from individual tissues of five to six animals for mRNA or tissues from five animals for binding measurement, respectively. Data expressed as variations over values from corresponding saline-treated animals. Statistical significance in the Student's *t* test is indicated. NS, non-significant.

So far, linkage studies of D_1, D_2, D_3, and D_4 receptor genes have been excluded in schizophrenia (Moises et al. 1991; Coon et al. 1993), indicating that these genes probably do not play a major role in the susceptibility to these diseases. Association studies may be more suitable than linkage studies to disclose discrete and minor effects of the genes, namely because no premise as regards the mode of inheritance of the disease has to be set. An association between the D_2 receptor gene and alcoholism, not revealed in linkage studies, was recently proposed (Blum et al. 1990; Uhl et al. 1992). Even though it is not established how sequence variations in the D_2 receptor gene and amog individuals contribute to vulnerability to alcohol abuse, these results constitute the very first example of involvement of a gene in a presumably polygenic psychiatric disease, for which environmental factors probably participate in the etiology.

No association was found between the DA receptor genes and schizophrenia if only allele frequencies are considered (Crocq et al. 1992; Jönsson et al. 1993; Nöthen et al. 1993). However, in the study by Crocq et al. (1992), gathering data of two independent association studies with schizophrenia, a significantly twofold higher risk of schizophrenia was found in homozygotes for the *Bal* I polymorphism in the D_3 receptor. The exact significance of this rather unexpected finding is not clear at the present time. Furthermore, the past history of false-positive results in psychiatric genetics suggests regarding this observation as preliminary, especially because similar studies were negative

(Jönsson et al. 1993; Nöthen et al. 1993), but it cannot be excluded that the D_3 receptor gene dysfunction makes a modest contribution to the etiology of schizophrenia. Replication and extension of this finding will be necessary to verify this hypothesis.

Conclusions

The emergence of three distinct D_2-like subtypes, D_2, D_3, and D_4, raises an important question: to which subtype can the various effects of neuroleptics be assigned, including beneficial therapeutic effects and unwanted side effects, both formerly attributed to the blockade of a single D_2 receptor? At this stage of our knowledge much caution is needed, but three clues point to the D_3 receptor as a key receptor in the antipsychotic effect: its recognition by neuroleptics, its selective expression in the limbic system, and its lack of upregulation under chronic blockade. Defects in limbic functions are likely to be involved in mood and cognitive disturbances. In addition, the peculiar mode of coupling of this receptor, whose stimulation may elicit long-lasting events, possibly affecting a variety of neuronal differentiation processes, also supports an implication in a disease in which developmental abnormalities are often assumed to contribute.

Nonetheless, the D_2 receptor is also expressed in limbic structures and is also (and even better) recognized by neuroleptics. Hence, it may also participate in the antipsychotic effect of these agents. However, its ubiquitous expression in the dopaminergic areas, especially the neostriatum and pituitary, suggests that its blockade is also responsible for the important motor and endocrine side effects of neuroleptics. In this respect, the marked upregulation of this receptor after chronic treatment by neuroleptics supports the idea that this receptor is also involved in tardive dyskinesia. The D_4 receptor is also blocked by some antipsychotics, e.g., clozapine, quite potently in some case, but it cannot be recognized as the common target for antipsychotics, since several compounds which are highly potent as antipsychotics are poorly recognized. To identify the molecular target(s) of antipsychotics in brain, it is obvious that more selective compounds than those presently available will be needed. Their design will be undoubtedly facilitated by the cloning and cellular expression of the novel DA receptor subtypes.

References

Autelitano DJ, Snyder L, Sealfon SC, Roberts JL (1989) Dopamine D_2-receptor messenger RNA is differentially regulated by dopaminergic agents in rat anterior and neurointermediate pituitary. Mol Cell Endocrinol 67:101–105

Baron JC, Martinot JL, Cambon H, Boulenger JP, Poirier MF, Caillard V, Blin J, Huret JD, Loc'h C, Maziere B (1989) Striatal dopamine receptor occupancy during and following withdrawal from neuroleptic treatment: correlative evaluation by positron emission tomography and plasma prolactin levels. Psychopharmacology 99:463–472

Benkert O, Gründer G, Wetzel H (1992) Dopamine autoreceptor agonists in the treatment of schizophrenia and major depression. Pharmacopsychiatry 25:254–260

Bernard V, Le Moine C, Bloch B (1991) Striatal neurons express increased level of dopamine D_2 receptor mRNA in response to haloperidol treatment: a quantitative in situ hybridization study. Neuroscience 45:117–126

Bertorello AM, Hopfield JE, Aperia A, Greengard P (1990) Inhibition by dopamine of (Na^+ + K^+) ATPase activity in neostriatal neurons through D_1 and D_2 dopamine receptor synergism. Nature 347:386–388

Björklund A, Lindvall O (1984) Dopamine-containing systems in the CNS. In: Björklund A, Hökfelt T (eds) Handbook of chemical neuroanatomy, vol 2. Classical transmitters in the CNS. Elsevier, Amsterdam, pp 55–122

Blum K, Noble EP, Sheridan PJ, Montgomery A, Ritchie T, Jagadeeswaran P, Nogami H, Briggs AH, Cohn JB (1990) Allelic association of human dopamine D_2 receptor gene in alcoholism. JAMA 263:2055–2060

Bouthenet ML, Martres MP, Sales N, Schwartz JC (1987) A detailed mapping of dopamine D_2 receptors in rat central nervous system by autoradiography with [^{125}I]iodosulpride. Neuroscience 20:117–155

Bouthenet ML, Souil E, Martres MP, Sokoloff P, Giros B, Schwartz JC (1991) Localization of dopamine D_3 receptor mRNA in the rat brain using in situ hybridization histochemistry: comparison with dopamine D_2 receptor mRNA. Brain Res 564:203–219

Bunzow JR, Van Tol HHM, Grandy DK, Albert P, Salon J, Christie McD, Machida CA, Neve KA, Civelli O (1988) Cloning and expression of a rat D_2 dopamine receptor cDNA. Nature 336:783–787

Cannon WB, Rosenblueth A (1949) The supersensitivity of denervated structures. Mac Millan, New York

Carlsson A (1988) The current status of the dopamine hypothesis of schizophrenia. Neuropsychopharmacology 1:179–186

Carlsson A, Lindqvist M (1963) Effect of chlorpromazine or haloperidol on formation of 3-methoxytyramine and normetanephrine in mouse brain. Acta Pharmacol Toxicol 20:140–144

Castro SW, Strange PG (1993) Differences in the ligand binding properties of the short and long versions of the D_2 dopamine receptor. J Neurochem 60:372–375

Clark D, White FJ (1987) Review: D_1 dopamine receptor. The search of a function: a critical evaluation of the D_1/D_2 dopamine receptor classification and its functional implication. Synapse 1:347–388

Cohen AI, Todd RD, Harmon S, O'Malley KL (1992) Photoreceptors of mouse retinas possess D_4 receptors coupled to adenylate cyclase. Proc Natl Acad Sci U S A 89:12093–12097

Coon H, Byerley W, Holik J, Hoff M, Myles-Worsley M, Lannfelt L, Sokoloff P, Schwartz J-C, Waldo M, Freedman R, Plaetke R (1993) Linkage analysis of schizophrenia with five dopamine receptor genes in nine pedigrees. Am J Hum Genet 52:327–334

Creese I, Sibley DR (1980) Receptor adaptations to centrally acting drugs. Annu Rev Pharmacol Toxicol 21:357–391

Creese I, Burt DR, Snyder SH (1976) Dopamine receptor binding predicts clinical and pharmacological potencies of antischizophrenic drugs. Science 192:481–483

Crocq MA, Mant R, Asherson P, Williams J, Hode Y, Mayerova A, Collier D, Lannfelt L, Sokoloff P, Schwartz J-C, Gill M, Macher JP, McGuffin P, Owen MJ (1992) Association between schizophrenia and homozygosity at the dopamine D_3 receptor gene. J Med Genet 29:858–860

Crow TJ (1982) The biology of schizophrenia. Experientia 38:1276–1282

Dal Toso R, Sommer B, Ewert M, Herb A, Pritchett DB, Bach A, Shivers BD, Seeburg PH (1989) The dopamine D_2 receptor: two mlecular forms generated by alternative splicing. EMBO J 8:4025–4034

Dawson TM, Barone P, Sidhu A, Wamsley JK, Chase TN (1988) The D_1 dopamine receptor in the rat brain: quantitative autoradiographic localization using an iodinated ligand. Neuroscience 26:83–100

Dearry A, Gingrich JA, Falardeau P, Fremeau RT, Bates MD, Caron MG (1990) Molecular cloning and expression of the gene for a human D_1 dopamine receptor. Nature 347:72–76

Delay J, Deniker P, Harl JM (1952) Utilisation en thérapeutique psychiatrique d'une phénothiazine d'action centrale élective (4560 RP). Ann Med Psychol 110:112–117

Drukarch B, Stoof JC (1990) D-2 dopamine autoreceptor selective drugs: do they really exist? Life Sci 47:361–376

Farde L, Wiesel FA, Hall H, Halldin C, Stone-Elander S, Sedvall G (1987) No D_2 receptor increase in PET study of schizophrenia. Arch Gen Psychiatry 44:671–672

Farde L, Wiesel FA, Nordström AL, Sedvall G (1989) D_1 and D_2-dopamine receptor occupancy during treatment with conventional and atypical neuroleptics. Psychopharmacology 99:S28–S31

Fishburn CS, Bellili D, David C, Carmon S, Fuchs S (1993) A novel short isoform of the D_3 dopamine receptor generated by alternative splicing in the third cytoplasmic loop. J Biol Chem 268:5872–5878

Fremeau RT, Duncan GE, Fornaretto MG, Dearry A, Gingrich JA, Brees GR, Caron MG (1991) Localisation of D_1 dopamine receptor mRNA in brain supports a role in cognitive, affective, and neuroendocrine aspects of dopaminergic neurotransmission. Proc Natl Acad Sci U S A 88:3772–3776

Gerfen CR, Engber TM, Mahan LC, Susel Z, Chase TN, Monsma FJ Jr, Sibley DR (1990) D_1 and D_2 dopamine receptor-regulated gene expression of striatonigral and striatopallidal neurons. Science 250:1429–1432

Gershon ES (1990) Genetics. In: Goodwin FK, Jamison KR (eds) Manic depressive illness. Oxford University Press, New York, pp 373–401

Giros B, Sokoloff P, Martres MP, Riou JF, Emorine LJ, Schwartz JC (1989) Alternative splicing directs the expression of two D_2 dopamine receptor isoforms. Nature 342:923–926

Giros B, Martres MP, Sokoloff P, Schwartz JC (1990) cDNA cloning of the human dopaminergic D_3 receptor and chromosome identification. CR Acad Sci [III] 311:501–508

Giros B, Martres MP, Pilon C, Sokoloff P, Schwartz JC (1991) Shorter variants of the D_3 dopamine receptor produced through various patterns of alternative splicing. Biochem Biophys Res Commun 176:1584–1592

Gottesman II (1991) Schizophrenia genesis: the origins of madness. Freedman, New York

Grandy DK, Marchionni MA, Makan H, Stofko RE, Alfano M, Prothingham L, Fischer JB, Burke-Howie KJ, Bunzow JR, Server AC, Civelli O (1989) Cloning of the cDNA and gene for a human D_2 dopamine receptor. Proc Natl Acad Sci USA 84:9762–9766

Grandy DK, Zhang Y, Bouvier C, Zhou Q, Johnson RA, Allen L, Buck K, Bunzow JR, Salon J, Civelli O (1991) Multiple human D_5 dopamine receptor genes: a functional receptor and two pseudogenes. Proc Natl Acad Sci USA 88:9175–9179

Heimer L, Wilson RD (1975) The subcortical projections of the allocortex: similarities in the neural associations of the hippocampus, the piriform cortex and the neocortex. In: Santini M (ed) Golgi Centennial symposium: perspectives in neurobiology. Raven, New York, pp 177–193

Henderson R, Unwin PN (1975) Three-dimensional model of purple membrane obtained by electron microscopy. Nature 257:28–32

Jaskiw GE, Weinberger DR (1992) Dopamine and schizophrenia – a cortically corrective perspective. Semin Neurosci 4:179–188

Jönsson E, Lannfelt L, Sokoloff P, Schwartz J-C, Sedvall G (1993) Lack of association between schizophrenia and alleles in the dopamine D_3 receptor gene. Acta Psychiatr Scand 87: 345–349

Kanterman RY, Mahan LC, Briley EM, Monsma FJ, Sibley DR, Axelrod J, Feldek CC (1991) Transfected D_2 dopamine receptors mediate the potentiation of arachidonic release in Chinese Hamster Ovary cells. Mol Pharmacol 39:364–369

Kebabian JW, Calne DB (1979) Multiple receptors for dopamine. Nature 277:93–96

Klawans HL (1973) The pharmacology of tardive dyskinesia. Am J Psychiatry 130:82–86

Kopp J, Lindefors N, Brene S, Hall H, Persson H, Sedvall G (1992) Effect of raclopride on dopamine D_2 receptor mRNA expression in rat brain. Neuroscience 47:771–779

Le Moine C, Normand E, Bloch B (1991) Phenotypical characterization of the rat striatal neurons expressing the D_1 dopamine receptor gene. Proc Natl Acad Sci USA 88:4205–4209

Mahan LC, Burch RM, Monsma FJ, Sibley DR (1990) Expression of striatal D_1 dopamine receptors coupled to inositol phosphate production and Ca^{2+} mobilization in Xenopus oocytes. Proc Natl Acad Sci USA 87:2196–2200

Malmberg A, Jackson DM, Eriksson A, Mohell N (1993) Unique binding characteristics of antipsychotic agents interacting with human dopamine D_{2A}, D_{2B} and D_3 receptors. Mol Pharmacol 43:749–754

Mansour A, Meador-Woodruff JH, Bunzow JR, Civelli O, Akil H, Watson SJ (1990) Localization of dopamine D_2 receptor mRNA and D_1 and D_2 receptor binding in the rat brain and pituitary. An in situ hybridization receptor autoradiographic analysis. J Neurosci 10:2587–2601

Martin GE, Williams M, Haubrich DR (1982) A pharmacological comparison of 6,7 dihydroxy-2 dimethylaminotetralin (TL 99) and N-n-propyl-3(3-hydroxyphenyl) piperidine (3-PPP) with selected dopamine agonists. J Pharmacol Exp Ther 223:298–304

Martinot JL, Paillere-Martinot ML, Loc'h C, Person Magnan P, Mazoyer B, Lecrubier Y, Hardy P, Beaufils B, Allilaire JF, Maziere B, Slama MF, Syrota A (1990) Central D_2 receptor blockade and antipsychotic effects of neuroleptics. Preliminary study with positon tomography. Psychiatr Psychobiol 5:231–240

Martres MP, Sokoloff P, Giros B, Schwartz JC (1992) Effects of dopaminergic transmission interruption on the D_2 receptor isoforms in various cerebral tissues. J Neurochem 58:673–679

Meador-Woodruff JH, Mansour A, Bunzow JR, Van Tol HHM, Watson SJ, Civelli O (1989) Distribution of D_2 dopamine receptor mRNA in rat brain. Proc Natl Acad Sci USA 86: 7625–7628

Moises HW, Gelernter J, Giuffra LA, Zarcone V, Wetterberg L, Civelli O, Kidd KK et al (1991) No linkage between D_2 dopamine receptor gene region and schizophrenia. Arch Gen Psychiatry 48:643–647

Monsma FJ, McVittie LD, Gerfen CR, Mahan LC, Sibley DR (1989) Multiple D_2 dopamine receptors produced by alternative RNA splicing. Nature 342:926–929

Monsma FJ, Mahan LC, McVittie LD, Gerfen CR, Sibley DR (1990) Molecular cloning and expression of a D_1 dopamine receptor linked to adenylyl cyclase activation. Proc Natl Acad Sci USA 87:6723–6727

Montmayeur JP, Guiramand J, Borelli E (1993) Preferential coupling between dopamine D_2 receptors and G-proteins. Mol Endocrinol 7:161–170

Morgan JI, Curran T (1989) Stimulus-transcription coupling in neurons: role of cellular immediate-early genes. Trends Neurosci 12:459–462

Neve KA, Henningen RA, Bunzow JR, Civelli O (1989) Functional characterization of a rat dopamine D_2 cDNA expressed in a mammalian cell line. Mol Pharmacol 36:446–451

Neve KA, Cox BA, Henningsen RA, Spanoyannis A, Neve RL (1991) Pivotal role for aspartate-80 in the regulation of dopamine D_2 receptor affinity for drugs and inhibition of adenylyl cyclase. Mol Pharmacol 39:733–739

Nöthen M, Körner J, Lannfelt L, Sokoloff P, Schwartz J-C, Lanczik M, Rietchel M, Cichon S, Kramer R, Fimmers R, Möller HJ, Beckman H, Propping P (1993) Lack of association between schizophrenia and alleles of the dopamine D_1, D_2, D_3 and D_4 receptor loci. Arch Gen Psychiatry (in press)

Pilon C, Lévesque D, Dimitriadou V, Griffon N, Martres M-P, Schwartz J-C, Sokoloff P (1994) Functional coupling of the human D_3 dopamine receptor in a transfected NG 108-15 neuroblastoma-glioma hybrid cell line. Eur J Pharmacol Mol Pharmacol Sect 268:129–139

Piomelli D, Greengard P (1990) Lipoxygenase metabolites of arachidonic acid in neuronal transmembrane signalling. Trends Pharmacol Sci 11:367–373

Piomelli D, Pilon C, Giros B, Sokoloff P, Martres MP, Schwartz JC (1991) Dopamine activation of the arachidonic acid cascade *via* a modulatory mechanism as a basis for D_1/D_2 receptor synergism. Nature 353:164–167

Roberts GW (1990) Schizophrenia: the cellular biology of a functional psychosis. Trends Neurosci 13:207–211

Robertson GS, Fibiger HC (1992) Neuroleptics increase C-fos expression in the forebrain: contrasting effects of haloperidol and clozapine. Neuroscience 46:315–328

Saibil H (1990) Cell and molecular biology of photoreceptors. Semin Neurosci 2:15–33
Schwartz JC, Llorens-Cortes C, Rose C, Quach TT, Pollard H (1980) Adaptative changes in neurotransmitter receptor mechanisms in the central nervous system. Prog Brain Res 58: 117–130
Seabrook GR, Patel S, Marwood R, Emms F, Knowles MR, Freedman SB, McAllister G (1992) Stable expression of human D_3 dopamine receptors in GH_4 Ci pituitary cells. FEBS Lett 312:123–126
Seeman P (1980) Brain dopamine receptors. Pharmacol Rev 32:229–313
Seeman P, Ulpian C, Bergeron C, Riederer P, Jellinger K, Gabriel E, Reynolds GP, Tourtelotte WW (1984) Bimodal distribution of dopamine receptor densities in brains of schizophrenics. Science 225:728–731
Seeman P, Niznik HB, Guan HC, Booth G, Ulpian C (1989) Link between D_1 and D_2 dopamine receptors is reduced in schizophrenia and Huntington diseased brain. Proc Natl Acad Sci USA 86:10156–10160
Siebert PD, Larrick JW (1992) Competitive PCR. Nature 359:557–558
Simon MI, Strathmann MP, Gautan N (1991) Diversity of G proteins in signal transduction. Science 252:802–808
Skirboll LR, Grace AA, Bunney BS (1979) Dopamine auto- and postsynaptic receptors: electrophysiological evidence for differential sensitivity to dopamine agonists. Science 206:80–82
Snyder LA, Roberts JL, Sealfon SC (1991) Alternative transcripts of the rat and human dopamine D_3 receptor. Biochem Biophys Res Commun 180:1031–1035
Sokoloff P, Giros B, Martres MP, Bouthenet ML, Schwartz JC (1990) Molecular cloning and characterization of a novel dopamine receptor (D_3) as a target for neuroleptics. Nature 347: 146–151
Sokoloff P, Andrieux M, Besançon R, Pilon C, Martres MP, Giros B, Schwartz JC (1992) Pharmacology of human D_3 dopamine receptor expressed in a mammalian cell line: comparison with D_2 receptor. Eur J Pharmacol Mol Pharmacol 225:331–337
Spano PF, Govoni S, Trabucchi M (1978) Studies on the pharmacological properties of dopamine receptors in various areas of the central nervous system. Adv Biochem Psychopharmacol 19:155–165
Sunahara RK, Niznik HB, Weiner DM, Stormann TM, Brann MR, Kennedy JL, Gelernter JE, Rozmahel R, Yang Y, Israel Y, Seeman P, O'Dowd BF (1990) Human dopamine D_1 receptor encoded by an intronless gene on chromosome 5. Nature 347:80–83
Sunahara RK, Guan HC, O'Dowd BF, Seeman P, Laurier LG, Ng G, George SR, Torchia J, Van Tol HHM, Niznik HB (1991) Cloning of the gene for a human dopamine D_5 receptor with higher affinity for dopamine than D_1. Nature 350:614–619
Svensson K, Johansson AM, Magnusson T, Carlsson A (1986) (+)-AJ 76 and (+)-UH 232: central stimulants acting as preferential dopamine autoreceptor antagonists. Naunyn Schmiedebergs Arch Pharmacol 334:234–245
Tamminga CA, Schaffer MH, Smith RC, Davis JM (1978) Schizophrenic symptoms improved with apomorphine. Science 200:567–568
Taylor CW (1990) The role of G proteins in transmembrane signalling. Biochem J 272:1–13
Tiberi M, Jarvie KR, Silvia C, Falardeau P, Gingrich JA, Godinot N, Bertrand L, Yang-Feng TL, Fremeau RT Jr, Caron MG (1991) Cloning, molecular characterization, and chromosomal assignment of a gene encoding a second D_1 dopamine receptor subtype: differential expression pattern in rat brain compared with the D_{1A} receptor. Proc Natl Acad Sci U S A 88:7491–7495
Todd RD, Tang L, Han S, Harmon S, Choi H, Heller A, O'Malley KO (1992) Characterization of transfected dopamine receptors in the immortalized mesencephalic cell line MN9D. Abstract of the 7th international catecholamine symposium
Uhl GR, Perisco AM, Smith SS (1992) Current excitement with D_2 dopamine receptor gene alleles in substance abuse. Arch Gen Psychiatry 49:157–160
Vallar L, Muca C, Magni M, Albert P, Bunzow J, Meldolesi J, Civelli O (1990) Differential coupling of dopaminergic D_2 receptors expressed in different cell types. J Biol Chem 265: 10320–10326

Vankova M, Arluison M, Leviel V, Tramu G (1992) Afferent connections of the rat substantia nigra pons lateralis with special reference to peptide-containing neurons of the amygdalo-nigral pathway. J Chem Neuroanat 5:39–50

Van Tol HHM, Bunzow JR, Guan HC, Sunahara RK, Seeman P, Niznik HB, Civelli O (1991) Cloning of the gene for a human dopamine D_4 receptor with high affinity for the antipsychotic clozapine. Nature 350:610–614

Van Tol HHM, Wu CM, Guan HC, Ohara K, Bunzow JR, Civelli O, Kennedy J, Seeman P, Niznik HB, Janovic V (1992) Multiple dopamine D_4 receptor variants in the human population. Nature 358:149–152

Wagner HN Jr, Weinberger DR, Kleinman JE, Casanova MF, Gibbs CJ Jr, Gur GE, Hornykiewicz O, Kuhar MJ, Pettegrew JW, Seeman P (1988) Neuroimaging and neuropathology. Schizophr Bull 14:383–397

Wolf ME, Roth RH (1987) Dopamine autoreceptors. In: Venter JC, Harrison LC (eds) Dopamine receptors. Receptor biochemistry and methodology, vol 8. Liss, New York, pp 45–96

Wong DF, Wagner HN Jr, Tune LE, Dannals RF, Pearlson GD, Links JM, Tamminga CA, Broussolle EP, Ravert HT, Wilson AA, Toung JKT, Malat J, Williams JA, O'Tuama LA, Snyder SH, Kuhar MH, Gjedde A (1986) Postron emission tomography reveals elevated D_2 dopamine receptors in drug-naive schizophrenics. Nature 234:1558–1563

Zhou QZ, Grandy DK, Thambi L, Kushner JA, Van Tol HHM, Cone R, Pribnow D, Salon J, Bunzow JR, Civelli O (1990) Cloning and expression of human and rat D_1 dopamine receptors. Nature 347:76–80

Increased Breakdown of Membrane Phospholipids in Schizophrenia: Implications for the Hypofrontality Hypothesis

W.F. Gattaz, J. Brunner, A. Schmitt, and A. Maras

Introduction

Phospholipase A_2 (PLA_2) is a key enzyme in the metabolism of phospholipids. The extracellular PLA_2 is secreted in the blood by the pancreas as a digestive enzyme (type I) or is released from activated platelets (type II). Intracellular PLA_2 has been found in all cells so far investigated: the enzyme is enriched in brain membranes, where it catalyzes the hydrolysis of membrane phospholipids to release free fatty acids and cytotoxic products such as lysophosphatidylcholine. PLA_2 plays an essential role in the regulation of the physicochemical properties of the neuronal membrane, which influences among others receptor function and signal transduction (for review see Farooqui et al. 1992). In brain membranes PLA_2 has been found to inhibit dopamine-sensitive adenylate cyclase (Anand-Srivastava and Johnson 1981) and to decrease [^{3}H] spiperone binding to dopamine receptors (Oliveira et al. 1984).

In two independent studies we found a significant increase in plasma and serum PLA_2 activity from drug-free schizophrenics as compared to healthy and to non-schizophrenic psychiatric controls (Gattaz et al. 1987, 1990). These results were recently confirmed by Noponen et al. (1993). However, in the latter increased serum PLA_2 activity was also observed in non-schizophrenic psychiatric patients, challenging thus the specificity of the findings for schizophrenia within the scope of psychiatric diagnoses investigated. Because extracellular PLA_2 is a heterogeneous class of enzymes, it is not unlikely that in both laboratories, at least in part, different enzyme activities were determined. In a third study with some methodological limitations (over 2 years difference in the storing time of the serum between patients and controls) no differences were found in serum PLA_2 activity between six schizophrenic patients and ten healthy controls (Albers et al. 1993).

In the first study described below we investigated in schizophrenia directly the *intracellular*, membrane-bound PLA_2 activity. We determined the enzyme activity in platelets, which are considered adequate peripheric models for neurons, as both have been shown to share some membrane and receptor properties (Pletscher 1986).

Neurobiology Unit, Central Institute of Mental Health, P. O. Box 12 21 20, 68072 Mannheim, Germany

H. Häfner and W.F. Gattaz (Eds.)
Search for the Causes of Schizophrenia, Vol. III

In the second study we investigated in an animal model the effects of stereotaxic intranigral injections of purified PLA_2 on dopamine (DA)-mediated behavior in rats.

Study I

Methods

Probands. The sample comprised 31 DSM-III-R paranoid schizophrenic patients (ten male, 21 female; mean age 31 ± 8 years); schizophrenics were paired to 31 healthy controls and to 31 non-schizophrenic psychiatric patients matched by age (±1 year) and gender. All schizophrenic patients and psychiatric controls were at least 1 week drug free. Probands with drugs and alcohol abuse and with medicaments known to influence PLA_2 were not included in the study. Fifteen schizophrenics were neuroleptic naive at the first episode of the disease. The DSM-III-R diagnoses of the psychiatric controls were as follows: 24 major depression, one bipolar disorder depressed, six personality disorder. After baseline determinations in all patients, in 20 schizophrenics the PLA_2 activity was again determined at weekly intervals during a 3-week standardized neuroleptic treatment with 10 mg haloperidol per day. During therapy the psychopathological changes were assessed weekly with the Brief Psychiatric Rating Scale (BPRS).

Laboratory Determinations. Platelets were obtained from 10 ml venous citrate blood, frozen immediately and stored at −70°C in a Tris-saccharose buffer (pH 7.4) until analyzed. Determination of PLA_2 activity was performed as described by Flesch et al. (1985). As enzyme substrate L-A-1-palmitoyl-2-arachidonyl-phosphatidylcholine was used, labeled in position 2 with [1-^{14}C]arachidonic acid (New England Nuclear). The assay mixture contained 0.1 *M* Tris-buffer (pH 9.0), 4 m*M* $CaCl_2$, 20 μg homogenate protein, and 0.06 μCi (1.035 nmol) phosphatidylcholine. After an incubation time of 30 min at 37°C, the reaction was stopped by adding a isopropanol–hydrochloric acid mixture. The liberated arachidonic acid was then extracted by the heptane-silica method with Dole's reagent followed by adsorption of phospholipids with a modified procedure according to Sundaram et al. (1978). The radioactivity of free fatty acid, measured in a Beckman liquid scintillation counter, was used for calculating the phospholipase activity, expressed as pmol/mg protein per minute. PLA_2 determinations were performed pairwise for each schizophrenic with the respective psychiatric and healthy controls. Data were analyzed by non-parametric tests (Wilcoxon test and Spearman correlation coefficients).

Results

Platelet PLA_2 activity was significantly increased in schizophrenics as compared to healthy controls ($p = 0.01$) and to non-schizophrenic psychiatric controls (p

= 0.03). The differences also remained when only the 15 first-onset schizophrenics were compared to their matched controls (Table 1). The enzyme activity did not differ between healthy and psychiatric controls. About two thirds of the schizophrenic patients had PLA_2 activity higher than their paired controls. In schizophrenics platelet PLA_2 activity did not correlate with the duration of the disease nor with number of previous psychiatric hospital treatments. There were no correlations between baseline enzyme activity and the psychopathological BPRS scores.

During 3 weeks on neuroleptic therapy, PLA_2 activity decreased significantly after the 1st week (15.3% ± 5.6% reduction from baseline, $p = 0.005$) and remained at this level throughout the treatment (14.6% ± 6.7% after 2 weeks and 15.0% ± 7.3% after 3 weeks).

The decrement in PLA_2 activity during neuroleptic therapy was not correlated with the improvement in psychopathology (BPRS total scores: baseline 58 ± 13, 1st week 46 ± 14, 2nd week 41 ± 13, 3rd week 38 ± 11).

Discussion

The present finding of increased platelet PLA_2 activity in schizophrenia is in line with our previous results in serum and in plasma (Gattaz et al. 1987, 1990). It is unlikely that the differences were secondary to a long-lasting disease or to the effects of previous drug treatment as they were also observed in the first-onset, neuroleptic-naive schizophrenics.

Also in agreement with our previous findings, no differences were found in the enzyme activity in platelets between non-schizophrenic psychiatric patients and healthy controls and no difference arose when only the depressed patients were compared to their controls (data not shown). This negative finding does not support the hypothesis advanced by Hibbeln et al. (1989) that disordered PLA_2 is a predisposing factor in affective illness. Thus, within the spectrum of psychiatric diagnoses investigated here, we can assume that increased platelet PLA_2 was not a result of non-specific stressors linked to mental disorder in general (such as arousal, anxiety, etc.). Increased PLA_2 in serum and plasma

Table 1. Platelet PLA_2 activity in schizophrenic patients and paired healthy and psychiatric controls

	Schizophrenics (pmol/mg protein/min) Means ± SD	Healthy controls (pmol/mg protein/min) Means ± SD	Psychiatric controls (pmol/mg protein/min) Means ± SD
All schizophrenics (n = 31)	28.4 ± 10.3[a]	24.1 ± 10.9	23.8 ± 8.3
First-onset schizophrenics (n = 15)	28.7 ± 11.3[b]	22.7 ± 8.2	22.0 ± 11.4

[a] $p = 0.01$ compared to healthy and $p = 0.03$ compared to psychiatric controls.
[b] $p = 0.06$ compared to healthy and $p = 0.04$ compared to psychiatric controls.

has been described in several medical conditions such as pancreatitis and inflammatory diseases (review in Chang et al. 1987). Concerning neuropsychiatric disorders, there is one report on increased brain PLA_2 activity in multiple sclerosis (Woelk and Peiler-Ichikawa 1974) and an activation of PLA_2 has been hypothesized in temporal lobe epilepsy (Simonato 1993). We have recently reported on decreased PLA_2 activity in the brains of Alzheimer's disease patients (Gattaz et al., in press).

Haloperidol treatment reduced the PLA_2 activity to the levels of controls. This finding is in line with our results in serum and plasma (Gattaz et al. 1987, 1990), but the inhibitory effect in platelets was less marked than in the extracellular enzyme activity. The inhibition of PLA_2 by different neuroleptics has been reported in several in vitro and animal experiments (Schröder et al. 1981; Aarsman et al. 1985; Taniguchi et al. 1988). The mechanism is ascribed to an interaction of neuroleptics with the cell membranes protecting them from the influence of PLA_2, probably through an interference with the substrate–enzyme interface or by changing the physical organization of the lipid bilayer in a way that promotes the desorption of the bound enzyme (Chang et al. 1987; Jain et al. 1989).

In an in vitro experiment Bennett et al. (1991) found no differences in the activity of PLA_2 in Epstein-Barr virus-transformed lymphoblasts of schizophrenic patients as compared with healthy controls. This finding makes it unlikely that a genetic defect in the structure of the enzyme itself would cause an increment in its activity in schizophrenia. The activity of the enzyme is modulated by a class of endogenous PLA_2-inhibiting proteins (PLIPs) that are under genetic control (review in Chang et al. 1987). Further studies should clarify whether a defect in the synthesis or release of PLIPs could disinhibit PLA_2 activity in schizophrenia.

Our results of increased platelet PLA_2 activity suggest an accelerated breakdown of membrane phospholipids in schizophrenia. This assumption is further supported by the findings in platelets of schizophrenics of decreased phosphatidylcholine and increased lysophosphatidylcholine concentrations, respectively, the precursor and breakdown product of membrane phospholipid metabolism by PLA_2 (Pangerl et al. 1991; Rotrosen and Wolkin 1987).

What would be the implications for schizophrenia if our findings in the blood are related to an increased PLA_2 activity in the brain? In three studies using ^{31}P-spectroscopy an accelerated breakdown of membrane phospholipids was demonstrated in the frontal cortex of drug-naive schizophrenics (Pettegrew et al. 1991; McClure et al., this volume; Williamson et al. 1991; Deicken et al. 1993). A similar acceleration of phospholipid metabolism in the frontal lobe was found in a post-mortem study in the brains of schizophrenics (Horrobin et al. 1991). Because a frontal dysfunction (hypofrontality) has been assumed in schizophrenia (Weinberger 1987; Robbins 1990), it is of interest to investigate whether an acceleration of phospholipid metabolism in the brain by increased PLA_2 may represent a possible mechanism leading to hypofrontality.

To clarify this question we investigated in a series of experiments in rats the effects of intracerebral injections of the enzyme on dopaminergic neurotransmission using the rotational model introduced by Ungerstedt and Arbuthnott (1970). The results are summarized below under study II.

Study II

Methods

Stereotaxic Surgery. Adult female Sprague Dawley rats (Charles River/Wiga, Sulzfeld) weighing 223 ± 10 g (mean ± SD) were anesthetized with 10 mg ketamine (Ketavet) s.c. per animal and 30 mg/kg pentobarbital (Nembutal) i.p. and mounted in a David Kopf stereotaxic frame with the incisor bar 5 mm above horizontal zero. The substances were injected unilaterally into the substantia nigra pars compacta (SNC) in a volume of 4 μl at a constant rate of 1 μl/min using a Hamilton microliter syringe (22 gauge). After completion of the injections, cannulas were left in place for additional 3 min. The stereotaxic coordinates of the SNC (AP: 2.2 mm, L: 1.6 mm, DV: 2.5 mm from interaural line) were taken from the atlas of De Groot (1959). To validate stereotaxic precision nine rats received a unilateral intranigral injection of 8 μg 6-hydroxydopamine (6-OHDA, Sigma) in 4 μl saline with 0.2 μg/μl ascorbic acid added. These animals were pretreated with 10 mg/kg desipramine (Thomae) p.o. 60 min prior to surgery to protect noradrenergic nigral neurons. With these stereotaxic coordinates the results of Ungerstedt and Arbuthnott (1970) could be replicated. Methamphetamine (5 mg/kg s.c.) induced an ipsilateral asymmetry ($p < 0.01$) with (mean ± SEM) 871 ± 69 ipsilateral and 4 ± 1 contralateral turns per hour, whereas apomorphine (APO) (0.5 mg/kg s.c.) induced a contralateral asymmetry with 133 ± 54 contralateral and 21 ± 9 ipsilateral turns per hour). In the following experiments we investigated the acute and subchronic effects of intranigral PLA_2 injections and the influence of neuroleptics on PLA_2-induced behavior. Control groups with saline-treated animals were introduced to clarify the effects of stereotaxic injections of vehicle itself on rotational behavior.

Experiment 1. PLA_2 isolated and purified from bovine pancreas (Sigma) with an enzymatic activity of 6.8 U/mg was unilaterally injected into the SNC in three different doses: 1 μg PLA_2 ($n = 12$), 3 μg PLA_2 ($n = 12$), and 5 μg PLA_2 ($n = 13$). One unit hydrolyzes 1 μmol L-α-phosphatidylcholine to L-α-lysophosphatidylcholine and a free fatty acid per minute at pH 8.0 and 37°C. The control group ($n = 12$) received unilateral intranigral vehicle injections (4 μl saline). Spontaneous rotational behavior was recorded for 30 min on days 1, 3, and 7 postoperatively. Seven days after stereotaxic surgery rotational behavior induced by 0.5 mg/kg APO s.c. was recorded in six animals from each group (the remaining animals were used for a histological investigation).

Experiment 2. Thirty-five rats received a unilateral intranigral injection of 5 μg PLA_2 (6.8 U/mg) dissolved in 4 μl saline. On the following day animals were randomly divided into the following groups: placebo (n = 12, 5 ml/kg saline i.p., latency: 30 min), clozapine [n = 12, 10 mg/kg clozapine (Sandoz, Basel) i.p., latency: 30 min] and haloperidol [n = 11, 0.25 mg/kg haloperidol (Haldol, Janssen) i.p., latency: 60 min]. Rotational behavior was recorded for 30 min.

Experiment 3. PLA_2 from bovine pancreas (Sigma) with a higher enzymatic activity (12 U/mg) was unilaterally injected into the SNC in two different doses: 20 μg PLA_2 (n = 11) and 100 μg PLA_2 (n = 12) dissolved in 4 μl saline. The control group (n = 11) received a unilateral intranigral injection of the vehicle (4 μl saline). Fourteen days after surgery spontaneous rotational behavior was recorded for 30 min. On days 21, 28, and 35 postoperatively rotational behavior was monitored for 60 min after stimulation with 0.5 mg/kg APO (Sigma) s.c., 5 mg/kg methamphetamine s.c., and 1 mg/kg APO s.c., respectively.

Rotational Assessments. Rotational behavior was recorded automatically in a rotometer. Only complete turns (360°) in each direction were counted.

Statistics. Data were analyzed by non-parametric tests. Mann-Whitney rank sum test was used for comparisons between different treatment groups. Rotational asymmetries (number of ipsilateral vs. contralateral turns) within the same group were analyzed using Wilcoxon matched pairs signed rank test.

Results

Experiment 1. On the first postoperative day a spontaneous contralateral asymmetry was observed in all groups with low-dose intranigral PLA_2 application as well as in saline-treated animals (range contralateral 34–43, ipsilateral 1–2 turns/30 min, $p < 0.001$ for all groups). The PLA_2 groups did not differ quantitatively from the saline group. The contralateral spontaneous asymmetry diminished in all treatment groups within the first postoperative week. One week after surgery 0.5 mg/kg APO induced an ipsilateral asymmetry only in PLA_2-treated rats. As no significant differences were found among the three PLA_2 dose groups (1, 3, and 5 μg/kg), data were pooled as presented in Fig. 1 (low PLA_2 group). No APO-induced asymmetry was observed in saline-treated animals.

Experiment 2. The finding of a spontaneous contralateral asymmetry ($p < 0.01$) 1 day after unilateral intranigral injection of 5 μg PLA_2 could be replicated in this experiment. Neuroleptic treatment reduced the number of contralateral turns by 78% after clozapine ($p < 0.001$) and 95% after haloperidol ($p < 0.001$, Fig. 2).

Experiment 3. There was no spontaneous rotational asymmetry 2 weeks after unilateral intranigral injection of 20 μg and 100 μg PLA_2. The administration of

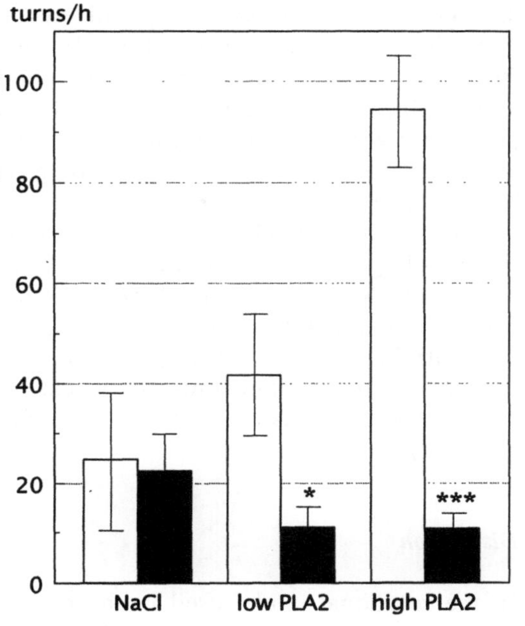

Fig. 1. Rotational behavior induced by 0.5 mg/kg apomorphine after intranigral injection of PLA_2 (1, 3, and 5 μg) and placebo (*NaCl*). Pooled data: *low PLA_2* group = 1, 3, and 5 μg; *high PLA_2* group = 20 and 100 μg. *Open columns*, ipsilateral turns; *solid columns*, contralateral turns. $^{*}p < 0.05$ and $^{***}p < 0.001$ compared to ipsilateral rotations

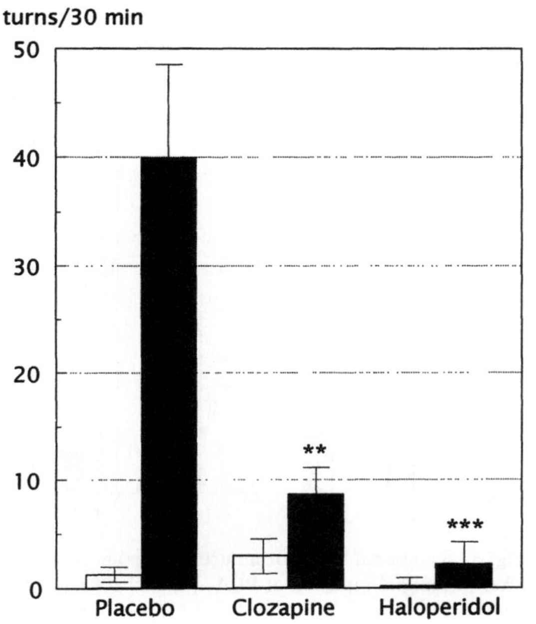

Fig. 2. Effects of clozapine (10 mg/kg) and haloperidol (0.25 mg/kg) on spontaneous contralateral asymmetry 1 day after intranigral injection of 5 μg PLA_2. *Open columns*, ipsilateral turns; *solid columns*, contralateral turns. $^{**}p < 0.01$ and $^{***}p < 0.001$ compared to placebo (NaCl)

0.5 mg/kg APO 3 weeks postoperatively caused a significant ipsilateral asymmetry in PLA_2-pretreated rats (Fig. 3), but not in the saline group (ipsilateral 56 ± 14, contralateral 29 ± 8 turns per hour, not significant). The APO-induced ipsilateral rotation counts in this experiment were similar in animals pretreated with 20 μg and 100 μg PLA_2, but were significantly higher than the ipsilateral counts in the low PLA_2 dose groups from experiment 1 ($p < 0.005$, Fig. 1). The administration of a higher APO dose (1.0 mg/kg) 5 weeks postoperatively in both PLA_2-pretreated groups caused an ipsilateral asymmetry that was larger than that induced by 0.5 mg/kg APO (Fig. 3). However, under this higher APO dose also rats pretreated with intranigral saline developed a smaller but still statistically significant ipsilateral asymmetry (ipsilateral 127 ± 23, contralateral 17 ± 7 turns per hour, $p < 0.05$). Similarly the administration of 5 mg/kg methamphetamine 4 weeks after surgery caused an ipsilateral asymmetry that was more marked in PLA_2 than in saline-pretreated rats (Table 2).

Discussion

In the rotational model used in the present study the direction of circling is ipsilateral to the hemisphere with the lower striatal dopaminergic activity

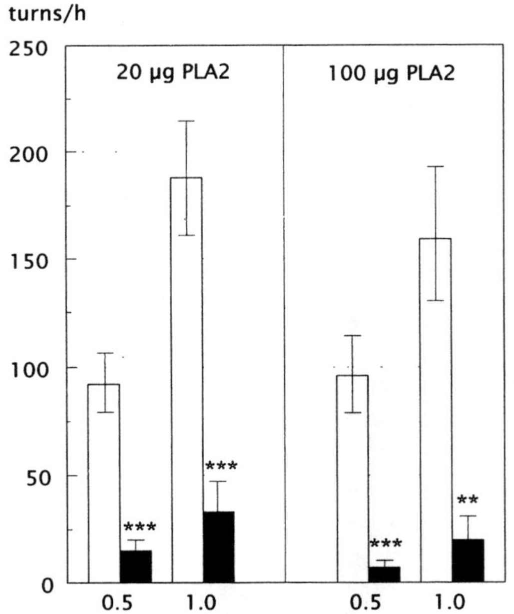

Fig. 3. Rotational behavior induced by apomorphine 3 weeks (0.5 mg/kg) and 5 weeks (1.0 mg/kg) after intranigral injection of PLA_2 (20 and 100 μg). *Open columns*, ipsilateral turns; *solid columns*, contralateral turns. $^{**}p < 0.01$ and $^{***}p < 0.001$ compared to ipsilateral rotations

Table 2. Ipsilateral asymmetry induced by 5 mg/kg methamphetamine 4 weeks after intranigral injections of PLA_2 or placebo

	Ipsilateral (Turns/h)	Contralateral (Turns/h)
NaCl ($n = 11$)	317 ± 30*	86 ± 38
20 μg PLA_2 ($n = 11$)	442 ± 38**	95 ± 27
100 μg PLA_2 ($n = 12$)	427 ± 61**	119 ± 23

*$p < 0.01$ and **$p < 0.001$ compared to contralateral rotations.

(Zetterström et al. 1986; Carman et al. 1991). One day after intranigral PLA_2 *and* saline injections there was a spontaneous contralateral rotational asymmetry that diminished within the first postoperative week. It can be assumed that this reversible acute effect resulted from a nonspecific dopaminergic stimulation caused by the fluid injections itself. In fact, the spontaneous contralateral circling 1 day after surgery could be antagonized by both typical (haloperidol) and atypical (clozapine) neuroleptics.

One week postoperatively APO at a dose of 0.5 mg/kg induced an ipsilateral asymmetry in rats treated with PLA_2. This effect was specific for PLA_2, as no asymmetry was induced by this APO dose in saline-treated animals. This asymmetry could be still observed 3 weeks after intranigral PLA_2 injection. The comparison between low and high PLA_2 dose groups showed that the ipsilateral asymmetry increased with the PLA_2 dose. These findings are in line with the results of Cadet et al. (1989) who observed an APO-induced ipsilateral circling behavior 10, 17, and 24 days after intranigral injection of PLA_2. The results from these animal experiments indicate that intracerebral injection of PLA_2 reduced dopaminergic activity. This assumption is supported by the findings of Cadet et al. (1989) who reported a reduction of the striatal concentrations of DA and its metabolites DOPAC and HVA on the side of the PLA_2 injection. An inhibitory effect of PLA_2 on dopaminergic neurotransmission has also been observed by in vitro experiments in which PLA_2 inhibited the activation of DA-sensitive adenylate cyclase in striatal tissue (Anand-Srivastava and Johnson 1981) and reduced the [^{3}H]spiperone binding to DA receptors (Oliveira et al. 1984).

In our experiments it is unlikely that intranigral PLA_2 injections caused extensive lesions of dopaminergic SNC neurons because PLA_2-treated rats showed a rotational behavior contrary to those with a selective dopaminergic SNC lesion induced by the neurotoxin 6-OHDA (see Methods). However, an alternative explanation for the APO-induced ipsilateral rotation after intranigral PLA_2 would be partial lesions of the substantia nigra pars reticulata (Olianas et al. 1978; Di Chiara et al. 1979; Reid et al. 1990). This alternative cannot be excluded without immunohistochemical morphological investigations.

In previous experiments from our laboratory no rotational asymmetry was induced by 2 mg/kg methamphetamine in rats pretreated with intranigral PLA_2 injections (unpublished data). In the present study the asymmetry induced by the higher methamphetamine dosis (5 mg/kg) was also ipsilateral, suggesting thus that PLA_2 affected the dopaminergic pathways at both pre- and post-synaptic levels. However, this interpretation must be seen with caution, as the use of higher doses of both methamphetamine and APO in our study induced ipsilateral asymmetry in animals treated with intranigral saline as well. Albeit smaller than in PLA_2-pretreated animals, this asymmetry suggests that the stereotaxic injection per se may cause discrete DA lesions that can be manifest in rotational behavior by the use of higher doses of DA agonists.

Conclusion

The results from our studies, together with data from the literature, suggest that (a) at least a subgroup of schizophrenic patients present in blood cells an accelerated breakdown of membrane phospholipids caused by increased PLA_2 activtity; and (b) in animal experiments a stimulation of the phospholipid breakdown by intracerebral PLA_2 injections reduced dopaminergic neurotransmission. How could these findings be involved in the pathophysiology of schizophrenia? A reduced dopaminergic activity in the prefrontal cortex (hypofrontality) has been postulated in schizophrenia (Weinberger 1987). An accelerated breakdown of membrane phospholipids in the prefrontal cortex of drug-naive schizophrenics was shown in vivo using ^{31}P NMR spectroscopy (Williamson et al. 1991; Deicken et al. 1993; Pettegrew et al. 1991; McClure et al., this volume). If this finding is caused by increased PLA_2 activity in the brain from schizophrenics, as observed in their peripheric cells (study I), then it is conceivable that increased PLA_2 activity may be related to the postulated hypodopaminergic activity in the mesocortico-prefrontal system in schizophrenia. This conclusion is highly speculative and can only be understood as a preliminary working hypothesis for the development of new studies.

It is prudent at the end to state that we, as many others, are also aware that schizophrenia is likely to be a heterogeneous disorder, so that a disordered phospholipid metabolism could at best characterize one subgroup of the disease. In an ongoing experiment we are investigating how far our findings on PLA_2 and phospholipids in platelet membranes can be used as a peripheric marker for the phospholipid metabolism in the brain. A relationship between central and peripheric enzyme activity is not unlikely in face of the endogenous, in part genetically determined mechanisms controlling the enzyme activity. In fact, after reporting on a significant decrease of PLA_2 activity in the brain of Alzheimer's disease patients (Gattaz et al., in press), we recently confirmed the same decrease in platelet membranes, suggesting thus that a relationship may exist between disordered brain and peripheric PLA_2 activity.

Acknowledgements. This work was supported by the Deutsche Forschungsgemeinschaft, SFB 258, Project S4. We are indebted to Prof. E. Ferber, Max Planck Institute for Immunobiology, Freiburg, and Prof. W.E. Müller, Department of Psychopharmacology at the CIMH, Mannheim, for the support and advice in the development of the laboratory assay. Thanks are due to Dr. Teschendorf and Dr. Gross (Knoll AG, Ludwigshafen) for valuable support with the animal experiments and to Mrs. Rosemarie Krämer for assistance in the experimental work.

References

Aarsman AJ, Roosenboom CFP, van Geffen GEW, van den Bosch H (1985) Some aspects of rat platelet and serum phospholipase A_2 activities. Biochim Biophys Acta 837:88–295

Albers M, Meurer H, Märki F, Klotz J (1993) Phospholipase A_2 activity in serum of neuroleptic-naive psychiatric inpatients. Pharmacopsychiatry 26:4–98

Anand-Srivastava MB, Johnson RA (1981) Role of phospholipids in coupling of adenosine and dopamine receptors to striatal adenylate cyclase. Neurochemistry 36:1819–1828

Bennett ER, Yedgar S, Lerer B, Ebstein RP (1991) Phospholipase A_2 activity in Epstein-Barr virus-transformed lymphoblast cells from schizophrenic patients. Biol Psychiatry 29:1058–1062

Cadet JL, Hu M, Jackson-Lewis V (1989) Behavioral and biochemical effects of intranigral injection of phospholipase-A_2. Biol Psychiatry 26:106–110

Carman LS, Gage FH, Shults CW (1991) Partial lesion of the substantia nigra: relation between extent of lesion and rotational behavior. Brain Res 553:275–283

Chang J, Musser JH, McGregor H (1987) Phospholipase A_2: function and pharmacological regulation. Biochem Pharmacol 36:2429–2436

De Groot J (1959) The rat forebrain in stereotaxic coordinates. Verhandeligen der koninklijke Nederlandse Akademie van Wetenschappen, afd. Natuurkunde. N.V. Noord-Hollandische Uitgevers Maatschappij, Amsterdam

Deicken RF, Merrin E, Calabrese G, Dillon W, Weiner MW, Fein G (1993) ^{31}Phosphorous MRSI of the frontal and parietal lobes in schizophrenia. Biol Psychiatry 33:46A

Di Chiara G, Porceddu ML, Morelli M, Mulas ML, Gessa GL (1979) Substantia nigra as an output station for striatal dopaminergic responses: role of a GABA-mediated inhibition of pars reticulata neurons. Naunyn Schmiedebergs Arch Pharmacol 306:153–159

Farooqui AA, Hirashima Y, Horrocks LA (1992) Brain phospholipases and their role in signal transduction. In: Bazan NG, Murphy MG, Tofano G (eds) Neurobiology of essential fatty acids. Plenum, New York, pp 11–26

Flesch I, Schmidt B, Ferber E (1985) Acylchain specifity and kinetic properties of phospholipase A_1 and A_2 of of bone marrow-derived macrophages. Z Naturforsch 40c:356–363

Gattaz WF, Köllisch M, Thuren T, Virtanen JA, Kinnunen PKJ (1987) Increased plasma phospholipase-A2 activity in schizophrenic patients: reduction after neuroleptic therapy. Biol Psychiatry 22:421–426

Gattaz WF, Hübner CK, Nevalainen TJ, Thuren T, Kinnunen PKJ (1990) Increased serum phospholipase A_2 activity in schizophrenia: a replication study. Biol Psychiatry 28:495–501

Gattaz WF, Maras A, Cairns NJ, Levy R, Förstl H (1994) Decreased phospholipase A_2 activity in Alzheimer brains. Biol Psychiatry (in press)

Hibbeln JR, Palmer JW, Davis JM (1989) Are disturbances in lipid-protein interactions by phospholipase A_2 a predisposing factor in affective illness? Biol Psychiatry 25:945–961

Horrobin DF, Manku MS, Hillman H, Iain A, Glen M (1991) Fatty acid levels in the brains of schizophrenics and normal controls. Biol Psychiatry 30:795–805

Jain MK, Yuan W, Gelb MH (1989) Competitive inhibition of phospholipase A_2 in vesicles. Biochemistry 28:4135–4139

Noponen M, Sanfilipo M, Samanich K, Ryer H, Ko G, Angrist B, Wolkin A, Rotrosen J (1993) Elevated PLA_2 activity in schizophrenics and other psychiatric patients. Biol Psychiatry 34: 641–649

Olianas MC, De Montis GM, Concu A, Tagliamonte A, Di Chiara G (1978) Intranigral kainic acid: evidence for nigral non-dopaminergic neurons controlling posture and behavior in a manner opposite to the dopaminergic ones. Eur J Pharmacol 49:223–232

Oliveira CR, Duarte EP, Carvalho AP (1984) Effect of phospholipase digestion and lysophosphatidylcholine on dopamine receptor binding. J Neurochem 43:455–465

Pangerl AM, Steudle A, Jaroni HW, Rüfer R, Gattaz WF (1991) Increased platelet membrane lysophosphatidylcholine in schizophrenia. Biol Psychiatry 30:837–840

Pettegrew JW, Keshavan MS, Panchalingam K, Strychor S, Kaplan DB, Tretta MG, Allen M (1991) Alterations in brain high-energy phosphate and membrane phospholipid metabolism in first-episode, drug naive schizophrenics. Arch Gen Psychiatry 48:563–568

Pletscher A (1986) Blood platelets as neuronal models: use and limitations. Clin Neuropharmacol 9[Suppl 4]:344–346

Reid MS, Herrera-Marschitz M, Hökfelt T, Lindefors N, Persson H, Ungerstedt U (1990) Striatonigral GABA, dynorphin, substance P and neurokinin A modulation of nigrostriatal dopamine release: evidence for direct regulatory mechanisms. Exp Brain Res 82:293–303

Robbins TW (1990) The case for frontostriatal dysfunction in schizophrenia. Schizophr Bull 16:391–402

Rotrosen J, Wolkin A (1987) Phospholipid and prostaglandin hypotheses of schizophrenia. In: Meltzer HY (ed) Psychopharmacology: the third generation of progress. Raven, New York, pp 759–764

Schröder T, Lempinen M, Nordling S, Kinnunen PKJ (1981) Chlorpromazine treatment of experimental acute fulminant pancreatitis in pigs. Eur Surg Res 13:143–151

Simonato M (1993) A pathogenetic hypothesis of temporal lobe epilepsy. Pharmacol Res 27: 217–225

Sundaram GS, Shakir KMM, Barnes G, Margolis S (1978) Release of phospholipase A_2 and triglyceride lipase from rat liver. J Biol Chem 253:7703–7710

Taniguchi K, Urakami M, Takanaka K (1988) Effects of various drugs on superoxide generation, arachidonic acid release and phospholipase A_2 in polymorphonuclear leukocytes. Jpn J Pharmacol 46:275–284

Ungerstedt U, Arbuthnott GW (1970) Quantitative recording of rotational behavior in rats after 6-hydroxydopamine lesions of the nigrostriatal dopamine system. Brain Res 24:485–493

Weinberger DR (1987) Implications of normal brain development for the pathogenesis of schizophrenia. Arch Gen Psychiatry 44:660–669

Williamson P, Drost D, Stanley J, Carr T, Morrison S, Merskey H (1991) Localized Phosphorus 31 Magnetic Resonance Spectroscopy in chronic schizophrenic patients and normal controls. Arch Gen Psychiatry 48:578

Woelk H, Peiler-Ichikava K (1974) Zur Aktivität der Phospholipase A_2 gegenüber verschiedenen 1-Alk-1′-enyl-2-acyl- and 1-Alkyl-2-acyl-Verbindungen wärend der multiplen Sklerose. J Neurol 207:319–326

Zetterström T, Herrera-Marschitz M, Ungerstedt U (1986) Simultaneous measurement of dopamine release and rotational behaviour in 6-hydroxy-dopamine denervated rats using intracerebral dialysis. Brain Res 376:1–7

^{31}P Magnetic Resonance Spectroscopy Study of Brain Metabolism in Schizophrenia

R. McClure, M. Keshavan, N.J. Minshew, K. Panchalingam, and J.W. Pettegrew

Introduction

The underlying neurobiological basis of schizophrenia remains poorly understood after over 30 years of intensive research. Although progress has been made in neuroanatomical, neurochemical, neurophysiolgic, and genetic research in schizophrenia, only a few integrative models have been proposed (for reviews see Bachneff 1991; Csernansky et al. 1991; Hoffman and McGlashan 1993; Keshavan and Ganguli 1990; Wyatt et al. 1989). Anatomical research has provided evidence that the frontal and limbic systems and their interconnections are involved in schizophrenia. However, structural changes in other brain areas have been reported and, taken together, the anatomical studies do not support a localized structural abnormality in schizophrenia. Cerebral metabolic functional imaging studies, discussed below, also provide evidence for frontal lobe involvement in schizophrenia. From a pathophysiological perspective, the dopaminergic system has been implicated (Meltzer and Stahl 1976; Seeman 1987), However, extensive research in this area has not provided unequivocal support for the classical dopamine hypothesis (Carlsson 1988; Crow 1987; Henley 1991; Reynolds 1989). The argument for the involvement of dopaminergic pathways in schizophrenia comes largely from the analysis of the mechanisms of action of antipsychotic drugs. The glutamatergic neurotransmitter system also has been implicated in schizophrenia (Kim et al. 1980). Findings of studies of the glutamatergic system in schizophrenia indicate that subtle imbalances of excitatory amino acid receptor types exist in schizophrenia which may influence synaptic events (Ulas and Cotman 1993). Reported membrane alterations in schizophrenia may impact membrane bound neurotransmitter receptors in both systems. Changes in membrane metabolite concentrations in schizophrenia provide valuable biochemical evidence of altered phospholipid metabolism, which may reflect abnormal neurodevelopment in schizophrenia. From an etiological perspective, genetic factors, abnormal neurodevelopment, and their interaction have been strongly implicated in schizophrenia (Weinberger 1987). Our search for the cause of schizophrenia utilizes ^{31}P magnetic resonance spectroscopy (MRS), a unique tool for

Neurophysics Laboratory, Department of Psychiatry, Western Psychiatric Institute and Clinic, 3811 O'Hara Street, Pittsburgh, PA 15213, USA

H. Häfner and W.F. Gattaz (Eds.)
Search for the Causes of Schizophrenia, Vol. III

obtaining in vivo quantitative information concerning membrane and energy metabolites (Pattegrew and Minshew 1992; Pettegrew et al. 1991, 1993a). In this chapter we describe the chemical information available from both in vitro and in vivo ^{31}P MRS and its relevance to neurodevelopment and aging. The findings of an in vivo ^{31}P MRS study of the frontal cortex of schizophrenic patients are described.

Membrane-Related Alterations in Schizophrenia

Neuronal cellular membranes, phospholipid bilayer structures, are responsible for sensing molecules and ions in the extracellular environment and modulating the activity of genetic and metabolic machinery in response to changes in the extracellular world. Also, neuronal intracellular metabolic and genetic changes are reflected in the molecular processing and activity modulation of transmembrane proteins that function as receptors and ion channels. The neuronal membrane also is responsible for communication between neurons through neurotransmitters and neuromodulators. Therefore, normal functioning of this membrane is especially important for normal brain function. Membrane abnormalities in schizophrenia may contribute to the pathophysiology of major psychoses. Evidence that schizophrenia is associated with membrane phospholipid alterations in peripheral cells (for a review see Rotrosen and Wolkin 1987) has stimulated our studies of brain membrane metabolites using ^{31}P MRS. For example, phospholipid alterations in peripheral cells such as erythrocytes have shown phosphatidylcholine (PtdC) to be reduced in some patients with schizophrenia (Henn 1980; Hitzemann et al. 1984; Stevens 1972).

Increases in phosphatidylserine (PtdS) and smaller decreases in phosphatidylethanolamine (PtdE) also have been noted, but less consistently replicated (Rotrosen and Wolkin 1987). Such abnormalities appear to be independent of previous drug treatment (Stevens 1972). Phospholipase A_2 activity in plasma, which alters phospholipid fatty acid composition through the deacylation–reacylation cycle, has been found to be increased in some schizophrenics (Gattaz et al. 1987). In addition, Kaiya et al. (1984) have described abnormalities in the phosphyatidylinositol (PtdI) cycle in a subgroup of schizophrenic patients. Support for drawing analogies between peripheral blood elements and brain tissue is derived from observed membrane abnormalities in peripheral cells in Huntington's chorea, myotonic dystrophy, affective illness, and Alzheimer's disease (Blass et al. 1985; Butterfield and Markesbery 1980; Butterfield et al. 1977, 1978, 1985; Diamond et al. 1983; Markesbery et al. 1980; Miller et al. 1989; Pettegrew et al. 1979a, b, c, 1981, 1982, 1983b, 1993b; Sherman et al. 1986; Zubenko et al. 1987). In vivo ^{31}P MRS is uniquely suited for studies of phospholipid metabolites of neural membranes.

Impaired Frontal Lobe Function in Schizophrenia

In our in vivo ^{31}P MRS study of schizophrenia we investigate energy and membrane metabolite changes in the dorsolateral prefrontal cortex area of the brain. This area is of importance since several lines of evidence indicate impaired frontal lobe function ("hypofrontality") in some schizophrenic patients (Buchsbaum et al. 1982; Weinberger et al. 1986). Cerebral blood flow studies using ^{133}xenon demonstrate a reduced ratio of frontal lobe to whole surface blood flow in schizophrenic patients under resting conditions (Ingvar and Franzen 1974) and during cortical activation with the Wisconsin Card Sort (Berman et al. 1988). Reduced frontal to occipital metabolic rate also is observed in some schizophrenics (Buchsbaum 1987). A recent anatomical study has reported an altered distribution of nicotinamide-adenine dinucleotide phosphate–diaphorase neurons in the white matter of the dorsolateral prefrontal area of schizophrenic brains (Akbarian et al. 1993). This could disrupt the normal pattern of cortical connections which could potentially lead to impaired frontal lobe function in schizophrenics. These studies suggest that frontal lobe metabolism is affected in schizophrenia. Normal brain function is dependent on normal high-energy phosphate and membrane phospholipid metabolism. We have focused our research on these aspects of brain function in schizophrenia to help elucidate the pathophysiology in schizophrenia.

^{31}P Magnetic Resonance Spectroscopy

The chemical information contained in both in vitro and in vivo ^{31}P MRS spectra is given below. In vitro MRS spectra provide the information base necessary to interpret in vivo spectra. In vitro animal studies of neurodevelopment are useful starting points to help interpret the in vivo results of the ^{31}P MRS schizophrenia and aging studies.

In Vitro ^{31}P MRS

Sample Preparation

The source of brain tissue for in vitro MRS analysis can be autopsy brain or, in the case of animal studies, freeze-clamped tissue. Autopsy tissue is an important source of brain tissue for measuring certain metabolite levels found to be stable after death. However, high-energy metabolites such as adenosine triphosphate (ATP) and phosphocreatine (PCr) are not stable after death. Immediate freeze clamping brain tissue of animals with liquid nitrogen at the time of harvesting minimizes the loss of labile high-energy metabolites. The frozen tissue is then extracted with either perchloric acid (PCA) (Pettegrew et al. 1990b) to obtain the water soluble metabolites or chloroform:methanol (2:1 v/v) to obtain lipid extracts. The identification of the major signals in the ^{31}P MRS spectra of PCA extracts of brain tissue is well documented (Glonek et al. 1982; Pettegrew 1991). The chemical shift of the metabolites may change with pH, so that the pH must be taken into account in identifying the metabolites.

^{31}P MRS Spectra

Signals from both membrane phospholipid and high-energy phosphate metabolites are present in the ^{31}P MRS spectrum shown in Fig. 1. This spectrum

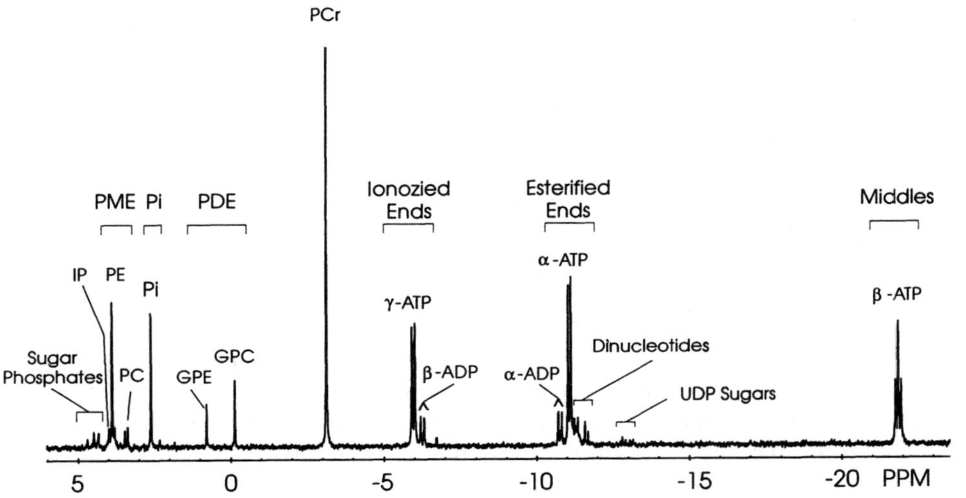

Fig. 1. ^{31}P MRS in vitro spectrum of perchloric acid extract of freeze-clamped rat brain tissue at a magnetic field strength of 11.7 Tesla, pH 9.6. *Ionized ends*, γATP and βADP; *esterified ends*, αATP and αADP; *middles*, βATP. *ADP, adenine diphosphate; ATP*, adenine triphosphate; *GPC*, glycerophosphocholine; *GPE*, glycerophosphoethanolamine; *IP*, phosphoinositol; *PC*, phosphocholine; *PCr*, phosphocreatine; *PDE*, phosphodiesters; *PE*, phosphoethanolamine; *Pi*, orthophosphate; *PME*, phosphomonoesters; *UDP*, uridine diphosphate

is from the PCA extract of freeze-clamped rat brain taken with a Bruker spectrometer at a magnetic field strength of 11.7 Tesla at pH 9.6. The resonances identified in the spectra and their chemical shifts (δ) include: (a) phosphomonoesters (PME), α-glycerol phosphate (α-GP, 4.35δ), inositol phosphate (IP, 4.05δ), phosphoethanolamine (PE, 3.94δ), and phosphocholine (PC, 3.43δ); (b) inorganic orthophosphate (Pi, 2.66δ); (c) the phsophodiesters (PDE), glycerol 3-phosphoethanolamine (GPE, 0.66δ) and glycerol 3-phosphocholine (GPC, -0.13δ); (d) PCr (-3.12δ); (e) the nucleotide triphosphates, predominantly ATP, γ -5.80δ, α -10.92δ, β -21.7δ; (f) the nucleotide diphosphates, predominantly adenosine diphosphate (ADP), β -6.11δ, α -10.618δ; (g) dinucleotides such as nicotinamide adenine dinucleotide (NAD) (-11.37δ); and (h) a complex resonance band centered around -12.89δ which is composed of nucleoside diphospho-derivatives such as uridine diphosphosugars and cytidine diphospho-derivatives such as cytidine diphosphocholine and cytidine diphosphoethanolamine.

Chemical Information

The PME are building blocks of phospholipids and the relative concentrations of these metabolites are a measure of the active synthesis of membranes (Vance 1991). Sources of PME are: (a) phosphorylation of their respective bases by kinases; (b) phospholipase C cleavage of their respective phospholipids; or (c) phosphodiesterase cleavage of their respective PDE such as GPC and GPE. The PME are broken down by phospholipase D to release Pi and the constituent base.

The PDE, such as GPE and GPC, are the major products of membrane phospholipid degradation (Vance 1988). The PDE are products of phospholipase $A_1 + A_2$ activity toward their phospholipids and are converted to their respective PME by PDE phosphodiesterase activity. In spite of their high abundance in brain, the physiological function of the PDE remains unknown. The PME/PDE ratio is an estimate of membrane phospholipid turnover.

Reliable information about energy metabolism can only be obtained from freeze-clamped tissue or in vivo studies. The resonances from the individual phosphate groups of di- and tri-phosphates have been characterized by their chemical identity which also determines their chemical shift. ATP and ADP are the predominant compounds. Since ADP has no middle phosphate, the middles region is the best measure of ATP levels. The ratio of PCr to Pi provides a convenient measure of the energy status of brain because it is the ratio of the most metabolically labile form of brain high-energy phosphates (PCr) to the ultimate breakdown product of all high-energy phosphates (Pi).

Quantitation of Spectra

Using appropriate routine instrumental conditions, the integrated area under an MRS peak corresponds to the number of atoms of that particular moiety in the sample. The quantitation of metabolites from brain tissue extracts can be accomplished by adding a known amount of an appropriate internal standard such as 3-aminopropylphosphoric acid to the weighed tissue at the time of extraction and reporting the results in absolute units of μmole/gram or by using tissue metabolites as their own internal reference and expressing the results in relative units such as mole percentage. These two approaches were investigated in a study of PCA extracts of brain tissue by ^{31}P MRS, ^{1}H MRS, and high-performance liquid chromatography in the same tissue sample (Klunk et al. 1994). This study found that the results were essentially identical for the mole percentage and μmol/gram methods. This supports reporting the results of in vivo MRS in mole percentage since adding an internal standard is not practical.

In Vivo ^{31}P MRS

In Vivo MRS Spectra

Resolution of in vivo MRS spectra of brain (Fig. 2) is diminished compared to in vitro MRS spectra of extracts. This loss of resolution of spectral peaks is due in part to the lower magnetic field strength (1.5 Tesla) of in vivo studies compared to in vitro studies (11.7 Tesla), since higher field strength gives greater dispersion of the chemical shift. The production of high field magnets sufficiently large to accommodate humans limits the field strength feasable for

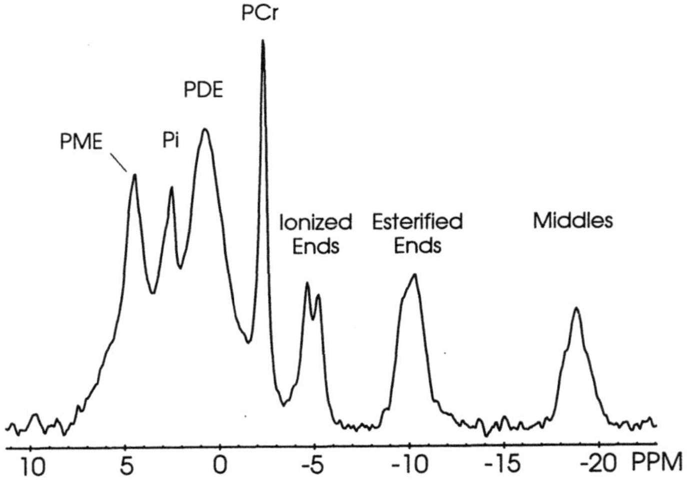

Fig. 2. ^{31}P MRS in vivo spectrum of human brain (dorsal prefrontal cortex) at a magnetic field strength of 1.5 Tesla. *Ionized ends*, γATP and βADP; *esterified ends*, αATP and αADP; *middles*, βATP. *PCr*, phosphocreatine; *PDE*, phosphodiesters; *Pi*, orthophosphate; *PME*, phosphomonoesters

in vivo MRS. Also, patient radio frequency power absorption, which also increases with field strength, limits the highest practical magnetic field strength for in vivo MRS to approximately 4 Tesla. In vivo MRS spectral peaks also are broadened by T_2 spin relaxation effects that reflect the environment of the molecules in living tissue versus the narrow in vitro spectra peaks of molecules in solution. Also, field inhomogeneity within the sample is an important contributor to line broadening. The identification of the origin of in vivo resonance signals requires detailed knowledge of in vitro spectra of brain extracts under conditions as close as possible to in vivo conditions.

Chemical Information

In vivo ^{31}P MRS of human brain provides spectral peaks for PME, Pi, PDE, PCR and the α, β, and γ phosphates of nucleotide diphosphates (predominantly ATP), and the α and β phosphates of nucleotide diphosphates such as ADP (Fig. 2). The individual PMEs and PDEs are not resolved. The intracellular pH is an important measurement that can be obtained from the chemical shift difference between the PCr and Pi peaks (Petroff et al. 1985).

Quantitation

Since in vivo MRS peaks are broad, of variable line width, and poorly resolved, areas of the peaks are usually obtained from curve-fitting software programs such as NMR1 of New Methods Research, Inc. or the GE/Nicolet curve analysis program. Figure 3 gives an example of curve-fitting an in vivo ^{31}P spectrum.

We report in vivo MRS data in relative units of mole percentage by using tissue metabolites as their own internal reference (Pettegrew et al. 1991). The best method of quantification would be based on comparison of sample signal to a reliable internal standard, however, no reliable internal standard exists for in vivo MRS. In order to express in vivo data in absolute units of μmoles/gram several studies have used external standards either attached in some fashion to the head and measured simultaneously (Bottomley et al. 1988, 1992) or in "phantoms" which are analyzed separately (Gruetter et al. 1992). Other studies obtain estimates of absolute concentration by assuming a known, unchanging concentration of one "reference" metabolite such as creatine (Frahm et al. 1989; Sappey-Marinier et al. 1992). This really measures a molar ratio despite the expression of data in absolute units. The external standard procedure introduces errors based on partial volume effects which are not present in the mole percentage values.

Signal Acquisition

A major attribute of in vivo MRS is its ability to measure levels of metabolites in the living human brain. However, the field of in vivo MRS is at an early, developing stage. Many technological hurdles remain to be overcome, including improvements in sensitivity, localization techniques, volume minimization, signal baseline correction, and quantitation. In our schizophrenia study, MRS

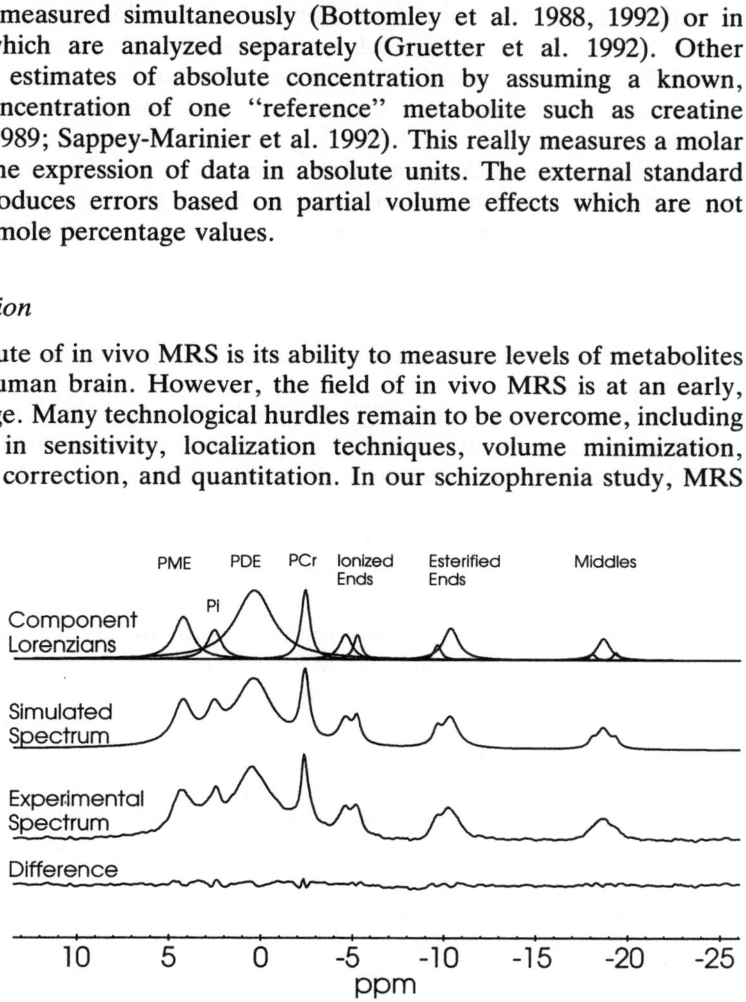

Fig. 3. Quantification of the ^{31}P MRS in vivo spectrum of brain. The experimental spectrum is shown after baseline correction in the trace labelled experimental spectrum. Component Lorentzians (*top trace*) are then fit to the individual peaks. Addition of the Lorentzians produces the trace-labeled simulated spectrum. The goodness of fit to the experimental data is controlled by insuring that the difference (*bottom trace*) between the experimental and simulated spectra is negligible. From Pettegrew et al. 1994

data was acquired using a straight-forward pulse-acquire sequence which minimizes the loss of spectral information due to delayed acquisition (Pettegrew et al. 1991). Localization of the brain area of interest was achieved by using a radio frequency field gradient produced by a surface coil as described by Bendall (1986), but without phase cycling. This localization technique is effective for studying 15–20 cc volumes of brain areas that are close to the surface of the skull, but it is not suitable for studying brain areas more distant from the surface coil. There are technical limitations present in all of the localization methods presently employed. More detailed information concerning the various in vivo MRS localization methods can be found in the literature (Aue 1986). Since MRS is safe and non-invasive, repeated measurements can be carried out in the same individual over a period of time making longitudinal studies possible.

Our in vivo ^{31}P MRS spectra are obtained using a General Electric Signa System with the spectroscopy research accessory (Pettegrew et al. 1991). The field strength is 1.5 Tesla, yielding a proton frequency of 63.970 MHz, and a phosphorus frequency of 25.895 MHz. The subject is placed supine on the MRS magnet transport table, and the head is positioned inside a plexiglass head holder containing a support bracket for securing a surface coil. The surface coil is positioned over the head to sample signals from the dorsal prefrontal cortex.

All subjects have routine T_1 weighted ^1H magnetic resonance imaging (MRI) scans in the sagittal, coronal, and axial planes immediately preceding or following the ^{31}P MRS spectroscopy studies. ^1H images, which are used to identify the location and volume of brain sampled by spectroscopy as well as the ^{31}P MRS spectra, are acquired with the same probe containing an 8 inch ^{31}P surface coil and a 3 inch surface coil dual-tuned at both the ^{31}P and ^1H frequencies. ^1H images for spectral localization are obtained by transmitting with a Helmholtz body coil, and receiving with the 3-inch surface coil. The localized ^1H images and the ^{31}P MRS spectra are from approximately 15–20 cm^3 of dorsal prefrontal cortex. The B_0 field is then shimmed on the H_2O ^1H signal to a line width of approximately 0.1 ppm.

^{31}P MRS spectra are obtained using a B_1 field gradient and a pulse-acquire sequence. A pulse width of 800 μs produces a spectrum with PCr/Pi and PME/PDE ratios typical of mammalian brain. The importance of these considerations have previously been pointed out (Pettegrew et al. 1983a). Other acquisition parameters are as follows: time to pulse repetition (TR) = 2 s, sweep width = 2 kHz, receiver filter = 1 kHz, number of complex data points = 2048. The integrated spectral areas are the same for TR values of 2, 4 and 6 s. Therefore, a TR of 2 s is used allowing a greater signal to noise for a fixed total acquisition time without signal saturation.

Our in vivo results compare quite favorably with previously published in vivo results (Cohen et al. 1984; Glonek et al. 1982; Pettegrew et al. 1987b, 1990b) verifying the validity of the in vivo method used in the present study. Test–retest and inter-rater reliability results give standard deviations of less than 5% of the mean for each parameter (n = 10).

Brain Development

Abnormal brain development has been implicated in the cerebral dysfunction demonstrated in schizophrenia (Feinberg 1982; Waddington 1993; Weinberger 1987). ^{31}P MRS is uniquely suited for tracking the energy and membrane metabolism of brain development. First, we will present in vitro ^{31}P MRS studies of the effect that development and aging have on rat brain metabolite levels. Also, the periods of human brain development will be briefly reviewed, followed by in vivo ^{31}P MRS studies of the effect of development and aging on human brain metabolite levels.

^{31}P MRS Study of Rat Brain Development

Brain maturation and development of the rat can be arbitrarily divided into four periods (McIlwain and Bachelard 1985). The age and anatomic events of these periods along with aging are shown in Table 1. Also, the in vitro ^{31}P MRS results are given for PCA freeze-clamped brain tissue extracts of animals within the defined age periods (Table 1).

There are striking effects of animal development and aging on the brain levels of the α-GP, PE, PC, GPE, GPC, PCr, and Pi (Pettegrew et al. 1990b). The PME are high in the immature brain and rapidly decrease from the newborn period to three months of age ($p = 0.0001$). The levels of α-GP and PC then remain relatively constant until 12 months of age, but the levels of PE increase slightly ($0.01 < p < 0.05$). From 12 to 24 months, the levels of PC decrease ($0.0001 < p < 0.001$). In contrast, the PDE (GPE and GPC) are very low or undetectable in the newborn period and then rapidly rise up to 3 months of age ($p = 0.0001$) after which the levels rise at a slower rate up to 24 months of age. The PME/PDE ratio, an estimate of membrane phospholipid anabolic/catabolic activity (Pettegrew et al. 1987a) is high in the newborn period, rapidly decreases up to 3 months of age ($p = 0.0001$) and then remains relatively constant up to 12 months of age. After 12 months of age the PME/PDE ratio decreases slightly ($0.01 < p < 0.05$) suggesting that membrane phospholipid breakdown is proceeding slightly faster than membrane phospholipid synthesis.

These ^{31}P MRS results tentatively suggest very high phospholipid anabolic activity without appreciable catabolic activity from 12 h to 10 days of age. This would coincide with an increase in the size of neuronal cells, the rapid outgrowth of axons and the rapid development of dendritic connections reported to occur in Period II. After 10 days of age, there is a rather rapid increase in PDE which are catabolic breakdown products of membrane phospholipids. The PME levels start to decrease at 10 days of age, but the PME/PDE ratio remains high (>10). The increasing PDE observed from 10 to 30 days of age could represent the metabolic correlates of naturally occurring cell death and the remodelling of neuritic connections. Cell death has been demonstrated to occur naturally during neural development and, in particular, occurs at about

Table 1. ^{31}P MRS correlates with rat brain development

Period	Age	Anatomic event	^{31}P MRS findings[a]	
			Phospholipid metabolites	High-energy metabolites
II	Birth-10 days	Neuronal Organization Rapid increase in brain lipid content Synaptogenesis Programmed cell death starts	PME levels high PDE levels low PME/PDE > 150	PCr levels low Pi levels high PCr/Pi ratio low
III	10–20 days	Neuronal Organization Synaptogenesis EEG activity appears Programmed cell death continues	PME levels decreasing PDE levels increasing	Rapid increase in PCr and PCr/Pi ratio Rapid decrease in Pi level, which correlates with development of glycolytic and oxidative pathways Increase in Na^+-K^+ ATPase and EEG activity
IV	20 days–12 months	Myelination	PME/PDE ≈ 1–2	Slight increase in PCr and Pi levels PCr/Pi: no change
V	>12 months	Aging and Senescence	PME/PDE < 1	Slight increase in Pi and PCr Slight increase in PCr/Pi suggests decrease in PCr utilization

[a] There were no changes in adenosine triphosphate with age.

PCr, phosphocreatine; PME, phosphomonoester; PDE, phosphodiester; Pi, inorganic orthophosphate; ATPase, adenosine triphosphatase; EEG, electroencephalographic.

the time of rapid synaptogenesis which is from birth to 20 days of age in the rat (Clarke 1985; Cowan et al. 1984; Pittman and Oppenheim 1979; Oppenheim 1985). From 1 to 3 months of age the PME continue to decrease and the PDE continue to increase, although not as rapidly as before. The PME/PDE ratio falls from ten to two indicating that the rate of phospholipid synthesis is decreasing. The changes in PME and PDE occurring from 1 to 3 months of age could reflect active myelination which is known to occur during this time in the rat. After 3 months of age, the turnover rate for phospholipids (PME/PDE) remains relatively constant until 12 months of age. From 12 to 24 months of age, the PME/PDE ratio decreases further ($0.01 < p < 0.05$) suggesting that membrane catabolism is proceeding slightly faster than anabolism.

Levels of PCr, a very labile brain high-energy phosphate, are quite low in the newborn period and appear to decrease even further up to 5 days of age. After 5 days of age, the PCr levels rapidly increase up to 1 month of age, with less rapid increases between 1 and 3 months of age (12-h versus 3-month $p = 0.0001$). After 3 months of age, the PCr levels undergo less dramatic but steady increases up to 24 months of age. In contrast, the levels of Pi are relatively high in the newborn, decrease up to 5 days of age followed by increases up to 10 days of age. After 10 days of age, the Pi levels drop rapidly until 3 months of age ($0.01 < p < 0.05$). From 3 months of age until 12 months of age, the Pi levels appear to increase slightly with more rapid increases from 12 to 24 months of age.

The PCr/Pi ratio is quite low in the newborn period until 5 days of age after which time the PCr/Pi ratio rapidly increases up to 3 months of age ($p = 0.0001$). After 3 months of age, the PCr/Pi ratio remains relatively constant up to 24 months of age. The rapid increase in the PCr/Pi ratio up to 1 month of age correlates with the development of the glycolytic and oxidative pathways, increasing Na^+-K^+ ATPase activity and the onset of electroencephalographic activity. The PCr/Pi ratio appears to increase slightly after 12 months of age suggesting decreased utilization of PCr. The brain levels of ATP do not undergo similar changes; there are no significant differences in brain ATP content comparing animals of 12 h versus 3 months, 3 months versus 12 months, or 12 months versus 24 months of age.

Neuronal Development of the Human Brain

A basic understanding of neuronal development in human brain is essential to interpret the biochemical changes monitored by ^{31}P MRS. The four major stages that characterize neuronal development of the human brain are: (a) neuronal proliferation; (b) migration of neurons to specific sites throughout the CNS; (c) organization of the neuronal circuitry; and (d) myelination of the neuronal circuitry (Volpe 1987).

The external form of the brain is established during the first 6 weeks of gestation. This is followed by neuronal proliferation which peaks during the 2nd to 4th month of gestation. All neurons and glia originate from the ventricular and subventricular zones of cellular membranes throughout the CNS. Glial multiplication occurs later throughout this stage, occurring from approximately 5 months of gestation to 1 year (or more) of age (Volpe 1987).

The second stage of neuronal development is characterized by the migration of millions of nerve cells from the ventricular and subventricular zones to loci within the CNS where they remain for life. The time period for this stage overlaps the neuronal proliferation stage, peaking in the 3rd to 5th month of gestation. Two major events are involved: (a) radial migration in the cerebral cortex to form the cerebral cortex and deep nuclear structures; and (b) radial migration in the cerebellum to form Purkinje cells, the dentate, and other roof nuclei and tangential migration in the cerebellum to form the internal granule cell layer of the cerebellar cortex. The migration of neurons in the cerebral cortex is completed by 20–24 weeks of gestation (Volpe 1987).

The third stage of neurodevelopment, organization of the neural circuitry, is most active from the 6th month of gestation to several years post natal. However, the timing of this organizational activity varies with the region of the brain under consideration. Of particular relevance to schizophrenia research is that the organization of neural circuitry in the dorsal prefrontal cortex can extend well into adolescence, overlapping with the timing of the onset of schizophrenia. The major events associated with neural organization include: (a) proper alignment, orientation, and layering of cortical neurons; (b) dendritic and axonal differentiation; (c) synaptic development; (d) synaptic pruning (cell death and selective elimination of neuronal processes); and (e) glial proliferation and differentiation (Volpe 1987). ^{31}P MRS is well suited to monitor both the process of synaptic development involving the proliferation of dendritic spines and neuronal cell death by measuring levels of metabolites implicated in membrane synthesis and degradation, respectively. The evaluation of the complex circuitry and synaptic connections that may be affected by disorders of neuronal organization is difficult. Perinatal insults are potential disturbances of disorders of neuronal organization (Volpe 1987). The increased incidence of schizophrenia among offspring who have had perinatal insults is one indication that schizophrenia may be a disorder of neuronal organization (Weinberger 1987).

Myelination is the fourth stage of neuronal development that involves the formation of a sheath of myelin membranes around axons. This process begins in the 2nd trimester of pregnancy and continues well into adult life. Myelination proceeds at different rates, dependent upon the brain region, but myelination of many areas of the brain is complete after 2–3 years. Perinatal insults also are potential disturbances in this stage (Volpe 1987).

Effect of Normal Human Development and Aging on ^{31}P MRS Metabolites

Knowledge of the changes in high-energy phosphate and membrane phospholipid metabolism associated with normal human brain development and aging is necessary in order to properly interpret the ^{31}P MRS findings in schizophrenia. In vivo ^{31}P MRS of the dorsal prefrontal cortex was performed on 62 normal volunteers aged 12–81 years who were recruited as controls for the autism, Down syndrome, schizophrenic, and Alzheimer's disease (AD) studies in our laboratory. All subjects had normal medical and neurological examinations, no subject met DSM-III-R criteria for any psychiatric disorder, and no subject had a positive family history for neuropsychiatric disorders. There was a significant decrease in the levels of PME with age ($p = 0.01$; $r = 0.52$) and a significant increase in the levels of PDE ($p = 0.01$; $r = 0.52$). There were no age-related changes in the levels of PCr, Pi, or ATP.

In vivo ^{31}P MRS spectra of these normal volunteers show similar metabolic effects of brain development and aging as demonstrated in the Fischer 344 rat. The PME increase between the ages of 10 and 20 years, remain relatively stable up to 40–50 years, and then slowly decline after 50 years of age (Fig. 4). The PDE decrease between the ages of 10 and 20 years, remain relatively stable up to 40–50 years, and then slowly increase over the age of 50 years (Fig. 4). The changes in brain levels of PME and PDE in individuals between 10 and 20 years of age probably reflect increasing membrane anabolic activity and decreasing membrane catabolic activity. The decreasing PME and increasing PDE that occur over the age of 50 presumably reflect the loss of dendritic spines and processes that occurs with aging.

In Vivo ^{31}P MRS Studies of Schizophrenics and Controls

Patient Characteristics

We have studied brain high-energy phosphate and membrane phospholipid metabolism in the dorsal prefrontal cortex of neuroleptic naive, first episode schizophrenic patients (Pettegrew et al. 1991). Neuroleptic naive patients were

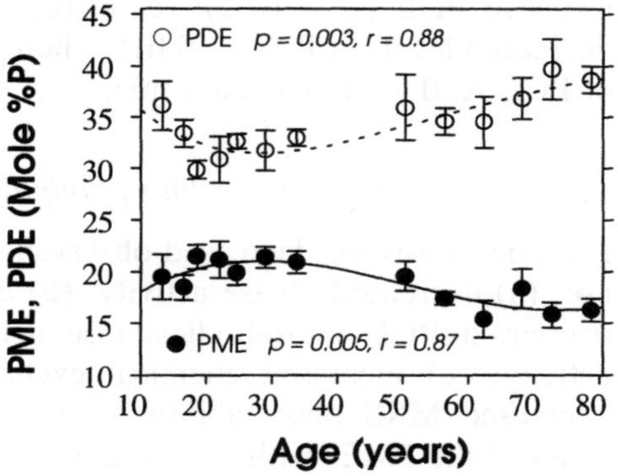

Fig. 4. In vivo ^{31}P MRS study of human brain (dorsal prefrontal cortex). Correlation of phosphomonoesters (*PME, solid circles*) and phosphodiesters (*PDE, open circles*) with age. Grouped average of 62 control subjects

chosen because of the known lasting effects of neuroleptics on membrane phospholipids. Also, patients at an early stage of the illness were chosen to exclude the possible effects of length of illness and hospitalization. The schizophrenic patients were compared to age, sex, race, education and parental education matched healthy controls under resting conditions. Eleven schizophrenic patients (seven males, four females; 24.4 ± 1.4 years, mean ± standard error of the mean, SEM); seven white, four black; 11.7 ± 0.8 years education) and 10 matched healthy controls (six males, four females; 24.1 ± 1.8 years; seven white, two black, one oriental; 12.7 ± 0.3 years education) were studied. Patients were assessed by the Schedule for Affective Disorders and Schizophrenia (SADS) (Endicott and Spitzer 1978) and met both DSM-III-R (American Psychiatric Association 1987) and Research Diagnostic Criteria (RDC) for schizophrenia (Spitzer et al. 1978). The DSM-III-R subclassification for the patients was six undifferentiated and five paranoid. None was previously treated with neuroleptics as reported by the patients' families or the referring physicians. The mean illness duration was 19.1 ± 5.9 months. None had a history of persistent substance abuse or met DSM-III-R criteria for it. Physical examination did not reveal abnormalities in any patient. The patients also were assessed using the Brief Psychiatric Rating Scale (BPRS) (Overall and Gorham 1962). All the patients have been followed for at least 6 months after entry into the study to ensure diagnostic stability. There are no significant group differences between the patients and controls for age, sex, race, education, or parental education.

Effect of Schizophrenia on Metabolites

The MRI results are normal in the schizophrenics and controls as reported by experienced neuroradiologists (Pettegrew et al. 1991). The following metabolic alterations are observed in the dorsal prefrontal cortex of the schizophrenic patients (see Fig. 5): (a) decreased levels of PME ($p = 0.002$); (b) increased levels of PDE ($p = 0.02$); (c) increased levels of ATP ($p = 0.02$); and (d) decreased levels of P_i ($p = 0.01$). There were no group differences in the levels of PCr, ADP or intracellular pH.

Effect of Schizophrenia on Phospholipid Metabolites

Possible causes for decreased observable levels of PME in schizophrenia brain are: (a) decreased kinase activity; (b) decreased phospholipase C activity; (c) decreased PDE phosphodiesterase activity; (d) increased phospholipase D activity; (e) increased chemical exchange with divalent cations leading to decreased MRS observability; or (f) increased NMR molecular correlation times of the PMEs with resulting decreased MRS observability.

Possible causes for increased PDE in schizophrenic brain include decreased PDE phosphodiesterase activity and increased phospholipase $A_1 + A_2$ activity.

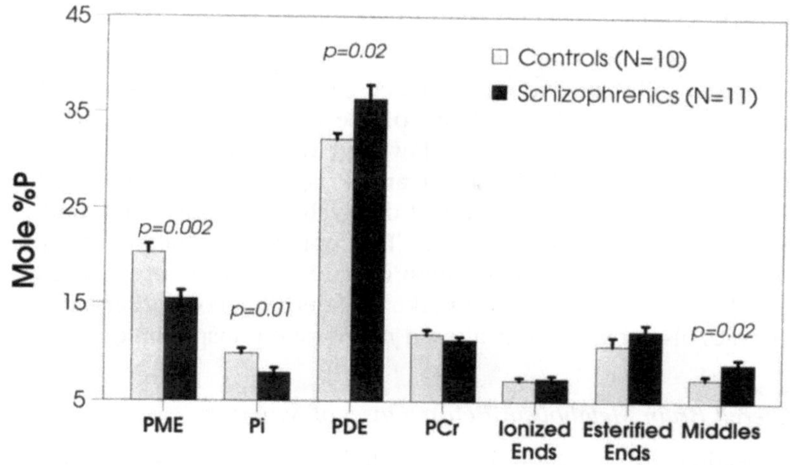

Fig. 5. Phospholipid metabolite levels in mole percentage P of schizophrenic subjects compared to controls. Relative mole percentage P obtained from in vivo ^{31}P MRS spectra of human brain (dorsal prefrontal cortex). *Ionized ends*, γATP and βADP; *esterified ends*, αATP and αADP; *middles*, βATP. *PCr*, phosphocreatine; *PDE*, phosphodiesters; *Pi*, orthophosphate; *PME*, phosphomonoesters

Gattaz et al. (1987) have reported increased activity of phospholipase A_2 in plasma from schizophrenic patients. Other possible causes of the increased observable PDE are decreased correlation times, an MRS parameter related to molecular motion, with increased MRS observability or decreased chemical exchange with divalent cations resulting in increased MRS observability. Decreased PDE phosphodiesterase activity could account for both the decreased PME and increased PDE levels observed in the schizophrenic patients studied.

The present findings of decreased PME and increased PDE levels also could suggest decreased synthesis and increased breakdown of membrane phospholipids in schizophrenia. Similar results are observed in normal aging in animals by in vitro ^{31}P MRS (Pettegrew et al. 1987b) and humans by in vivo ^{31}P MRS spectroscopy (Cady et al. 1983; Pettegrew et al. 1990a). Demisch et al. (1987) have reported a significantly decreased incorporation of 14-C labelled arachidonic acid in PtdC, PtdE and PtdI in patients with schizoaffective and schizophreniform disorders which could suggest decreased membrane phospholipid synthesis. As previously mentioned, decreased levels of membrane PtdC have been observed in erythrocytes in some schizophrenics (Henn 1980; Hitzemann et al. 1984; Stevens 1972). All the schizophrenic patients in the present study had normal MRI findings (as reported by neuroradiologists) at the time of their ^{31}P MRS studies. It is possible, therefore, that the PME and PDE alterations could represent metabolic alterations which antedate the onset of anatomical changes.

Effect of Schizophrenia on High-Energy Metabolites

The increased ATP and decreased Pi observed in the present study suggests decreased ATP utilization with decreased production of Pi. These findings are consistent with hypoactivity of the dorsal prefrontal cortex as suggested by decreased cerebral blood flow and decreased uptake of 2-deoxyglucose (Buchsbaum et al. 1982; Ingvar and Franzen 1974; Weinberger et al. 1986). Both cerebral blood flow and 2-deoxyglucose uptake are indirect measures of cerebral energy metabolism. The uptake of glucose and presumably 2-deoxyglucose are dependent upon normal membrane properties (Carruthers et al. 1989) and the decreased uptake of 2-deoxyglucose observed in some schizophrenics also could reflect altered membrane phospholipids.

Altered Brain Metabolism Before Onset of Symptoms

Alterations in phospholipid metabolites similar to those in schizophrenia were seen in a "healthy" control subject who was studied 2 years before a psychotic episode. This suggests that the phospholipid alterations seen in schizophrenia may represent "trait" markers (Keshavan et al. 1991). The ^{31}P MRS findings appear to persist with neuroleptic treatment (J.W. Pettegrew, unpublished findings). Longitudinal studies of subjects at risk for schizophrenia (e.g., first-degree relatives of schizophrenic probands) and of schizophrenic subjects in episode as well as untreated remission are needed to clarify state-trait issues.

Comparison of the Effect of Schizophrenia with Other ^{31}P MRS Studies

Other ^{31}P MRS studies of schizophrenic patients (Fukuzako et al. 1992; O'Callaghan et al. 1991; Williamson et al. 1991) have appeared in the literature since this original study (Table 2) (Pettegrew et al. 1991). With the exception of one study (O'Callaghan et al. 1991), which examined the temporal lobe, these studies have generally replicated the phospholipid findings of our study but have been inconsistent with respect to alterations in energy metabolism. The differences across these studies may lie in the MRS methodology or may relate to variability in the clinical status of the schizophrenic patients, for example, medicated or unmedicated, acute or chronic.

In Vivo ^{31}P MRS Studies of Autism

Autism is the prototype for a large group of disorders which form a clinical spectrum between autism and schizophrenia. To help determine the specificity of the ^{31}P MRS results to schizophrenia, autistic patients were studied by in vivo ^{31}P MRS using the identical protocol used for the schizophrenic patients. Normal IQ, adult autistic patients, who are similar in age to the schizophrenic patients in this study and share the negative symptoms of schizophrenia, do not have altered levels of brain PME or PDE by in vivo ^{31}P MRS (Minshew et al. 1993). A number of significant correlations were observed with the metabolites,

Table 2. ^{31}P MRS studies in schizophrenia

Reference	Patient population	Localization/technique	MRS findings	Remarks
Pettegrew et al. 1991	First-episode neuroleptic-naive SCZ (n = 11); controls (n = 10) matched for age, sex, race, and parental education	DLPFC/depth resolved pulse sequence	Decreased PME Increased PDE Increased ATP Decreased Pi	
O'Callaghan et al. 1991	Outpatient chronic SCZ, all but 1 on medication (n = 18); and agematched controls (n = 10)	Left temporoparietal	Increased pH	Controls somewhat older and not matched for handedness; medication effects cannot be excluded
Williamson et al. 1991	Chronic, medicated SCZ (n = 10); controls matched for age, sex, and handedness	Left DLPFC	Decreased PME Increased PCr Increased Pi	Increase in Mg^{2+} also seen; significance unclear
Fukuzako et al. 1992	Treatment-resistant chronic SCZ (n = 12)	Different brain regions/3-D CSI	Increased PDE	Medication effects cannot be excluded

ATP, adenosine triphosphate; DLPFC, dorsolateral prefrontal cortex; 3-D CSI, three-dimensional chemical shift imaging; PCr, phosphocreatine; PDE, phosphodiester; PME, phosphomonoester; Pi, inorganic orthophosphate; SCZ, schizophrenic subjects.

PCr, PME, and PDE and neuropsychology and language test scores for the autistic group, but no such correlations for the carefully matched control group. As test scores declined, the levels of PCr and PME decreased and the levels of PDE increased. Accordingly, the levels of PCr and esterified ends (αATP and αADP) were decreased in the autistic subjects compared to controls, but the phospholipid correlations were not reflected in statistically significant differences in the autistic subjects compared to controls. The correlation of a decrease in PME and increase in PDE with declining test scores is consistent with under synthesis and enhanced degradation of brain membranes. These findings are consistent with other recent research which, taken together, suggest that autism may be the consequence of maldevelopment of a distributed neural network responsible for processing complex information. The phospholipid findings in autism are similar to those in schizophrenia with the exception of the high-energy metabolites. The autistic patients have decreased levels of ATP which is the opposite of that observed in the schizophrenic patients. The decreased levels of ATP suggests increased ATP utilization in the autistics. This is consistent with a PET study of 12 young adult autistic men which reported an increase in metabolic rate for glucose in frontal, parietal, temporal, and occipital cortex, hippocampus, thalamus, and basal ganglia (Rumsey et al. 1985).

Discussion

Schizophrenia, Normal Aging, and Alzheimer's Disease

It is of interest that the changes in PME and PDE observed in schizophrenics are distinct and in the opposite direction from those observed with Alzheimer's disease (AD) which is associated with increases in brain PME early in the course of the disease. Elevated brain levels of PDE occur later in the course of AD and probably reflect the onset of membrane degenerative changes (Brown et al. 1989; Pettegrew et al. 1984, 1987b, 1988a,b, 1990b, 1994). As previously mentioned, the changes observed in schizophrenia are quite similar to those observed in normal aging in which the brain levels of PME are decreased and the PDE are increased (Pettegrew et al. 1987b). From this perspective, schizophrenia could be considered a form of premature brain aging, perhaps involving only certain neural systems. This formulation would be in keeping with the concept of dementia praecox put forth many years ago (Kraepelin 1919; Morel 1860). All the schizophrenic patients in the present study had normal MRI findings, as reported by neuroradiologists, at the time of their ^{31}P MRS studies. It is possible, therefore, that the PME and PDE alterations could represent metabolic alterations which antedate the onset of anatomical changes. This could provide a possible metabolic explanation for the development of cerebral atrophy observed in some schizophrenics (Andreasen et al. 1986; Andreasen 1988) including the affected twin of monozygotic twins discordant for schizophrenia (Suddath et al. 1990).

Schizophrenia and Programmed Synaptic Pruning

Decreased PME and increased PDE in first-episode, drug-naive schizophrenic patients also could be due to abnormal brain development (Feinberg 1982; Weinberger 1987). There is now substantial evidence for programmed neuronal cell loss, loss of callosal axons and synaptic elimination in normal brain development in animals and humans (Cowan et al. 1984; Huttenlocher 1979; Huttenlocher et al. 1982; Purves and Lichtman 1980; Rakic and Riley 1983). The present in vivo ^{31}P MRS findings indicate that the cell bodies, processes and terminals in the sampled area of the dorsal prefrontal cortex of the schizophrenic patients contain depressed levels of PMEs and elevated levels of PDEs as compared to the control subjects. In vivo ^{31}P MRS studies of normal volunteers between the ages of 12 and 85 years also reveal a transient decrease in the levels of PME and an increase in the levels of PDE during adolescence (10–20 years) (Panchalingam et al. 1990). However, the decreased PME and increased PDE observed in the schizophrenic patients appear to be more pronounced than in the normal volunteers of comparable age. This suggests decreased synthesis and enhanced breakdown of membranes occurring in the schizophrenic patients. This could be due to altered timing or abnormal enhancement of normal programmed synaptic pruning during adolescence as initially suggested by Feinberg (1982) and demonstrated to occur in the prefrontal cortex of humans during adolescence by Huttenlocher (1979). An exaggeration of these regressive synaptic events involving apical and basal dendrites could produce brain structural changes similar to those observed in schizophrenia in which the predominant reduction is in grey but not white matter. An exaggeration of normal neuronal cell death should produce decreased numbers of neurons and their projection axons and, therefore, reduce the volume of both grey and white matter.

Schizophrenia and Prefrontal Glutamatergic Pathways

The descending prefrontal glutamatergic pathway is a major input to both the striatum and the limbic cortex. Exaggerated synaptic pruning of prefrontal-limbic glutamatergic axonal terminals could produce a reduced neuropil volume resulting in what might appear as an enhanced clustering of cells as observed for pre-α-cells in the parahippocampal gyrus (Falkai et al. 1988). Excessive pruning of prefrontal–striatal glutamatergic axonal terminals ending on striatal dopaminergic terminals could decrease the tonic release of dopamine from the dopaminergic terminals resulting in secondary up-regulation of the postsynaptic dopaminergic receptors. A number of postmorten studies have demonstrated increased numbers of D_2 receptors in the caudate and putamen of patients with schizophrenia (Jaskiw and Kleinman 1988; Seeman et al. 1984) but a recent study suggests that this is secondary to neuroleptic treatment within 3 months of death (Kornhuber et al. 1989). Two in vivo positron emission tomography studies of brain D_2 receptors in schizophrenic patients have appeared. One study reported increased numbers of D_2 receptors in the striatum of drug-naive schizophrenic patients (Wong et al. 1986) but the other study could not replicate this finding (Farde et al. 1987). The phasic release of dopamine is thought to be responsive to stress. With up-regulation of postsynaptic dopaminergic receptors, stress could result in exaggerated mesolimbic activity resulting in positive symptoms. Neuroleptics which block postsynaptic D_2 receptors could modulate the mesolimbic activity back toward normal. This formulation has been suggested recently by Grace (1991).

Future Directions

The alterations of phospholipid metabolite levels observed in the schizophrenic patients could be accounted for by either an altered timing of the normal synaptic pruning of the prefrontal cortex, which occurs in adolescence or premature aging of the prefrontal cortex. Consideration should be given to the possible use of brain trophic factors early in the course of schizophrenia if decreased brain PME and increased PDE are found in a much larger sample of schizophrenic patients. Also, ^{31}P MRS might be used to identify individuals who demonstrate PME and PDE changes early in the course of the illness or, possibly, prior to the onset of symptoms. These individuals would then be candidates for neurotrophic therapy to slow, stop or, hopefully, reverse the membrane alterations. This is highly speculative, and even if this approach should prove beneficial, its implementation is far into the future. However, by recognizing the potential of such an approach and planning now, perhaps we can reduce the time required to implement approaches utilizing ^{31}P MRS for diagnostic and monitoring applications.

The ability of ^{31}P MRS to monitor high-energy phosphates as well as membrane phospholipid metabolites could provide new insights into the causes of schizophrenia. ^{31}P MRS also appears to be a viable noninvasive procedure for longitudinal studies which has the potential to provide new information on the possible evolution of regional biochemical changes in schizophrenia.

References

Akbarian S, Bunney WE, Potkin SG, Wigal SB, Hagman JO, Sandman CA, Jones EG (1993) Altered distribution of nicotinamide-adenine dinucleotide phosphate-diaphorase cells in frontal lobe of schizophrenics implies disturbances of cortical development. Arch Gen Psychiatry 50:169–177

American Psychiatric Association (1987) DSM-III-R diagnostic and statistical manual of mental disorders, 3rd edn. Workgroup to revise DSM-III, Washington DC

Andreasen NC (1988) Brain Imaging: Applications in psychiatry. Science 239:1381–1388

Andreasen N, Nasrallah HA, Dunn V, Olson SC, Grove WM, Ehrhardt JC, Coffman JA, Crossett JHW (1986) Structural abnormalities the frontal system in schizophrenia. Arch Gen Psychiatry 43:136–144

Aue WP (1986) Localization methods for in vivo nuclear magnetic resonace spectroscopy. Magn Reson Med 1:21–72

Bachneff SA (1991) Position emission tomography and magnetic resonance imaging: a review and a local circuit neurons hypo (dys) function hypothesis of schizophrenia. Biol Psychiatry 30:857–886

Bendall MR (1986) Surface coil techniques for in vivo NMR. Bull Magn Reson 8:17–44

Berman KF, Illowsky BP, Weinberger DR (1988) Physiological dysfunction of dorsolateral prefontal cortex in schizophrenia. Arch Gen Psychiatry 45:616–622

Blass JP, Hanin I, Barclay L, Kopp U, Reding MJ (1985) Red blood cell abnormalities in Alzheimer's disease. J Am Geriatr Soc 33:401–405

Bottomley PA, Charles HC, Roemer PB, Flaming D, Engeseth H, Edelstein WA, Mueller OM (1988) Human in vivo phosphate metabolite imaging with ^{31}P NMR. Magn Reson Med 7:319–336

Bottomley PA, Cousins JP, Pendrey DL, Wagle WA, Hardy CJ, Eames FA, McCaffrey RJ, Thompson DA (1992) Alzheimer dementia: quantification of energy metabolism and mobile phosphoesters with P-31 NMR spectroscopy. Radiology 183:695–699

Brown GG, Levine SR, Gorell JM, Pettegrew JW, Gdowski JW, Bueri JA, Helpern JA, Welch KM (1989) In vivo ^{31}P NMR profiles of Alzheimer's disease and multiple subcortical infarct dementia. Neurology 39:1423–1427

Buchsbaum MS (1987) Positron emission tomography in schizophrenia. In: Meltzer HY (ed) Psychopharmacology, the third generation of progress. Raven, New York, pp 783–792

Buchsbaum MS, Ingvar DH, Kessler R, Waters RN, Cappelette J, van Kammen DP, King C, Johnson JL, Manning RB, Flynn RW, Mann LS, Bunney WE, Sokoloff L (1982) Cerebral glucography with positron tomography. Arch Gen Psychiatry 39:251–259

Butterfield DA, Markesbery WR (1980) Specificity of biophysical and biochemical alterations in erythrocyte membranes with neurological disorders. J Neurol Sci 97:261–271

Butterfield DA, Oeswein JW, Markesbery WR (1977) Electron spin resonance study of membrane protein alterations in erythrocytes in Huntington's disease (letter). Nature 267:453–455

Butterfield DA, Oeswein JW, Prunty ME, Hisle KC, Markesbery WR (1978) Increased sodium plus potassium adenosine triphosphatase activity in erythrocyte membranes in Huntington's disease. Ann Neurol 4:60–62

Butterfield DA, Nicholas MM, Markesbery WR (1985) Evidence for an increased rate of choline efflux across erythrocyte membranes in Alzheimer's disease. Neurochem Res 10:909–918

Cady EB, Dawson MJ, Hope PL, Tofts PS, Costello AM, Delpy DT, Reynolds EOR, Wilkie DR (1983) Non-invasive investigation of cerebral metabolism in newborn infants by phosphorus nuclear magnetic resonance spectroscopy. Lancet 1:1059–1062

Carlsson A (1988) The current status of the dopamine hypothesis of schizophrenia. Neuropsychopharmacology 1:179–186

Carruthers A, Helgerson AL, Herbert DN, Tefft RE Jr, Naderi S, Melchior DL (1989) Effects of calcium, ATP and lipids on human erythrocyte sugar transport. Ann N Y Acad Sci 568:52–67

Clarke PG (1985) Neuronal death in the development of the vertebrate nervous system. Trends Neurosci 8:345–349

Cohen MM, Pettegrew JW, Kopp SJ, Minshew N, Glonek T (1984) P-31 nuclear magnetic resonance analysis of brain: normoxic and anoxic brain slices. Neurochem Res 9:785–801

Cowan WM, Fawcett JW, O'Leary DD, Stanfield BB (1984) Regressive events in neurogenesis. Science 225:1258–1265

Crow RJ (1987) The dopamine hypothesis survives, but there must be a way ahead. Br J Psychiatry 151:460–465

Csernansky JG, Murphy GM, Faustman WO (1991) Limbic/mesolimbic connections and the pathogenesis of schizophrenia. Biol Psychiatry 30:383–400

Demisch L, Gerbaldo H, Heinz K, Kirsten R (1987) Transmembranal signalling in schizophrenic and affective disorders: studies on arachidonic acid and phospholipids. Schizophr Res 22:275–282

Diamond JM, Matsuyama SS, Meier K, Jarvik LF (1983) Elevation of erythrocyte countertransport rates in Alzheimer's dementia (letter). N Engl J Med 309:1061–1062

Endicott J, Spitzer RL (1978) A diagnostic interview: the schedule for affective disorders and schizophrenia. Arch Gen Psychiatry 35:837–844

Falkai P, Bogerts B, Rozumek M (1988) Cell loss and volume reduction in the entorhinal cortex of schizophrenics. Eur Arch Psychiatry Neurol Sci 24:515–521
Farde L, Wiesel FA, Hall H, Halldin C, Stone-Elander S, Sedvall G (1987) No D2 receptor increase in PET study of schizophrenia (letter). Arch Gen Psychiatry 44:671–672
Feinberg I (1982) Schizophrenia: caused by a fault in programmed synaptic elimination during adolescence? J Psychiatr Res 17[Suppl 4]:319–334
Frahm J, Bruhn H, Gyngell ML, Merboldt KD, Hanicke W, Sauter R (1989) Localized proton NMR spectroscopy in different regions of the human brain in vivo. Relaxation times and concentrations of cerebral metabolites. Magn Reson Med 11:47–63
Fukuzako H, Takeuchi K, Fujimoto T, Hokazono Y, Hirakawa K, VeYama K, Matsumoto K, Fukuzako T (1992) 31P magnetic resonance spectroscopy of schizophrenic patients with neuroleptic resistant positive and negative symptoms. Biol Psychiatry 31[Suppl]:204A–205A
Gattaz WF, Kolisch M, Thuren T, Virtanen JA, Kinnunen PKJ (1987) Increased plasma phospholipase A2 activity in schizophrenic patients: reduction after neuroleptic therapy. Biol Psychiatry 22:421–426
Glonek T, Kopp SJ, Kot E, Pettegrew JW, Harrison WH, Cohen MM (1982) P-31 nuclear magnetic resonance analysis of brain: the perchloric acid extract spectrum. J Neurochem 39:1210–1219
Grace AA (1991) Phasic versus tonic dopamine release and the modulation of dopamine system responsivity: a hypothesis for the etiology of schizophrenia. Neuroscience 41:1–24
Gruetter R, Rothman DL, Novotny EJ, Shulman RG (1992) Localized 13C NMR spectroscopy of myo-inositol in the human brain in vivo. Magn Reson Med 25:204–210
Healey D (1991) D_1 and D_2 and D_3. Br J Psychiatry 159:319–324
Henn F (1980) Biological concepts of schizophrenia. In: Baxter C, Melnachuk T (eds) Perspectives in schizophrenia research. Raven, New York, pp 209–223
Hitzemann R, Hirschowitz D, Garver D (1984) Membrane abnormalities in the psychoses and affective disorders. J Psychiatr Res 18:319–326
Hoffman RE, McGlashan TH (1993) Neurodynamics and schizophrenia research: editors' introduction. Schizophr Bull 19:15–19
Huttenlocher PR (1979) Synaptic density in human frontal cortex. Developmental changes and effects of aging. Brain Res 163:195–205
Huttenlocher PR, deCourten C, Garey LJ, Van Der Loos H (1982) Synaptogenesis in human visual cortex-evidence for synapse elimination during normal development. Neurosci Lett 33:247–252
Ingvar DH, Franzen G (1974) Abnormalities of cerebral blood flow distribution in patients with chronic schizophrenics. Acta Psychiatr Scand 50:425–462
Jaskiw G, Kleinman J (1988) Postmortem neurochemistry studies in schizophrenia. In: Schulz SC, Tamminga HK (eds) Schizophrenia: a scientific focus. Oxford University Press, New York, pp 264–273
Kaiya H, Takeuchi K, Namba M, Imai A, Nakashima S, Nozawa Y (1984) Abnormal phosphatidylinositol-cycle of platelet membrane in schizophrenia – a preliminary study. Folia Psychiatr Neurol Jpn 38:437–444
Keshavan MS, Ganguli R (1990) Biology of schizophrenia. In: Gershon S, Pohl R (eds) Progress in basic and clinical pharmacology 3: biological basis of psychiatric treatment. Karger, Basel, pp 1–33
Keshavan MS, Pettegrew JW, Panchalingam KS, Kaplan D, Bozik E (1991) Phosphorus 31 magnetic resonance spectroscopy detects altered brain metabolism before onset of schizophrenia. Arch Gen Psychiatry 48:1112–1113
Kim JS, Kornhuber HH, Schid-Burgk W et al (1980) Low cerebrospinal fluid glutamate in schizophrenic patients and a new hypothesis on schizophrenia. Neurosci Lett 20:379–392
Klunk WE, Xu CJ, Panchalingam K, McClure RJ, Pettegrew JW (1994) Analysis of magnetic resonance spectra by mole percent: comparison to absolute units. Neurobiol Aging 15:133–140
Kornhuber J, Riederer P, Reynolds GP, Beckmann H, Jellinger K, Gabriel E (1989) 3H-spiperone binding sites in post-mortem brains from schizophrenia patients: relationship to neuroleptic drug treatment, abnormal movements, and positive symptoms. J Neural Transm 75:1–10

Kraepelin E (1919) Dementia praecox and paraphrenia. Churchill Livingston, Edinburgh
Markesbery WR, Leung PK, Butterfield DA (1980) Spin label and biochemical studies of erythrocyte membranes in Alzheimer's disease. J Neurol Sci 45:323–330
McIlwain H, Bachelard HS (1985) Biochemistry and the central nervous system, 5th edn. Churchill Livingstone, Edinburgh
Meltzer T, Stahl SM (1976) Dopamine hypothesis of schizophrenia: a review. Schizophr Bull 2:19–76
Miller BL, Jenden D, Tang C, Read S (1989) Choline and choline-bound phospholipids in aging and Alzheimer's disease (abstract). Neurology 39[Suppl 1]:254
Minshew NJ, Goldstein G, Dombrowski SM, Panchalingam K, Pettegrew JW (1993) A preliminary ^{31}P MRS study of autism: evidence for undersynthesis and increased degradation of brain membranes. Biol Psychiatry 33:762–773
Morel BA (1860) Traitement des maladies mentales. Masson, Paris
O'Callaghan EO, Redmond O, Ennis R, Stack J, Kinsella A, Ennis JT, Conall L, Waddington JL (1991) Initial investigation of the left temporoparietal region in schizophrenia by ^{31}P magnetic resonance spectroscopy. Biol Psychiatry 29:1149–1152
Oppenheim RW (1985) Naturally occurring cell death during neural development. Trends Neurosci 8:487–493
Overall JE, Gorham DR (1962) The brief psychiatric rating scale. Psychol Rep 10:799–812
Panchalingam K, Pettegrew JW, Strychor S, Tretta M (1990) Effect of normal aging on membrane phospholipid metabolism by ^{31}P in vivo NMP spectroscopy (abstract). Soc Neurosci Abstr 16:843
Petroff OAC, Prichard JW, Behar KL, Alger JR, den Hollander JA, Shulman RG (1985) Cerebral intracellular pH by ^{31}P nuclear magnetic resonance spectroscopy. Neurology 35:781–788
Pettegrew JW (1991) Nuclear magnetic resonance: principles and applications to neuroscience research. In: Boller F, Grafman J (eds) Handbook of neuropsychology, 5th edn. Elsevier Science, New York, pp 39–56
Pettegrew JW, Minshew NJ (1992) Molecular insights into schizophrenia. J Neural Transm 36:23–40
Pettegrew JW, Glonek T, Baskin F, Rosenberg RN (1979a) Phosphorus-31 NMR of neuroblastoma clonal lines: effect of cell confluency state and dibutyryl cyclic AMP. Neurochem Res 4:795–801
Pettegrew JW, Nichols JS, Stewart RM (1979b) Studies of the fluorescence of fibroblasts from Huntington's disease: evidence of a membrane abnormality. N Engl J Med 300:678
Pettegrew JW, Nichols JS, Stewart RM (1979c) Fluorescence spectroscopy on Huntington's fibroblasts. J Neurochem 33:905–911
Pettegrew JW, Nichols JS, Stewart RM (1981) Membrane studies in Huntington's disease: steady-state and time-dependent fluorescence spectroscopy of intact lymphocytes. J Neurochem 36: 1966–1976
Pettegrew JW, Nichols JS, Minshew NJ, Rush AJ, Stewart RM (1982) Membrane biophysical studies of lymphocytes and erythrocytes in manic-depressive illness. J Affective Disord 4:237–247
Pettegrew JW, Minshew NJ, Diehl J, Smith T, Kopp SJ, Glonek T (1983a) Anatomical considerations for interpreting topical P-31 NMR (letter). Lancet 2:913
Pettegrew JW, Minshew NJ, Stewart RM (1983b) Dynamic membrane studies in individuals at risk for Huntington's disease. Life Sci 32:1207–1212
Pettegrew JW, Minshew NJ, Cohen MM, Kopp SJ, Glonek T (1984) P-31 NMR changes in Alzheimer's and Huntington's disease brain (abstract). Neurology 34[Suppl 1]:281
Pettegrew JW, Kopp SJ, Minshew NJ, Glonek T, Feliksik JM, Tow JP, Cohen MM (1987a) ^{31}P nuclear magnetic resonance studies of phosphoglyceride metabolism in developing and degenerating brain: preliminary observations. J Neuropathol Exp Neurol 46:419–430
Pettegrew JW, Withers G, Panchalingam K, Post JF (1987b) ^{31}P nuclear magnetic resonance (NMR) spectroscopy of brain in aging and Alzheimer's disease. J Neural Transm Suppl 24:261–268

Pettegrew JW, Moossy J, Withers G, McKeag D, Panchalingam K (1988a) ^{31}P nuclear magnetic resonance study of the brain in Alzheimer's disease. J Neuropathol Exp Neurol 47:235–248

Pettegrew JW, Panchalingam K, Moossy J, Martinez J, Rao G, Boller F (1988b) Correlation of phosphorus-31 magnetic resonance spectroscopy and morphologic findings in Alzheimer's disease. Arch Neurol 45:1093–1096

Pettegrew JW, Panchalingam K, Strychor S, Branthoover G (1990a) Analysis of membrane phospholipids in Alzheimer's disease brain by ^{31}P NMR (abstract). Soc Neurosci Abstr 16:498

Pettegrew JW, Panchalingam K, Withers G, McKeag D, Strychor S (1990b) Changes in brain energy and phospholipid metabolism during development and aging in the Fischer 344 rat. J Neuropathol Exp Neurol 49:237–249

Pettegrew JW, Keshavan MS, Panchalingam K, Strychor S, Kaplan DB, Tretta MG, Allen M (1991) Alterations in brain high-energy phosphate and phospholipid metabolism in first episode, drug-naive schizophrenia. A pilot study of the dorsal prefrontal cortex by in vivo ^{31}P NMR spectroscopy. Arch Gen Psychiatry 48:563–568

Pettegrew JW, Keshavan MS, Minshew NJ (1993a) ^{31}P Nuclear magnetic resonance spectroscopy: neurodevelopment and schizophrenia. Schizophr Bull 19:35–53

Pettegrew JW, Minshew NJ, Spiker D, Tretta M, Strychor S, McKeag D, Munez LR, Miller GM, Carbone D, McClure RJ (1993b) Alterations in membrane molecular dynamics in erythrocytes of patients with affective illness. Depress 1:88–100

Pettegrew JW, Panchalingam K, Klunk WE, McClure RJ, Muenz LR (1994) Alterations of cerebral metabolism in probable Alzheimer's disease; a preliminary study. Neurobiol Aging 15:117–132

Pittman R, Oppenheim RW (1979) Cell death of motoneurons in the chick embryo spinal cord. IV. Evidence that a functional neuromuscular interaction is involved in the regulation of naturally occurring cell death and the stabilization of synapses. J Comp Neurol 187:425–446

Purves D, Lichtman JW (1980) Elimination of synapses in the developing nervous system. Science 210:153–157

Rakic P, Riley KP (1983) Overproduction and elimination of retinal axons in the fetal rhesus monkey. Science 219:1441–1444

Reynolds GP (1989) Beyond the dopamine hypothesis; the neurochemical pathology of schizophrenia. Br J Psychiatry 155:305–316

Rotrosen J, Wolkin A (1987) Phospholipid and prostaglandin hypothesis in schizophrenia. In: Meltzer HY (ed) Psychopharmacology: the third generation of progress. Raven, New York, pp 759–764

Rumsey JM, Duara R, Grady C, Rapoport JL, Margolin RA, Rapoport SI, Cutler NR (1985) Brain metabolism in autism. Resting cerebral glucose utilization rates as measured with positron emission tomography. Arch Gen Psychiatry 42:448–455

Sappey-Marinier D, Calabrese G, Hetherington HP, Fisher SN, Deicken R, Van Dyke C, Fein G, Weiner MW (1992) Proton magnetic resonance spectroscopy of human brain: applications to normal white matter, chronic infarction, and MRI white matter signal hyperintensities. Magn Reson Med 26:313–327

Seeman P (1987) Dopamine receptors and the dopamine hypothesis of schizophrenia. Synapse 1:133–152

Seeman P, Ulpian C, Bergeron C, Riederer P, Jellinger K, Gabriel E, Reynolds GP, Tourtellotte WW (1984) Bimodal distribution of dopamine receptor densities in brains of schizophrenics. Science 225:728–731

Sherman KA, Gibson SE, Blass JP (1986) Human red blood cell choline uptake with age and Alzheimer's disease. Neurobiol Aging 7:205–209

Spitzer RL, Endicott J, Robins E (1978) Research diagnostic criteria (RDC) for a selected group of functional disorders, 3rd edn. New York State Psychiatric Institute Biometric Division, New York

Stevens JD (1972) The distribution of phospholipid fractions in the red cell membrane of schizophrenics. Schizophr Bull 6:60–61

Suddath RL, Christison GW, Torrey EF, Casanova MF, Weinberger DR (1990) Anatomical abnormalities in the brains of monozygotic twins discordant for schizophrenia. N Engl J Med 322:789–794

Ulas J, Cotman CW (1993) Excitatory amino acid receptors in schizophrenia. Schizophr Bull 19:105–117

Vance DE (1991) Phospholipid metabolism and cell signalling in eucaryotes. In: Vance DE, Vance J (eds) Biochemistry of lipids, lipoproteins and membranes, vol 20. Elsevier, New York, pp 205–240

Volpe JJ (1987) Neuronal proliferation, migration, organization, and myelination. In: Volpe JJ (ed) Neurology of the newborn, 2nd edn. Saunders, Philadelphia, pp 33–68

Waddington JL (1993) Neurodynamics of abnormalities in cerebral metabolism and structure in schizophrenia. Schizophr Bull 19:55–69

Weinberger DR (1987) Implications of normal brain development for the pathogenesis of schizophrenia. Arch Gen Psychiatry 44:464–469

Weinberger DR, Berman KF, Zec DF (1986) Physiological dysfunction of the dorsolateral prefrontal cortex in schizophrenia. Arch Gen Psychiatry 43:114–124

Williamson P, Drost D, Stanley J, Carr T, Morrison S, Merskey H (1991) Localized phosphorus 31 Magnetic Resonance Spectroscopy in chronic schizophrenic patients and normal controls (letter). Arch Gen Psychiatry 48:578

Wong DF, Wagner HN, Tune LE, Dannals RF, Pearlson GD, Links JM, Tamminga CA, Broussolle EP, Ravert HT, Wilson AA, Toung JKT, Malat J, Williams JA, Lorcan A, O'Tuama O, Snyder SH, Kuhar MJ, Gjedde A (1986) Positron emission tomography reveals elevated D2 dopamine receptors in drug-naive schizophrenics. Science 234:1558–1563

Wyatt RJ, Alexander RC, Egan MF, Kirch DG (1988) Schizophrenia just the facts; What do we know; How well do we know it? Schizophr Res 1:3–18

Zubenko GS, Cohen BM, Reynolds CF, Boller F, Malinakova I, Keefe MA (1987) Platelet membrane fluidity in Alzheimer's disease and major depression. Am J Psychiatry 144:860–868

Positron Emission Tomography, Receptors, and Schizophrenia

F.-A. Wiesel

The monoamines have been extensively studied in relation to the mechanism of action of antipsychotics and the pathophysiology of schizophrenia. Dopamine was first thought to play only a precursor role in norepinephrine synthesis, but was later suggested to have an independent function in the brain when it was discovered that the distribution of norepinephrine and dopamine differed (Bertler and Rosengren 1959; Ehringer and Hornykiewicz 1960). In 1963, Carlsson and Lindqvist suggested that antipsychotic drugs block postsynaptic catecholamine receptors and thereby induce a compensatory activation of the presynaptic neurons. Later it was demonstrated that most antipsychotic drugs preferentially accelerated the synthesis and turnover of dopamine (Nybäck and Sedvall 1968, 1970). The dopamine hypothesis in schizophrenia was initially based on findings demonstrating that neuroleptics antagonized dopamine transmission probably by a blockade of its receptors and that compounds which increased the dopaminergic transmission were psychotomimetic. In humans it was demonstrated that the major dopamine metabolite homovanillic acid increased in the spinal fluid of patients treated with neuroleptics, a finding in line with the proposed negative feedback mechanism regulating activity of dopamine neurons (Carlsson and Lindqvist 1963; Sedvall et al. 1975). However, untreated patients with schizophrenia did not have increased HVA levels in the spinal fluid which would be expected if schizophrenia were to be caused by an increased dopaminergic transmission (Bjerkenstedt et al. 1985).

During the 1970s major methodological progress was made with the development of radioactive ligands with highly specific activities. It allowed the study of receptor distribution and numbers and receptor effects of psychotropical compounds. Thus it was possible to demonstrate that neuroleptics blocked D_2 dopamine receptors with a potency which was positively correlated with the antipsychotic potency of the compounds (Creese et al. 1975; Seeman et al. 1975; Peroutka and Snyder 1980). This observation, together with the finding that deceased schizophrenic patients had increased levels of D_2 dopamine receptors put the focus on the D_2 dopamine receptor for the mechanism of action of antipsychotics and the pathophysiology of schizophrenia (Seeman et al. 1984). Moreover, if the patients had an increased number of D_2 dopamine

Department of Psychiatry, University of Uppsala, Ulleråker Hospital, 750 17 Uppsala, Sweden

H. Häfner and W.F. Gattaz (Eds.)
Search for the Causes of Schizophrenia, Vol. III

receptors, the proposed negative feedback mechanism of dopamine neurons could explain the reduced HVA levels in the spinal fluid of schizophrenics. However, the relevance of the increased D_2 dopamine receptors in postmorem brains of schizophrenics was questioned since neuroleptic treatment is known to upregulate these receptors (Mackay et al. 1982; Kornhuber et al. 1989). Besides problems with previous neuroleptic treatment of deceased patients, most of them had suffered from their disease for decades and it is quite possible that an increase in receptor numbers could be a consequence of the disease rather than the pathogenetic mechanism.

Positron Emission Tomography

Positron emission tomography (PET) has made it possible to test some aspects of the dopamine hypothesis in schizophrenia. The use of radioactively labeled ligands makes PET the principle receptor autoradiography in vivo. However, the resolution is relatively low even though it has improved and is about 4–5 mm (FWMH) in today's system (the theoretical resolution is 2 mm). This means that important anatomical structures cannot be studied in detail. Another limiting factor is the development of suitable ligands which, within a relatively short period of time, must obtain a high ratio between specific and non-specific binding (Sedvall et al. 1986). However, these obstacles are not present in testing the D_2 dopamine hypothesis in schizophrenia using PET. Appropriate ligands have been developed and the structures of interest are big enough to be correctly analyzed. (Elevations of D_2 dopamine receptors in deceased schizophrenics were found in the putamen, accumbens, and caudate.) Densities of D_2 dopamine receptors have been determined in neuroleptic-naive schizophrenic patients in several studies. In the study of the mechanism of action of neuroleptics both D_1 and D_2 dopamine receptors have been investigated and to some extent also serotonin$_2$ receptors.

D_2 Dopamine Receptors and Schizophrenia

The first ligand to be used in this context was [^{11}C]N-methylspiperone ([^{11}C] NMSP). This ligand has a high affinity for both D_2 dopamine and serotonin$_2$ ($5HT_2$) receptors and was used by Wong et al. (1986) to examine D_2 dopamine receptor densities in vivo in the caudate nucleus of ten drug-naive schizophrenic patients. To determine receptor numbers, a kinetic compartment model was used in which k_3, the rate constant for the association of free [^{11}C]NMSP to receptors in the brain, was determined. Receptor numbers (B_{max}) was calculated in principle from the difference in k_3 before and after the oral administration of 7.5 mg haloperidol. In a linear graph the reciprocals of the two k_3 values were plotted versus the brain concentrations of haloperidol which were estimated from plasma values. Wong et al. (1986) reported a two- to threefold elevation in D_2 dopamine densities of the patients over healthy controls (Table 1). In our

Table 1. D_2 dopamine receptor in schizophrenic patients

Author	Ligand	B_{max}	Striatum / Cerebellum
Wong et al. 1986[a]	^{11}C-NMSP		
Controls = 11 (23)		17 (16)	No difference
Patients = 10 (15)		42 (34)	
Farde et al. 1990	^{11}C-raclopride		
Controls = 20		28	No difference
Patients = 18		30	
Nordström et al. 1993a	^{11}C-NMSP		
Controls = 7		24	
Patients = 7		32	
Hietala et al. 1992	^{11}C-raclopride		
Controls = 13		35	No difference
Patients = 10		39	
Martinot et al. 1990	^{76}Br-SP		
Controls = 12			2.1
Patients = 12			2.1
Martinot et al. 1991	^{76}Br-lisuride		
Controls = 14			3.9
Patients = 19			4.0

Patients were neuroleptic naive or drug free for at least 6 months.
[a] B_{max} was elevated only in patients with a disease duration >2 years. (Tune et al. 1992). Figures in parentheses represent a greater patient sample (Tune et al. 1992).

study of D_2 dopamine receptor densities in schizophrenics, we used the selective D_2 dopamine receptor antagonist [^{11}C]raclopride (Farde et al. 1990). It is a competitive antagonist allowing an in vivo saturation analysis to determine D_2 dopamine receptor densities and affinities (K_D) in the caudate and the putamen of young neuroleptic-naive patients with schizophrenia (n = 18) and healthy controls (n = 20). We did not find any differences in B_{max} or K_d values between the two groups. Our results did not support the hypothesis of schizophrenia as due to generally elevated D_2 dopamine receptor densities in the basal ganglia. The finding contradicts the results of the PET study by Wong et al. (1986). The discrepancy in results could have several explanations. It has been proposed that the patients should have an increased dopaminergic activity causing elevated concentrations of dopamine in the synapse. This elevated concentration of dopamine was suggested to interfere with [^{11}C]raclopride binding, since raclopride is a competetive antagonist. However, if this were the case, then the patients should have a reduced apparent K_D, which was not the case. Moreover, determination of homovanillic acid in the spinal fluid does not support the hypothesis that most schizophrenic patients should have an increased turnover of dopamine in the brain (Bjerkenstedt et al. 1985; Lindström

1985; Wik and Wiesel 1991). To find out if kinetic analysis and saturation analysis would give similar results, Farde et al. (1989) determined receptor numbers in healthy controls with both type of analyses with [^{11}C]raclopride as a ligand. The two methods fave similar D_2 dopamine densities. It has also been proposed that the discrepancy in results is due to differences in patient samples between the two groups. The patients examined by Wong et al. (1986) had a higher mean age (31.4 ± 4 years) and a longer duration of illness (5 ± 3 years) than our patients who had a mean age of 24.2 ± 3.3 with a median duration of illness of 10 months and a range of 1–72 months. If increased D_2 dopamine receptor densities develop with the time course of the disease, there should be a correlation between receptor numbers and duration of schizophrenia. However, we did not find such a relationship.

Other groups studying D_2 dopamine receptor densities in drug-free schizophrenics have also not found elevations in the patients (Table 1). In two different studies by Martinot et al. (1990, 1991) with either [^{76}Br]bromospiperone or [^{76}Br]bromolisuride as ligands, no differences between patients and controls were found. [The ratio of striatal to cerebellar radioactivity was taken as an index of striatal D_2 dopamine densities (Table 1)]. However, some support was obtained for D_2 dopamine receptor densities to be state dependent. This possibility is further strengthened by Wong et al. (1989) who found increases in D_2 dopamine receptor densities in psychotic manic depressive patients and Tourette's syndrome. It may be that the ligand and method used by Wong et al. (1986) is more sensitive to patients' characteristics than raclopride since they later reported that B_{max} elevations were found only in patients with a duration of illness of more than 2 years (Tune et al. 1992).

In order to reproduce the findings by Wong et al. (1986), the group at the Karolinska Institute investigated neuroleptic-naive schizophrenic patients (n = 7) and healthy controls (n = 7) according to the protocol described by Wong et al. (1986) (Nordström et al. 1993a). However, a difference between normals and patients with schizophrenia in D_2 dopamine receptor densities could also not be demonstrated in this replication study (Table 1). The results obtained from different research groups together with the replication study strongly suggest that patients with schizophrenia do not have increased D_2 dopamine receptor densities. The pathogenetic role of the D_2 dopamine receptor is also questioned by negative findings of the D_2 dopamine receptor gene and schizophrenia (Moises et al. 1991). However, this does not exclude the possibility that dopaminergic transmission may be disturbed at another level or receptor subtype.

D_1 Dopamine Receptors and Schizophrenia

The D_1 dopamine receptor, the first demonstrated subtype of the dopamine receptor, has become somewhat less significant than the D_2 dopamine receptor. However, changes in D_1 dopamine receptors have been reported. Thus, Hess et al. (1987) found decreases in D_1 dopamine receptor densities in deceased schizophrenics. There are PET methods available to study the D_1 dopamine receptor in vivo (Farde et al. 1987). With [^{11}C]-labeled SCH 23390, a selective D_1 dopamine antagonist, Karlsson et al. (1993) studied D_1 receptor densities in a small sample of schizophrenics ($n = 5$) and normals ($n = 5$). No major differences between the two groups were found in D_1 dopamine receptor binding in the putamen (normals 48 ± 9 nmol/l, patients 61 ± 9 nmol/l) and neocortex (normals 18 ± 7 nmol/l, patients 30 ± 14 nmol/l).

Mechanisms of Action of Antipsychotics

Even if changes in the D_2 dopamine receptor may not be a pathogenetic mechanism to schizophrenia, this receptor subtype may be of importance in mediating the antipsychotic effect of neuroleptics. Neuroleptics are effective in both toxic and so-called functional psychoses and are in principal suitable for all types of psychoses. The use of the term "antischizophrenic drugs" is therefore inappropriate. It is also recognized that patients with schizophrenia treated with neuroleptics are not cured but have a symptomatic improvement of their disorder allowing the combination of drug treatment with other types of psychiatric and psychological treatment (Hogarty et al. 1974; Goldstein et al. 1978; Fallon et al. 1985; Leff et al. 1985).

As discussed above, strong support for the importance of a blockade of the D_2 dopamine receptor for the antipsychotic effect is obtained from the positive correlation between the potency of the D_2 dopamine antagonistic activity and the antipsychotic potency of neuroleptics (Peroutka and Snyder 1980). Such correlations have not been demonstrated in other receptor types. It should be pointed out that the correlation demonstrated was calculated from in vitro data of the antagonistic potency of neuroleptics and from general recommendations of dosage of compounds. However, optimal dosage seems to vary almost tenfold (Baldessarini et al. 1988). The PET technique makes it possible to study in vivo how neuroleptics block the receptors. In studies to determine D_2 dopamine receptor occupancy of neuroleptics, the highly selective D_2 dopamine antagonist raclopride was used. After the i.v. injection of [^{11}C]raclopride there is a high accumulation of radioactivity in the putamen and caudate in both normals and untreated patients. However, in neuroleptic-treated patients radioactivity will not accumulate because the D_2 dopamine-binding sites are occupied or blocked by the neuroleptic compound. In our studies the patients had all been under treatment with conventional doses of one antipsychotic compound (monotherapy) for at least 1 month and had responded well to the treatment. The occurrence of extrapyramidal side effects was rated according to the Simpson and Angus (1970) scale and Barnes' (1989) scale for akathisia. Receptor occupancy was determined in a PET experiment 6 h after the morning (last) dose or in the case of depot neuroleptics 1 week after the last injection. Similar methodology was used to determine D_1 dopamine receptor occupancy but with another ligand, [^{11}C]-labeled SCH-23390, a D_1 dopamine antagonist (Farde et al. 1992).

D_2 Receptor Occupancy

In patients treated with conventional neuroleptics, levels of receptor occupancy varied between 70%–89% (Table 2) (Farde et al. 1992). Patients with an occupancy above 80% had a high frequency of extrapyramidal side effects. Haloperidol blocked D_2 dopamine receptors with a high potency. Thus

Table 2. Central D_2 and D_1 dopamine receptor occupancy in neuroleptic-treated patients with schizophrenia. (Modified from Farde et al. 1992)

Drug	Dosage (mg)	Occupancy		EPS[a]
		D_2 (%)	D_1 (%)	
Phenothiazines				
Chlorpromazine	100 × 2	78	–	–
Thioridazine	150 × 2	74	–	–
Thioridazine	200 × 2	81	30	–
Trifluoperazine	5 × 2	75	–	–
Perphenazine enantate (7 days)	100	76	–	Parkinsonism
Perphenazine	8 × 2	–	0	–
Thioxanthenes				
Flupentixol	3 × 2	71	44	–
Flupentixol	3 × 2	70	–	–
Flupentixol decanoate (7 days)	40	81	36	Parkinsonism
Zuclopenthixol decanoate (14 days)	200	81	16	Dystonia
Butyrophenones				
Haloperidol	6 × 2	84	–	Akathisia
Haloperidol	3 × 2	89	–	Parkinsonism
Haloperidol	3 × 2	84	–	Parkinsonism
Haloperidol	2 × 2	75	–	Akathisia
Haloperidol	2 × 2	84	–	Akathisia, parkinsonism
Haloperidol	3 × 2	86	3	Akathisia
Haloperidol decanoate (28 days)	50	85	–	Parkinsonism
Haloperidol decanoate (28 days)	70	74	5	–
Melperone	125 × 2	71	–	–
Melperone	100 × 3	70	–	–
Diphenylbutyls				
Pimozide	4 × 2	79	–	Akathisia
Substituted benzamides				
Remoxipride	200 × 2	71	–	–
Sulpiride	400 × 2	78	0	–
Dibenzodiazepines				
Clozapine	300 × 2	63	–	–
Clozapine	150 × 2	40	38	–
Clozapine	250 × 2	38	42	–
Clozapine	250 × 2	46	36	–
Clozapine	250 × 2	55	52	–

[a] EPS, extrapyramidal syndromes.

moderate doses of the drug (4–12 mg) resulted in high occupancy values. Clozapine, an atypical antipsychotic compound, distinguished among the antipsychotic drugs with lower occupancy values varying between 38%–63% (Table 2).

The relationship between clinical drug effects and D_2 receptor occupancy was examined in a prospective study with a controlled design (Nordström et al. 1993c). The patients were treated with raclopride for 4 weeks. They were randomly assigned to one of three parallel dose groups. Daily doses of raclopride were 2, 6 or 12 mg, respectively. Patients and staff involved in the study were blind to the dosage. Central D_2 receptor occupancy was determined in a PET experiment during the 3rd or 4th week of raclopride treatment (Nordström et al. 1993c). The occurrence and intensity of psychotic symptoms were rated with the Brief Psychiatric Rating Scale (BPRS) before the start of the treatment and weekly during the treatment period (Overall and Gorham 1962). Extrapyramidal symptoms were also rated on the same occasions. Seventeen patients were recruited for the study and 13 participated in the PET experiments for determination of D_2 receptor occupancy. Ratings were also made on the day of the PET experiment.

A curvilinear relationship between raclopride concentrations in plasma and D_2 dopamine receptor occupancy in the brain was obtained confirming in vivo the expected hyperbolic relationship (Farde et al. 1988). A linear relationship between D_2 dopamine receptor occupancy both for the antipsychotic effect and extrapyramidal side effects was obtained. The results of this prospective controlled study give strong support for the hypothesis that D_2 dopamine receptor blockade mediates the antipsychotic effect and extrapyramidal side effects. However, this does not exclude the fact that other subtypes of the dopamine receptor or other neuronal systems could also mediate therapeutic and extrapyramidal side effects. One obvious explanation of the fact that clozapine does not induce extrapyramidal side effects is its low occupancy values obtained during clinical treatment (Table 2).

D_1 Receptor Occupancy

The degree of receptor occupancy of D_1 dopamine receptors was dependent on the chemical structure of the neuroleptic compound (Table 2) (Farde et al. 1992). Perphenazine and sulpiride did not seem to interact with D_1 dopamine receptors, whereas thioridazine and zuclopenthixol occupied D_1 receptors to some degree but much less than D_2 receptors. Clozapine had the same degree of D_1 receptor occupancy as for D_2 receptors. Thus, clozapine was distinguished from the other compounds by having a similar receptor occupancy for both D_1 and D_2 receptors. The lower but similar occupancy of D_1 and D_2 receptors may result in a synergistic effect and therefore result in a full antidopaminergic effect with a pronounced antipsychotic effect in clinical treatment (Longoni et al. 1987).

With the advance in molecular genetics a dopamine receptor family has been identified: the D_3 dopamine receptor (Sokoloff et al. 1990), the D_4 dopamine receptor (van Tol et al. 1991), and the D_5 dopamine receptor (Sunahara et al. 1991). The functional significance of these receptor subtypes for therapeutic effects has not been demonstrated. It may be that the D_4 dopamine receptor is of special interest since clozapine has about a tenfold great affinity to that subtype than to the other dopamine receptors.

5-HT$_2$ Receptors

Serotoninergic mechanisms have been discussed both in relation to therapeutic effects and extrapyramidal side effects. Ritanserin is a selective 5-HT_2 and 5-HT_{2C} receptor antagonist which has been shown to reduce extrapyramidal side effects and negative symptoms in neuroleptic-treated schizophrenic patients (Bersani et al. 1990). In a 4-week study with an open design of ritanserin treatment in acute schizophrenic patients, both positive and negative symptoms were found to be reduced (Wiesel et al. 1993). It was also demonstrated that this effect was not due to a D_2 dopamine antagonistic effect. It has been suggested that the atypical effect of clozapine could be due to its antagonistic effects on 5-HT_2 receptors (Melzer 1989). In a study of three clozapine-treated patients, Nordström et al. (1993b) demonstrated with [^{11}C]NMSP (this ligand binds to both D_2 dopamine and 5-HT_2 receptors) and PET that 5-HT_2 receptor occupancy in the frontal cortex was above 80%. D_2 dopamine receptor occupancy in the basal ganglia was determined in two patients and was about 50%. Thus, in clinical treatment clozapine results in a moderate occupancy of D_1 and D_2 dopamine receptors and a high occupancy of 5-HT_2 receptors. The broader clinical effects of clozapine and the low frequency of extrapyramidal side effects may be due to the combination of effects on both dopamine and serotonin systems.

Comments

The dopamine hypothesis could be divided into two parts: (a) D_2 dopamine receptors are elevated in patients with schizophrenia and are of pathogenetic importance to the disease; (b) the antipsychotic effect of neuroleptic treatment is mediated by a blockade of D_2 dopamine receptors.

The first hypothesis could not be confirmed by PET studies of young neuroleptic-naive patients with schizophrenia. The second part of the hypothesis was strongly supported in conventional neuroleptics. However, the clozapine results also demonstrated that the antipsychotic effect could be mediated by receptor mechanisms other than a high D_2 dopamine receptor blockade. It should be pointed out that, even if clozapine has a broader clinical spectrum than conventional neuroleptics, a relatively high proportion of the schizophrenic population is still only partly helped by clozapine treatment. About 20% of patients with schizophrenia will have an unsatisfactory response to clozapine treatment making it difficult to initiate rehabilitation programs in these patients. It is evident that too little is known about the pathophysiological mechanisms leading to psychotic disorders. In this context the exploration of the glutamate systems seems to be of special interest in order to understand how psychotic symptoms develop and to find new treatment strategies (Kornhuber et al. 1984; Carlsson and Carlsson 1990; Javitt and Zukin 1991).

Acknowledgements. Ms. A. Liberg is gratefully acknowledged for preparing the manuscript. This study was supported by grants from the Swedish Medical Research Council (Nr 8313) and the Söderström-Königska Foundation.

References

Baldessarini RJ, Cohen BM, Teicher MH (1988) Significance of neuroleptic dose and plasma level in the pharmacological treatment of psychoses. Arch Gen Psychiatry 45:79–91

Barnes TRE (1989) A rating scale for drug-induced akathisia. Br J Psychiatry 154:672–676

Bersani G, Grispini A, Marini S, Pasini A, Valducci M, Ciani N (1990) $5HT_2$ antagonist ritanserin in neuroleptic-induced parkinsonism: a double-blind comparison with orphenadrine and placebo. Clin Neuropharmacol 13:500–506

Bertler Å, Rosengren E (1959) Occurrence and distribution of catecholamines in brain. Acta Physiol Scand 47:350–361

Bjerkenstedt L, Edman G, Hagenfeldt L, Sedvall G, Wiesel FA (1985) Plasma amino acids in relation to monoamine metabolites in schizophrenic patients and healthy controls. Br J Psychiatry 147:276–282

Carlsson A, Lindqvist M (1963) Effect of chlorpromazine or haloperidol on formation of 3-methoxytyramine and norm etanephrine in mouse brain. Acta Pharmacol Toxicol 20:140–144

Carlsson M, Carlsson A (1990) Interactions between gluamatergic and monoaminergic systems within the basal ganglia – implications for schizophrenia and Parkinson's disease. TINS 13:272–276

Creese I, Burt DR, Snyder SH (1975) Dopamine receptor binding: differentiation of agonist and antagonist states with ^3H-dopamine and ^3H-haloperidol. Life Sci 17:993–1002

Ehringer H, Hornykiewicz O (1960) Verteilung von Noradrenalin und Dopamin (3-Hydroxytyramin) im Gehirn des Menschen und ihr Verhalten bei Erkrankungen des extrapyramidalen Systems. Klin Wochenschr 38:1236–1239

Falloon JRH, Boyd JL, McGill CW et al (1985) Family management in the prevention of morbidity of schizophrenia. Arch Gen Psychiatry 42:887–896

Farde L, Halldin H, Stone-Elander C, Sedvall G (1987) PET-analysis of human dopamine receptor subtypes using [^{11}C]-SCH23390 and [^{11}C]-raclopride. Psychopharmacology 92:278–284

Farde L, Wiesel F-A, Halldin C, Sedvall G (1988) Central D_2- dopamine receptor occfupancy in schizophrenic patients treated with antipsychotic drugs. Arch Gen Psychiatry 45:71–76

Farde L, Eriksson L, Blomquist G, Halldin C (1989) Kinetic analysis of central [11C]raclopride binding to D2-dopamine receptors studied with PET: a comparison to the equilibrium analysis. J Cereb Blood Flow Metab 9:696–708

Farde L, Wiesel F-A, Stone-Elander S, Halldin C, Nordström A-L, Hall H, Sedvall G (1990) D_2 dopamine receptors in neuroleptic-naive schizophrenic patients. Arch Gen Psychiatry 47:213–219

Farde L, Nordström A-L, Wiesel F-A, Pauli S, Halldin C, Sedvall G (1992) Positron emission tomographic analysis of central D_1- and D_2-dopamine receptor occupancy in patients treated with classical neuroleptics and clozapine – relation to extrapyramidal side effects. Arch Gen Psychiatry 49:538–544

Goldstein MJ, Rodnick EH, Evans JR et al (1978) Drug and family therapy in the aftercare of acute schizophrenics. Arch Gen Psychiatry 35:1169–1177

Hess EJ, Bracha HS, Kleinman JE et al (1987) Dopamine receptor subtype imbalance in schizophrenia. Life Sci 40:1487–1497

Hietala J, Syvälahti E, Vuorio K, Någren K, Lehikoinen P, Ruotsalainen U, Räkköläinen V, Lehtinen V, Wegelius U (1991) Striatal dopamine D_2 receptor density in neuroleptic-naive schizophrenics studied with positron emission tomography. In: Racagni G, Brunello N, Fukuda T (eds) Biological psychiatry, vol 2. Excerpta Medica, Amsterdam, pp 386–387

Hogarty GE, Goldberg SC, Schooler NR (1974) Drug and sociotherapy in the aftercare of schizophrenic patients. Arch Gen Psychiatry 31:609–618

Javitt DC, Zukin SR (1991) Recent advances in the phencyclidine model of schizophrenia. Am J Psychiatry 148:1301

Karlsson P, Farde L, Halldin C, Nordström A-L, Sedvall G (1993) D_1-dopamine receptor binding in drug naive schizophrenic patients measured by PET. J Cereb Blood Flod Metab 13[Suppl 1]:S556

Kornhuber HH, Kornhuber J, Kim JS, Kornhuber ME (1984) Zur biochemischen Theorie der Schizophrenie. Nervenarzt 55:602–606

Kornhuber J, Rieder P, Reynolds GP, Beckman H, Jellinger K, Gabriel E (1989) 3H-spiperone binding in post-mortem brains from schizophrenic patients: relationship to neuroleptic drug treatment, abnormal movements, and positive symptoms. J Neural Transm 75:1–10

Leff J, Kuipers L, Berkowitz R et al (1985) A controlled trial of social intervention in the families of schizophrenic patients: two year follow-up. Br J Psychiatry 146:594–600

Lindström L (1985) Low HVA and normal 5HIAA CSF levels in drug-free schizophrenic patients compared to healthy volunteers: correlations to symptomatology and family history. Psychiatry Res 14:265–273

Longoni R, Spina L, Dichiara G (1987) Permissive role of D-1 receptor stimulation for the expression of D-2 mediated behavioral responses: a quantitative phenomenological study in rats. Life Sci 41:2135–2145

Mackay AVP, Iversen LL, Rossor M, Spokes E, Bird E, Arregui A, Creese I, Snyder SH (1982) Increased brain dopamine and dopamine receptors in schizophrenia. Arch Gen Psychiatry 39:991–997

Martinot J-L, Peron-; agnan P, Hure J-D, Mazoyer B, Baron J-C, Boulenger J-P, Loc'h C, Maziere B, Caillard V, Loo H, Syrota A (1990) Striatal D_2 dopamine receptors assessed with positron emission tomography and [^{76}Br]Bromospiperone in untreated schizophrenic patients. Am J Psychiatry 147:44–50

Martinot J-L, Paillere-Martinot ML, Loc'h C, Hardy P, Poirier MF, Mazoyer B, Beaufils B, Maziere B, Allilaire JF, Syrota A (1991) The estimated density of D_2 striatal receptors in schizophrenia – a study with positron emission tomography and ^{76}Br-Bromolisuride. Br J Psychiatry 158:346–350

Melzer HY (1989) Clinical studies on the mechanism of action of clozapine: the dopamine-serotonin hypothesis of schizophrenia. Psychopharmacology 99:18–27

Moises HW, Gelernter J, Giuffra LA, Zarcone V, Wetterberg L, Civelli O, Kidd KK, Cavalli-Sforza LL (1991) No linkage between D_2 dopamine receptor gene region and schizophrenia. Arch Gen Psychiatry 48:643–647

Nordström A-L, Farde L, Eriksson L, Halldin C (1993a) No elevated D_2 dopamine receptors in neuroleptic naive schizophrenic patients revealed by PET and [^{11}C]NMSP. Manuscript. In: Nordström AL (ed) PET evaluation of dopamine hypotheses for antipsychotic drugs and schizophrenia. Thesis from the Department of Psychiatry and Psychology, Karolinska Institute, 1993

Norström A-L, Farde L, Halldin C (1993b) High 5-HT_2 receptor occupancy in clozapine treated patients demonstrated by PET. Psychopharmacology 110:365–367

Nordström A-L, Farde L, Wiesel F-A, Forslund K, Pauli S, Halldin C, Uppfeldt G (1993c) Central D2-dopamine receptor occupancy in relation to antipsychotic drug effects: a double-blind PET study of schizophrenic patients. Biol Psychiatry 33:227–235

Nybäck H, Sedvall G (1968) Effect of chlorpromazine on accumulation and disappearance of catecholamine formed from tyrosine ^{14}C in brain. J Pharmacol Exp Ther 162:294–301

Nybäck H, Sedvall G (1970) Further studies on the accumulation and disappearance of catecholamines formed from tyrosine-^{14}C in mouse brain. Effect of some phenothiazine analogues. Eur J Pharmacol 10:193–205

Overall JE, Gorham DR (1962) The brief psychiatric rating scale. Psychol Rep 10:799–812

Peroutka SJ, Snyder SH (1980) Relationship of neuroleptic drug effects at brain dopamine, serotonin, adrenergic and histamine receptors to clinical potency. Am J Psychiatry 137:1518–1522

Sedvall G, Alfredsson G, Bjerkenstedt L et al (1975) Selective effects of psychoactive drugs on levels of monoamine metabolites and prolactin in cerebrospinal fluid of psychiatric patients. In: Airaksinen H (ed) Proceedings of the 6th international congress of pharmacology. Forssan, Kirjapaino, Oy, pp 255–267

Sedvall G, Farde L, Persson A, Wiesel FA (1986) Imaging of neurotransmitter receptors in the living human brain. Arch Gen Psychiatry 43:995–1005

Seeman P, Chau-Wong M, Tedesco J et al (1975) Brain receptors for antipsychotic drugs and dopamine binding assay. Proc Natl Acad Sci USA 72:4376–4380

Seeman P, Ulpian P, Bergeron C, Riederer P, Jellinger K, Gabriel E, Reynolds GP, Tourtelotte WW (1984) Bimodal distribution of dopamine receptor densities in brain of schizophrenics. Science 225:728–730

Simpson GM, Angus JWS (1970) A rating scale for extrapyramidal side effects. Acta Psychiatr Scand 45[Suppl 212]:11–19

Sokoloff P, Giros B, Martres MP, Bouthenet ML, Schwartz JC (1990) Molecular cloning and characterization of a novel dopamine receptor (D3) as a target for neuroleptics. Nature 347:146–151

Sunahara RK, Guan H-C, O'Dowd BF et al (1991) Cloning of the gene for a human dopamine D_5 with higher affinity for dopamine than D_1. Nature 350:614–619

Tune LE, Wong DF, Pearlson GD (1992) Elevated dopamine 2 receptor density in 23 schizophrenic patients: a positron emission tomography study with [^{11}C]N-methylspiperone. Schizophrenia Res (special issue) 6:147

Van Tol H, Bunzow J, Guan H, Sunahara R, Seeman P, Niznik H, Civelli O, (1991) Cloning of the gene for a human dopamine D_4 receptor with high affinity for the antipsychotic clozapine. Nature 350:610–614

Wiesel F-A, Nordström A-L, Farde L, Eriksson B (1993) An open clinical and biochemical study of ritanserin in acute patients with schizophrenia. Psychopharmacol (in press)

Wik G, Wiesel F-A (1991) Regional brain glucose metabolism: correlations to biochemical measures and anxiety in patients with schizophrenia. Psychiatry Res Neuroimaging 40:101–114

Wong DF, Wagner HN, Tune LE, Dannals RF, Pearlsson GD, Links JM, Tamminga CA, Broussolle EP, Ravert HT, Wilson AA, Toung JKT, Malat J, Williams FA, O'Touma LA, Snyder SH, Kuhar MJ, Gjedde A (1986) Positron emission tomography reveals elevated D2 dopamine receptors in drug-naive schizophrenics. Science 234:1558–1563

Wong DF, Pearlson GD, Young LT, Singer H, Willemagne V, Tune L, Ross C, Dannals RF, Links JM, Chan B, Wilson SS, Ravert HT, Wagner HN Jr, Gjedde A (1989) D_2 dopamine receptors are elevated in neuropsychiatric disorders other than schizophrenia. J Cereb Blood Flow Metab 9[Suppl 1]:S593

Brain, Receptors, and Schizophrenia: Discussion

F.A. Henn

In considering the neurotransmitters associated with schizophrenia one normally begins with dopamine. This will be no exception, in fact we will begin and end this discussion with a consideration of dopamine. The initial paper in this section by Sokoloff is a superb review of the recent work on DA receptors identified by cloning techniques. This methodology at once promises the possibility of greater specificity in the design of drugs and almost ununderstandable levels of complexity in the analysis of transmitter function. This is illustrated by serotonin, where, up until today, 14 clearly different receptors have been identified using cloning methodology. In the case of DA there are apparently six major forms of receptor falling into two classes. Sokoloff clearly reviews this information and presents some of his laboratories extensive data arguing for a special role for the D_3 receptor in anti-psychotic effects. His argument rests on three points: (a) the distribution of the receptor is appropriate for mediating psychotic symptoms; (b) the receptor recognizes a wide variety of neuroleptic receptors in the correct order of clinical potency; and (c) long-term blockade does not result in up regulation of the receptor. This last point is important because clinically nothing like tardive psychosis is seen while there are clear clinical signs of up regulation in the motor system leading to tardive dyskinesia. Thus, if the D_3 receptor is the site of action for anti-psychotic activity this receptor will not be up regulated by long-term anti-psychotic drug usage, accounting for the clinical evidence that increasing psychotic symptoms are not generally seen with long-term treatment.

The data gathered over the last 5 years using molecular approaches have confirmed the complexity of the DA receptor family. It has also opened up the possibility of much more specific treatments; if a specific receptor is involved in either the etiology of the illness or the final common pathway involved in the expression of psychotic symptoms, it should be possible to target these symptoms specifically with fewer side effects of the correct receptor sub-type is identified. In addition, molecular sub-typing prompts new etiological hypotheses. One which has recently been put forth by Seeman et al. (1993) suggests that schizophrenia involves the specific up regulation of the D_4 receptor. Another is the interesting idea Sokoloff raised in this volume suggesting D_3 activation leads to c-fos involvement with the potential to stimulate synaptic

Department of Psychiatry, SUNY Stony Brook, Stony Brook, NY 11794, USA

H. Häfner and W.F. Gattaz (Eds.)
Search for the Causes of Schizophrenia, Vol. III

outgrowth. This suggests a mechanism whereby structural changes could be a result of alterations in receptor activity, uniting two divergent areas of schizophrenia research. If a single receptor is the source of pathology or the final pathway in the expression of symptoms, then we may be close to more specific therapeutic approaches. However, the inability to find a clear linkage with any cloned receptor and the multiplicity of reasonable non-dopaminergic hypotheses suggest this simple solution may not be the case. Over the years hypotheses concerning a transmitter defect involving serotonin, glutamate, GABA, CCK, and somatostatin have been proposed. Currently PCP psychosis is focusing attention on the glutamate system, post-mortem studies are pointing toward GABA and serotonin interactions with DA are favored in explaining the action of clozapine. The point is that all of these systems have reasonable evidence suggesting their involvement in the disease process. This points to the fact that the brain is an intensely interactive system, alterations in any one part will be widely reflected in multiple systems as the brain attempts to compensate. A reasonable assumption is that the disease will ultimately not be understood as a disorder of a given neurotransmitter but as a complex response to a group of yet undefined metabolic defects.

Given my bias, the next two papers provide an interesting possible paradigm. They point to the possibility of a defect in membrane structure secondary to a change in phospholipid composition. The phospholipid story goes back to the work of Stevens (1972) who reported that RBC membranes had altered phospholipid composition. Subsequent to this many laboratories have attempted to replicate these findings with mixed results (Rotrosen and Wolkin 1987). The work of Gattaz and his collaborators presented here attempts to look at this issue by examining an enzyme which might be responsible for alterations in phospholipid composition. Gattaz has now replicated his finding that phospholipase A_2 is higher in schizophrenic patients, including first break neuroleptic-free patients, compared to well controls and patients with affective disorder. What this study very nicely shows is a correlation between enzyme activity and changes in substrate and product concentration in the thrombocyte. Gattaz goes on to show in a series of animal experiments that the infusion of the phospholipase A_2 into the substantia nigra effects the DA system. These data do suggest that changes in enzyme levels may effect DA but say little about the possible mechanisms which may be operative in schizophrenic patients. In view of the difficulty in replicating membrane composition changes throughout the world, some caution prior to further replication is warranted in assessing these results. Some independent evidence on this issue was provided in the paper by McClure et al. presented in this volume.

The work described by McClure et al. involves the use of in vivo NMR spectroscopy to measure phosphate compounds in the brains of patients and controls. The development of techniques which allow spectroscopy in living tissue is a major advance. In principle both phosphate and hydrogen spectra can be utilized to measure specific compounds. The technique involves no radiation exposure and coupled with MRI should allow the measurement of specific groups of compounds in well defined anatomical regions. McClure et al. have used the phosphate spectra to define monophosphate compounds and confirms a decrease in this pool as would be predicted by Gattaz's results. The spectroscopic approach is new for intact brains and again caution is necessary in interpreting the data, however this clearly is a developing area of research with great promise. Taken together these papers point to an emerging area of research in schizophrenia which may, if we are fortunate, be getting us closer to major etiological factors contributing to the disease or some forms of the disease.

The final paper returns us to the dopamine hypothesis, which was initially put forth as an etiological theory and has been modified substantially to account for new findings. We suggested (Henn 1982) over 10 years ago that DA might clearly play a role in the final common pathway which leads to the expression of positive symptoms but was unlikely to play an important role in the expression of negative symptoms. Recent PET studies summarized by Wiesel in the final paper of this session have tended to support and strengthen this view. Several groups have found a good correlation between receptor occupancy and therapeutic effect. Evidence that increased receptor levels are present in untreated schizophrenic patients is considerably more confused. The PET findings have been much discussed and disputed in the last several years following the report of Wong et al. (1986) that dopamine receptors were elevated consistent with the postmortem findings of Seeman and his group. The inability of the Swedish group to duplicate these findings and the problems with the various ligands used to measure dopamine receptors has fostered significant work in this area including the contribution to this volume by Wiesel. This work suggests that DA may be central to understanding the anti-psychotic activity of various drugs but may not be central to understanding the causes of schizophrenia itself. This conclusion is also reached when considering the work with HVA and psychosis. Plasma HVA appears to roughly correlate with severity of psychosis in three laboratories (Bowers 1974; Pickar et al. 1986; Davis et al. 1985) while it is never found to correlate with the motivational and affective features of the disease. There remain many cloudy issues including an explanation of the apparently drug-free cases Seeman reports with elevated D_2 receptors, and the inability to find correlations with CSF HVA and psychosis.

In spite of the somewhat inconsistent and incomplete data on the question of the etiological role of DA the broad picture is relatively clear. It suggests that specific pathology in DA neurotransmission can not be the major etiological factor in the majority of cases of schizophrenia. There might in fact be a subset of schizophrenic patients who have a primary defect in DA neurotransmission, but this has not as yet been clearly demonstrated. Just as clear is the fact that psychotic symptoms appear to be mediated by a pathway which involves DA neurotransmission and that this pathway appears to play a role in the formation of psychotic symptoms whether the etiology is a toxin, mania, depression or schizophrenia. My view of this is that there are only a few ways the human brain can react to subtle forms of dysfunction, which may predominantly reside in altered physiological function instead of the more gross pathology seen in neurological disease, where structural causes can reliably be identified in most cases. Changes in the way information is processed can result from imbalances betwen modulating systems or between alterations in the balance of inhibitory and excitatory activity. This appears to either lead to changes in mood or thinking and can often result in false perception of sensory information (hallucinations) or false beliefs not based on sensory input (delusions). These latter symptoms may be analogous to generalized responses in other body dysfunctions represented by fever. Fever is the response to a variety of dysfunctions including infection and malignancy and signals a general metabolic response to a variety of dysfunctions. So too may psychosis be a generalized response to a variety of brain dysfunctions, which could have quite different etiologies. As fever responds to aspirin, which effects the final common pathway in the production of fever so psychosis responds to neuroleptics which effects the final common pathway in its formation. Such an argument is in direct opposition to a continuum viewpoint and suggests that in fact it may be likely that the psychoses encompass a wide variety of disorders with a variety of etiologies. The resolution lies in the discovery of etiologies and I would argue that we already have some. For example, we understand some of the pathophysiology of Huntington's disease, and recently the gene responsible has been localized. Whatever that gene does appears to trigger at some time in mid-life the death of predominantly GABAergic neurons. This disequilibrium very often leads to typical delusions, often of a paranoid type and at times mood alterations are seen which appear phenomenologically exactly like mania or depression. The psychotic symptoms in this condition clearly respond to neuroleptic medications, and it appears that they arise secondary to a functionally overactive DA system reacting to the loss of GABAergic input. Since this is a progressive and fatal disease the structural pathophysiology was easier to track than less progressive disorders such as schizophrenia or affective disorders. However the model of neurotransmitter disequilibrium resulting in some changes in DA function, realized in this disease, could be realized in a variety of ways in what we refer to as the functional psychoses. As other causes are found categories will be cleanly pulled out of what appears to be a psychotic continuum, just as in mental retardation, a continuum of low IQ is giving way to specifically identified diseases, all of which effect IQ among other signs and symptoms.

The point I want to make with the above is that almost any functional change in GABAergic, glutaminergic or peptidergic activity could in some way ultimately alter activity in the dopamine system and so lead to psychosis. All these conditions would in principle respond to DA antagonists. This response tells us nothing about etiology. An appreciation of these issues emphasizes that psychiatric diagnostic formulations are in all likelihood syndromic formulations, not categorical formulations of etiologically distinct diseases. This makes the argument of a continuation of psychosis irrelevant for diagnostic purposes, just as a continuation of fever has little utility in defining medical diseases. Thus, this volume has let us look at three powerful new ways of looking into brain function, molecular neurobiology, NMR in vivo spectroscopy and PET scanning, and a different focus for pathology, phospholipase A_2. While these contributions have not led us to the causes of schizophrenia, the efforts have moved us down the path towards that end and may provide us with clues to the etiologies which will allow us to better treat and classify these difficult disorders in the future.

References

Bowers MB Jr (1974) Central dopamine turnover in schizophrenic syndromes. Arch Gen Psychiatry 31:50–54

Davis KL, Davidson M, Mohs RC, Kendler KS, Davis BM, Johns CA, DeNigris Y, Horvath TH (1985) Plasma homovanillic acid concentration and the severity of schizophrenic illness. Science 227:1601–1602

Henn FA (1982) Dopamine: a role in psychosis or schizophrenia. In: Henn FA, Nosrallah HN (eds) Schizophrenia as a brain disease. Oxford University Press, New York, pp 176–195

Pichar D, Labarca R, Doran AR, Wolkowitz OM, Roy A, Brier A, Linonoila M, Paul SM (1986) Longitudinal measurement of plasma homovanillic acid in schizophrenic patients: correlation with psychosis and response to neuroleptic treatment. Arch Gen Psychiatry 43:669–676

Rotrosen J, Wolkin A (1987) Phospholipid and prostaglandin hypothesis of schizophrenia. In: Meltzer Hy (ed) Psychopharmacology, the third generation of progress. Raven, New York, pp 759–764

Seeman P, Guan HC, VanTol HH (1993) Dopamine D_4 receptors elevated in schizophrenia. Nature 365:441–455

Stevens JD (1972) The distribution of the phospholipid fractions in the red cell membrane of schizophrenics. Schizophr Bull 6:60–61

Wong DF, Wagner HN, Tune LE, Dannals RF, Pearlson GD, Links JM, Tamminga CA, Broussolle EP, Ravert HT, Wilson AA, Toung JKT, Malat J, Williams FA, O'Toung LA, Snyder SH, Kuhar MJ, Gjedde A (1986) Positron emission tomography reveals elevated D_2 dopamine receptors in drug-naive schizophrenics. Science 234:1558–1563

Subject Index

Springer-Verlag and the Environment

We at Springer-Verlag firmly believe that an international science publisher has a special obligation to the environment, and our corporate policies consistently reflect this conviction.

We also expect our business partners – paper mills, printers, packaging manufacturers, etc. – to commit themselves to using environmentally friendly materials and production processes.

The paper in this book is made from low- or no-chlorine pulp and is acid free, in conformance with international standards for paper permanency.

MIX
Papier aus verantwortungsvollen Quellen
Paper from responsible sources
FSC® C105338

If you have any concerns about our products,
you can contact us on
ProductSafety@springernature.com

In case Publisher is established outside the EU,
the EU authorized representative is:
Springer Nature Customer Service Center GmbH
Europaplatz 3, 69115 Heidelberg, Germany

Printed by Libri Plureos GmbH
in Hamburg, Germany